Information Geometry and Its Applications

Information Geometry and Its Applications

Editor: Brayan Cox

NY RESEARCH PRESS
New York

Published by NY Research Press
118-35 Queens Blvd., Suite 400,
Forest Hills, NY 11375, USA
www.nyresearchpress.com

Information Geometry and Its Applications
Edited by Brayan Cox

International Standard Book Number: 978-1-64725-463-6 (Hardback)

Cataloging-in-Publication Data

Information geometry and its applications / edited by Brayan Cox.
p. cm.
Includes bibliographical references and index.
ISBN 978-1-64725-463-6
1. Geometry, Differential. 2. Mathematical statistics. 3. Information theory in mathematics.
4. Information theory--Mathematics. I. Cox, Brayan.
QA641 .I54 2023
516.36--dc23

Contents

Preface

Information geometry refers to an interdisciplinary field that studies statistics and probability theory by applying the techniques of differential geometry. It investigates statistical manifolds, which are Riemannian manifolds, where each point is a probability distribution. Information geometry offers a differential-geometric structure on manifolds that help in designing and studying statistical decision rules. It has been utilized in a variety of applications such as machine learning, quantum systems, mathematical finance, statistical inference, and neural networks. The amount of information integration within various terminals of a causal dynamical system is measured using information geometry. The amount of information lost can be measured using integrated information when a system is divided into parts and information transmission between the parts is stopped. This book provides comprehensive insights into information geometry and its applications. It presents researches and studies performed by experts across the globe. The book aims to equip students and experts with the advanced topics and upcoming concepts in this area of study.

This book has been the outcome of endless efforts put in by authors and researchers on various issues and topics within the field. The book is a comprehensive collection of significant researches that are addressed in a variety of chapters. It will surely enhance the knowledge of the field among readers across the globe.

It gives us an immense pleasure to thank our researchers and authors for their efforts to submit their piece of writing before the deadlines. Finally in the end, I would like to thank my family and colleagues who have been a great source of inspiration and support.

Editor

1

An Elementary Introduction to Information Geometry

Frank Nielsen

Sony Computer Science Laboratories, Tokyo 141-0022, Japan; Frank.Nielsen@acm.org

Abstract: In this survey, we describe the fundamental differential-geometric structures of information manifolds, state the fundamental theorem of information geometry, and illustrate some use cases of these information manifolds in information sciences. The exposition is self-contained by concisely introducing the necessary concepts of differential geometry. Proofs are omitted for brevity.

Keywords: differential geometry; metric tensor; affine connection; metric compatibility; conjugate connections; dual metric-compatible parallel transport; information manifold; statistical manifold; curvature and flatness; dually flat manifolds; Hessian manifolds; exponential family; mixture family; statistical divergence; parameter divergence; separable divergence; Fisher–Rao distance; statistical invariance; Bayesian hypothesis testing; mixture clustering; α-embeddings; mixed parameterization; gauge freedom

---

1. Introduction

1.1. Overview of Information Geometry

We present a concise and modern view of the basic structures lying at the heart of Information Geometry (IG), and report some applications of those information-geometric manifolds (herein termed "information manifolds") in statistics (Bayesian hypothesis testing) and machine learning (statistical mixture clustering).

By analogy to Information Theory (IT) (pioneered by Claude Shannon in his celebrated 1948 paper [1]) which considers primarily the communication of messages over noisy transmission channels, we may define Information Sciences (IS) as the fields that study "communication" between (noisy/imperfect) data and families of models (postulated as a priori knowledge). In short, information sciences seek methods to distill information from data to models. Thus information sciences encompass information theory but also include the fields of Probability and Statistics, Machine Learning (ML), Artificial Intelligence (AI), Mathematical Programming, just to name a few.

We review some key milestones of information geometry and report some definitions of the field by its pioneers in Section 5.2. Professor Shun-ichi Amari, the founder of modern information geometry, defined information geometry in the preface of his latest textbook [2] as follows: "Information geometry is a method of exploring the world of information by means of modern geometry." In short, information geometry geometrically investigates information sciences. It is a mathematical endeavour to define and bound the term geometry itself as geometry is open-ended. Often, we start by studying the invariance of a problem (e.g., invariance of distance between probability distributions) and get as a result a novel geometric structure (e.g., a "statistical manifold"). However, a geometric structure is "pure" and thus may be applied to other application areas beyond the scope of the original problem (e.g., use of the dualistic structure of statistical manifolds in mathematical programming [3]): the method of geometry [4] thus yields a pattern of abduction [5,6].

A narrower definition of information geometry can be stated as the field that studies the geometry of decision making. This definition also includes model fitting (inference) which can be interpreted as a decision problem as illustrated in Figure 1; namely, deciding which model parameter to choose from a family of parametric models. This framework was advocated by Abraham Wald [7–9] who considered all statistical problems as statistical decision problems. Dissimilarities (also loosely called distances among others) play a crucial role not only for measuring the goodness-of-fit of data to model (say, likelihood in statistics, classifier loss functions in ML, objective functions in mathematical programming or operations research, etc.) but also for measuring the discrepancy (or deviance) between models.

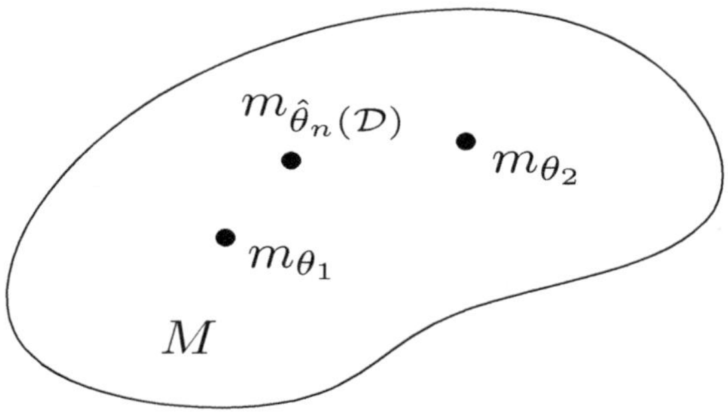

Figure 1. The parameter inference $\hat{\theta}$ of a model from data $\mathcal{D}$ can also be interpreted as a decision making problem: decide which parameter of a parametric family of models $M = \{m_\theta\}_{\theta\in\Theta}$ suits the "best" the data. Information geometry provides a differential-geometric structure on manifold M which useful for designing and studying statistical decision rules.

One may ponder why adopting a geometric approach? Geometry allows one to study invariance of "figures" in a coordinate-free framework. The geometric language (e.g., line, ball or projection) also provides affordances that help us reason intuitively about problems. Note that although figures can be visualized (i.e., plotted in coordinate charts), they should be thought of as purely abstract objects, namely, geometric figures.

Geometry also allows one to study equivariance: For example, the centroid $c(T)$ of a triangle is equivariant under any affine transformation A: $c(A.T) = A.c(T)$. In Statistics, the Maximum Likelihood Estimator (MLE) is equivariant under a monotonic transformation g of the model parameter θ: $\widehat{(g(\theta))} = g(\hat{\theta})$, where the MLE of θ is denoted by $\hat{\theta}$.

1.2. Rationale and Outline of the Survey

The goal of this survey on information geometry [2] is to describe the core dualistic structures on manifolds without assuming any prior background on differential geometry [10], and explain several important related principles and concepts like invariance, covariance, projections, flatness and curvature, information monotonicity, etc. In doing so, we shall illustrate the basic underlying concepts with selected examples and applications, and shall make clear of some potential sources of confusion (e.g., a geometric statistical structure can be used in non-statistical applications [3], untangle the meaning of α in the α-connections, the α-divergences, and the α-representations, etc.). In particular, we shall name and state the fundamental theorem of information geometry in Section 3.5. We refer the reader to the books [2,4,11–17] for an indepth treatment of the field with its applications in information sciences.

This survey is organized as follows:

In the first part (Section 2), we start by concisely introducing the necessary background on differential geometry in order to define a manifold structure (M, g, ∇), i.e., a manifold M equipped with a metric tensor field g and an affine connection ∇. We explain how this framework generalizes the Riemannian manifolds (M, g) by stating the fundamental theorem of Riemannian geometry that defines a unique torsion-free metric-compatible Levi–Civita connection which can be derived from the metric tensor.

In the second part (Section 3), we explain the dualistic structures of information manifolds: we present the conjugate connection manifolds (M, g, ∇, ∇^*), the statistical manifolds (M, g, C) where C denotes a cubic tensor, and show how to derive a family of information manifolds $(M, g, \nabla^{-\alpha}, \nabla^{\alpha})$ for $\alpha \in \mathbb{R}$ provided any given pair $(\nabla = \nabla^{-1}, \nabla^* = \nabla^1)$ of conjugate connections. We explain how to get conjugate connections ∇ and ∇^* coupled to the metric g from any smooth (potentially asymmetric) distances (called divergences), present the dually flat manifolds obtained when considering Bregman divergences, and define, when dealing with parametric family of probability models, the exponential connection ${}^e\nabla$ and the mixture connection ${}^m\nabla$ that are dual connections coupled to the Fisher information metric. We discuss the concept of statistical invariance for the metric tensor and the notion of information monotonicity for statistical divergences [2,18]. It follows that the Fisher information metric is the unique invariant metric (up to a scaling factor), and that the f-divergences are the unique separable invariant divergences.

In the third part (Section 4), we illustrate how to use these information-geometric structures in simple applications: First, we described the natural gradient descent method in Section 4.1 and its relationships with the Riemannian gradient descent and the Bregman mirror descent. Second, we consider two applications in dually flat spaces in Section 4.2: In the first application, we consider the problem of Bayesian hypothesis testing and show how Chernoff information (which defines the best error exponent) can be geometrically characterized on the dually flat structure of an exponential family manifold. In the second application, we show how to cluster statistical mixtures sharing the same component distributions on the dually flat mixture family manifold.

Finally, we conclude in Section 5 by summarizing the important concepts and structures of information geometry, and by providing further references and textbooks [2,16] for further readings to more advanced structures and applications of information geometry. We also mention recent studies of generic classes of principled distances and divergences.

In Appendix A, we show how to estimate the statistical f-divergences between two probability distributions in order to ensure that the estimates are non-negative in Appendix B, and report the canonical decomposition of the multivariate Gaussian family, an example of exponential family which admits a dually flat structure.

At the beginning of each part, we start by outlining its contents. A summary of the notations used throughout this survey is provided in Appendix C.

2. Prerequisite: Basics of Differential Geometry

In Section 2.1, we review the very basics of Differential Geometry (DG) for defining a manifold (M, g, ∇) equipped with both a metric tensor field g and an affine connection ∇. We explain these two independent metric/connection structures in Sections 2.2 and 2.3, respectively. From an affine connection ∇, we show how to derive the notion of covariant derivative in Section 2.3.1, parallel transport in Section 2.3.2 and geodesics in Section 2.3.3. We further explain the intrinsic curvature and torsion of manifolds induced by the connection in Section 2.3.4, and state the fundamental theorem of Riemannian geometry in Section 2.4: the existence of a unique torsion-free Levi–Civita connection ${}^{\mathrm{LC}}\nabla$ compatible with the metric (metric connection) that can be derived from the metric tensor g. Thus the Riemannian geometry (M, g) is obtained as a special case of the more general manifold structure $(M, g, {}^{\mathrm{LC}}\nabla)$: $(M, g) \equiv (M, g, {}^{\mathrm{LC}}\nabla)$. Information geometry shall further consider a dual structure (M, g, ∇^*) associated to (M, g, ∇), and the pair of dual structures shall form an information manifold (M, g, ∇, ∇^*).

2.1. Overview of Differential Geometry: Manifold (M, g, ∇)

Informally speaking, a smooth D-dimensional manifold M is a topological space that locally behaves like the D-dimensional Euclidean space $\mathbb{R}^D$. Geometric objects (e.g., points, balls, and vector fields) and entities (e.g., functions and differential operators) live on M, and are coordinate-free but can conveniently be expressed in any local coordinate system of an atlas $\mathcal{A} = \{(\mathcal{U}_i, x_i)\}_i$ of charts$(\mathcal{U}_i, x_i)$'s (fully covering the manifold) for calculations. Historically, René Descartes (1596–1650) allegedly invented the global Cartesian coordinate system while wondering how to locate a fly on the ceiling from his bed. In practice, we shall use the most expedient coordinate system to facilitate calculations. In information geometry, we usually handle a single chart fully covering the manifold.

A C^k manifold is obtained when the change of chart transformations are C^k. The manifold is said smooth when it is C^∞. At each point $p \in M$, a tangent plane T_p locally best linearizes the manifold. On any smooth manifold M, we can define two independent structures:

1. A metric tensor g, and
2. An affine connection ∇.

The metric tensor g induces on each tangent plane T_p an inner product space that allows one to measure vector magnitudes (vector "lengths") and angles/orthogonality between vectors. The affine connection ∇ is a differential operator that allows one to define:

1. The covariant derivative operator which provides a way to calculate differentials of a vector field Y with respect to another vector field X: namely, the covariant derivative $\nabla_X Y$,
2. The parallel transport $\prod_c^\nabla$ which defines a way to transport vectors between tangent planes along any smooth curve c,
3. The notion of ∇-geodesics γ_∇ which are defined as autoparallel curves, thus extending the ordinary notion of Euclidean straightness,
4. The intrinsic curvature and torsion of the manifold.

2.2. *Metric Tensor Fields g*

The tangent bundle of M is defined as the "union" of all tangent spaces:

$$TM := \sqcup_p T_p := \{(p,v), \quad p \in M, v \in T_p\}. \tag{1}$$

Thus the tangent bundle TM of a D-dimensional manifold M is of dimension $2D$ (the tangent bundle is a particular example of a fiber bundle with base manifold M).

Informally speaking, a tangent vector v plays the role of a directional derivative, with vf informally meaning the derivative of a smooth function f (belonging to the space of smooth functions $\mathfrak{F}(M)$) along the direction v. Since the manifolds are abstract and not embedded in some Euclidean space, we do not view a vector as an "arrow" anchored on the manifold. Rather, vectors can be understood in several ways in differential geometry like directional derivatives or equivalent class of smooth curves at a point. That is, tangent spaces shall be considered as the manifold abstract too.

A smooth vector field X is defined as a "cross-section" of the tangent bundle: $X \in \mathfrak{X}(M) = \Gamma(TM)$, where $\mathfrak{X}(M)$ or $\Gamma(TM)$ denote the space of smooth vector fields. A basis $B = \{b_1, \ldots, b_D\}$ of a finite D-dimensional vector space is a maximal linearly independent set of vectors: A set of vectors $B = \{b_1, \ldots, b_D\}$ is linearly independent if and only if $\sum_{i=1}^D \lambda_i b_i = 0$ iff $\lambda_i = 0$ for all $i \in [D]$. That is, in a linearly independent vector set, no vector of the set can be represented as a linear combination of the remaining vectors. A vector set is linearly independent maximal when we cannot add another linearly independent vector. Tangent spaces carry algebraic structures of vector spaces. Furthermore, to any vector space V, we can associate a dual covector space V^* which is the vector space of real-valued linear mappings. We do not enter into details here to preserve this gentle introduction to information geometry with as little intricacy as possible. Using local coordinates on a chart $(\mathcal{U}, x)$, the vector field X can be expressed as $X = \sum_{i=1}^D X^i e_i \stackrel{\Sigma}{=} X^i e_i$ using Einstein summation convention on dummy indices (using notation $\stackrel{\Sigma}{=}$), where $(X)_B := (X^i)$ denotes the contravariant vector components (manipulated as "column vectors" in algebra) in the natural basis $B = \{e_1 = \partial_1, \ldots, e_D = \partial_D\}$ with $\partial_i :=: \frac{\partial}{\partial x_i}$. A tangent plane (vector space) equipped with an inner product $\langle \cdot, \cdot \rangle$ yields an inner product space. We define a reciprocal basis $B^* = \{e^{*i} = \partial^i\}_i$ of $B = \{e_i = \partial_i\}_i$ so that vectors can also be expressed using the covariant vector components in the natural reciprocal basis. The primal and reciprocal basis are mutually orthogonal by construction as illustrated in Figure 2.

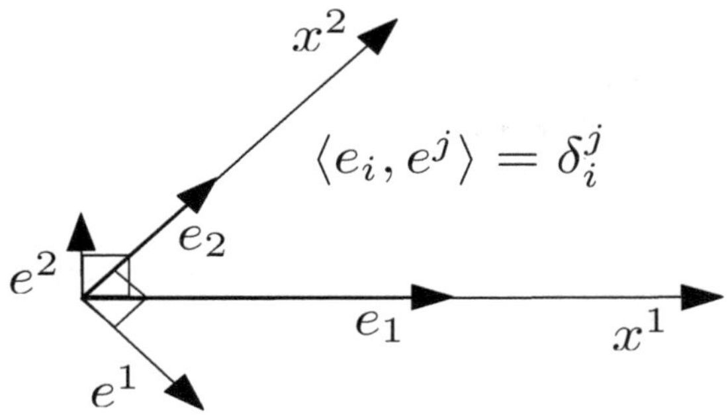

Figure 2. Primal basis (red) and reciprocal basis (blue) of an inner product $\langle \cdot, \cdot \rangle$ space. The primal/reciprocal basis are mutually orthogonal: e^1 is orthogonal to e_2, and e_1 is orthogonal to e^2.

For any vector v, its contravariant components v^i's (superscript notation) and its covariant components v_i's (subscript notation) can be retrieved from v using the inner product with the use of the reciprocal and primal basis, respectively:

$$v^i = \langle v, e^{*i} \rangle, \tag{2}$$

$$v_i = \langle v, e_i \rangle. \tag{3}$$

The inner product defines a metric tensor g and a dual metric tensor g^*:

$$g_{ij} := \langle e_i, e_j \rangle, \tag{4}$$

$$g^{*ij} := \langle e^{*i}, e^{*j} \rangle. \tag{5}$$

Technically speaking, the metric tensor $g_p : T_pM \times T_pM \to \mathbb{R}$ is a 2-covariant tensor field:

$$g \stackrel{\Sigma}{=} g_{ij}\, \mathrm{d}x_i \otimes \mathrm{d}x_j, \tag{6}$$

where $\otimes$ is the dyadic tensor product performed on pairwise covector basis $\{\mathrm{d}x_i\}_i$ (the covectors corresponding to the reciprocal vector basis). We do not describe tensors in details for sake of brevity. A tensor is a geometric entity of a tensor space that can also be interpreted as a multilinear map. A contravariant vector lives in a vector space while a covariant vector lives in the dual covector space. We recommend the textbook [19] for a concise and well-explained description of tensors.

Let $G = [g_{ij}]$ and $G^* = [g^{*ij}]$ denote the $D \times D$ matrices. It follows by construction of the reciprocal basis that $G^* = G^{-1}$. The reciprocal basis vectors e^{*i}'s and primal basis vectors e_i's can be expressed using the dual metric g^* and metric g on the primal basis vectors e_j's and reciprocal basis vectors e^{*j}'s, respectively:

$$e^{*i} \stackrel{\Sigma}{=} g^{*ij} e_j, \tag{7}$$

$$e_i \stackrel{\Sigma}{=} g_{ij} e^{*j}. \tag{8}$$

The metric tensor field g ("metric tensor" or "metric" for short) defines a smooth symmetric positive-definite bilinear form on the tangent bundle so that for $u, v \in T_p$, $\quad g(u,v) \geq 0 \in \mathbb{R}$. We can also write equivalently $g_p(u,v) {:=:} \langle u, v \rangle_p {:=:} \langle u, v \rangle_{g(p)} {:=:} \langle u, v \rangle$. Two vectors u and v are said orthogonal, denoted by $u \perp v$, iff $\langle u, v \rangle = 0$. The length of a vector is induced from the norm $\|u\|_p {:=:} \|u\|_{g(p)} = \sqrt{\langle u, u \rangle_{g(p)}}$. Using local coordinates of a chart $(\mathcal{U}, x)$, we get the vector contravariant/covariant components, and compute the metric tensor using matrix algebra (with column vectors by convention) as follows:

$$g(u,v) = (u)_B^\top \times G_{x(p)} \times (v)_B = (u)_{B^*}^\top \times G_{x(p)}^{-1} \times (v)_{B^*}, \tag{9}$$

since it follows from the primal/reciprocal basis that $G \times G^* = I$, the identity matrix. Thus on any tangent plane T_p, we get a Mahalanobis distance:

$$M_G(u,v) := \|u - v\|_G = \sqrt{\sum_{i=1}^{D}\sum_{j=1}^{D} G_{ij}(u^i - v^i)(u^j - v^j)}. \tag{10}$$

The inner product of two vectors u and v is a scalar (a 0-tensor) that can be equivalently calculated as:

$$\langle u, v\rangle := g(u,v) \stackrel{\Sigma}{=} u^i v_i \stackrel{\Sigma}{=} u_i v^i. \tag{11}$$

A metric tensor g of manifold M is said conformal when $\langle\cdot,\cdot\rangle_p = \kappa(p)\langle\cdot,\cdot\rangle_{\mathrm{Euclidean}}$. That is, when the inner product is a scalar function $\kappa(\cdot)$ of the Euclidean dot product. More precisely, we define the notion of a metric g' conformal to another metric g when these metrics define the same angles between vectors u and v of a tangent plane T_p:

$$\frac{g'_p(u,v)}{\sqrt{g'_p(u,u)}\sqrt{g'_p(v,v)}} = \frac{g_p(u,v)}{\sqrt{g_p(u,u)}\sqrt{g_p(v,v)}}. \tag{12}$$

Usually g' is chosen as the Euclidean metric. In conformal geometry, we can measure angles between vectors in tangent planes as if we were in an Euclidean space, without any deformation. This is handy for checking orthogonality in charts. For example, the Poincaré disk model of hyperbolic geometry is conformal but Klein disk model is not conformal (except at the origin), see [20].

2.3. *Affine Connections* ∇

An affine connection ∇ is a differential operator defined on a manifold that allows us to define (1) a covariant derivative of vector fields, (2) a parallel transport of vectors on tangent planes along a smooth curve, and (3) geodesics. Furthermore, an affine connection fully characterizes the curvature and torsion of a manifold.

2.3.1. Covariant Derivatives $\nabla_X Y$ of Vector Fields

A connection defines a covariant derivative operator that tells us how to differentiate a vector field Y according to another vector field X. The covariant derivative operator is denoted using the traditional gradient symbol ∇. Thus a covariate derivative ∇ is a function:

$$\nabla : \mathfrak{X}(M) \times \mathfrak{X}(M) \to \mathfrak{X}(M), \tag{13}$$

that has its own special subscript notation $\nabla_X Y{:=:}\nabla(X,Y)$ for indicating that it is differentiating a vector field Y according to another vector field X.

By prescribing D^3 smooth functions $\Gamma_{ij}^k = \Gamma_{ij}^k(p)$, called the Christoffel symbols of the second kind, we define the unique affine connection ∇ that satisfies in local coordinates of chart $(\mathcal{U}, x)$ the following equations:

$$\nabla_{\partial_i}\partial_j = \Gamma_{ij}^k \partial_k. \tag{14}$$

The Christoffel symbols can also be written as $\Gamma_{ij}^k := (\nabla_{\partial_i}\partial_j)^k$, where $(\cdot)^k$ denotes the k-th coordinate. The k-th component $(\nabla_X Y)^k$ of the covariant derivative of vector field Y with respect to vector field X is given by:

$$(\nabla_X Y)^k \stackrel{\Sigma}{=} X^i(\nabla_i Y)^k \stackrel{\Sigma}{=} X^i\left(\frac{\partial Y^k}{\partial x^i} + \Gamma_{ij}^k Y^j\right). \tag{15}$$

The Christoffel symbols are not tensors (fields) because the transformation rules induced by a change of basis do not obey the tensor contravariant/covariant rules.

2.3.2. Parallel Transport $\prod_c^\nabla$ along a Smooth Curve c

Since the manifold is not embedded in a Euclidean space, we cannot add a vector $v \in T_p$ to a vector $v' \in T_{p'}$ as the tangent vector spaces are unrelated to each others without a connection (the Whitney embedding theorem [21] states that any D-dimensional Riemannian manifold can be embedded into $\mathbb{R}^{2D}$; when embedded, we can implicitly use the ambient Euclidean connection ${}^{\mathrm{Euc}}\nabla$ on the manifold, see [22]). Thus a connection ∇ defines how to associate vectors between infinitesimally close tangent planes T_p and $T_{p+\mathrm{d}p}$. Then the connection allows us to smoothly transport a vector $v \in T_p$ by sliding it (with infinitesimal moves) along a smooth curve $c(t)$ (with $c(0) = p$ and $c(1) = q$), so that the vector $v_p \in T_p$ "corresponds" to a vector $v_q \in T_q$: this is called the parallel transport. This mathematical prescription is necessary in order to study dynamics on manifolds (e.g., study the motion of a particle on the manifold). We can express the parallel transport along the smooth curve c as:

$$\forall v \in T_p, \forall t \in [0,1], \quad v_{c(t)} = \prod_{c(0)\to c(t)}^{\nabla} v \in T_{c(t)} \tag{16}$$

The parallel transport is schematically illustrated in Figure 3.

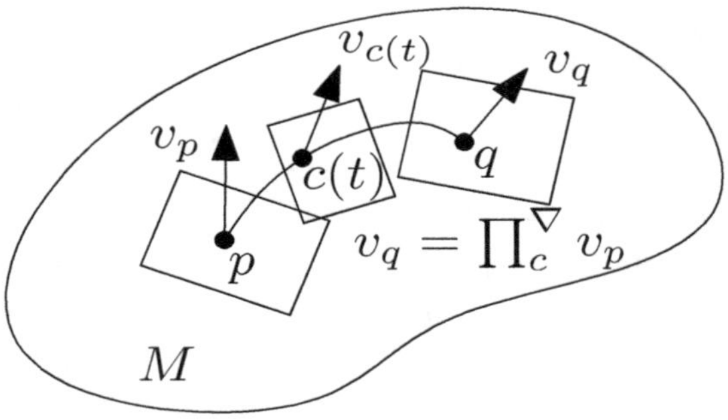

Figure 3. Illustration of the parallel transport of vectors on tangent planes along a smooth curve. For a smooth curve c, with $c(0) = p$ and $c(1) = q$, a vector $v_p \in T_p$ is parallel transported smoothly to a vector $v_q \in T_q$ such that for any $t \in [0,1]$, we have $v_{c(t)} \in T_{c(t)}$.

Elie Cartan introduced the notion of affine connections [23,24] in the 1920s motivated by the principle of inertia in mechanics: a point particle, without any force acting on it, shall move along a straight line with constant velocity.

2.3.3. ∇-Geodesics γ_∇: Autoparallel Curves

A connection ∇ allows one to define ∇-geodesics as autoparallel curves, that are curves γ such that we have:

$$\nabla_{\dot\gamma}\dot\gamma = 0. \tag{17}$$

That is, the velocity vector $\dot\gamma$ is moving along the curve parallel to itself (and all tangent vectors on the geodesics are mutually parallel): In other words, ∇-geodesics generalize the notion of "straight Euclidean" lines. In local coordinates $(\mathcal{U}, x)$, $\gamma(t) = (\gamma^k(t))_k$, the autoparallelism amounts to solve the following second-order Ordinary Differential Equations (ODEs):

$$\ddot{\gamma}(t) + \Gamma^k_{ij}\dot{\gamma}(t)\dot{\gamma}(t) = 0, \quad \gamma^l(t) = x^l \circ \gamma(t), \tag{18}$$

where Γ^k_{ij} are the Christoffel symbols of the second kind, with:

$$\Gamma^k_{ij} \overset{\Sigma}{=} \Gamma_{ij,l}g^{lk}, \quad \Gamma_{ij,k} \overset{\Sigma}{=} g_{lk}\Gamma^l_{ij}, \tag{19}$$

where $\Gamma_{ij,l}$ the Christoffel symbols of the first kind. Geodesics are 1D autoparallel submanifolds and ∇-hyperplanes are defined similarly as autoparallel submanifolds of dimension $D-1$. We may specify in subscript the connection that yields the geodesic γ: γ_∇.

The geodesic equation $\nabla_{\dot{\gamma}(t)}\dot{\gamma}(t) = 0$ may be either solved as an Initial Value Problem (IVP) or as a Boundary Value Problem (BVP):

- Initial Value Problem (IVP): fix the conditions $\gamma(0) = p$ and $\dot{\gamma}(0) = v$ for some vector $v \in T_p$.
- Boundary Value Problem (BVP): fix the geodesic extremities $\gamma(0) = p$ and $\gamma(1) = q$.

2.3.4. Curvature and Torsion of a Manifold

An affine connection ∇ defines a 4D curvature tensor R (expressed using components R^i_{jkl} of a $(1,3)$-tensor). The coordinate-free equation of the curvature tensor is given by:

$$R(X,Y)Z := \nabla_X\nabla_Y X - \nabla_Y\nabla_X Z - \nabla_{[X,Y]}Z, \tag{20}$$

where $[X,Y](f) = X(Y(f)) - Y(X(f))$ $(\forall f \in \mathfrak{F}(M))$ is the Lie bracket of vector fields. When the connection is the metric Levi–Civita, the curvature is called Riemann–Christoffel curvature tensor. In a local coordinate system, we have:

$$R(\partial_j, \partial_k)\partial_i \overset{\Sigma}{=} R^l_{jki}\partial_l. \tag{21}$$

Informally speaking, the curvature tensor as defined in Equation (20) quantifies the amount of non-commutativity of the covariant derivative. It follows from symmetry constraints that the number of independent components of the Riemann tensor is $\frac{D^2(D^2-1)}{12}$ in D dimensions.

A manifold M equipped with a connection ∇ is said flat (meaning ∇-flat) when $R = 0$. This holds in particular when finding a particular coordinate system x of a chart $(\mathcal{U}, x)$ such that $\Gamma^k_{ij} = 0$, i.e., when all connection coefficients vanish. For example, the Christoffel symbols vanish in a rectangular coordinate system of a plane but not in the polar coordinate system of it.

A manifold is torsion-free when the connection is symmetric. A symmetric connection satisfies the following coordinate-free equation:

$$\nabla_X Y - \nabla_Y X = [X,Y]. \tag{22}$$

Using local chart coordinates, this amounts to check that $\Gamma^k_{ij} = \Gamma^k_{ji}$. The torsion tensor is a $(1,2)$-tensor defined by:

$$T(X,Y) := \nabla_X Y - \nabla_Y X - [X,Y]. \tag{23}$$

For a torsion-free connection, we have the first Bianchi identity:

$$R(X,Y)Z + R(Z,X)Y + R(Y,Z)X = 0, \tag{24}$$

and the second Bianchi identity:

$$(\nabla_V R)(X,Y)Z + (\nabla_X R)(Y,V)Z + (\nabla_Y R)(V,X)Z = 0. \tag{25}$$

In general, the parallel transport is path-dependent. The angle defect of a vector transported on an infinitesimal closed loop (a smooth curve with coinciding extremities) is related to the curvature. However for a flat connection, the parallel transport does not depend on the path, and yields absolute parallelism geometry [25]. Figure 4 illustrates the parallel transport along a loop curve for a curved manifold (the sphere manifold) and a flat manifold (the cylinder manifold).

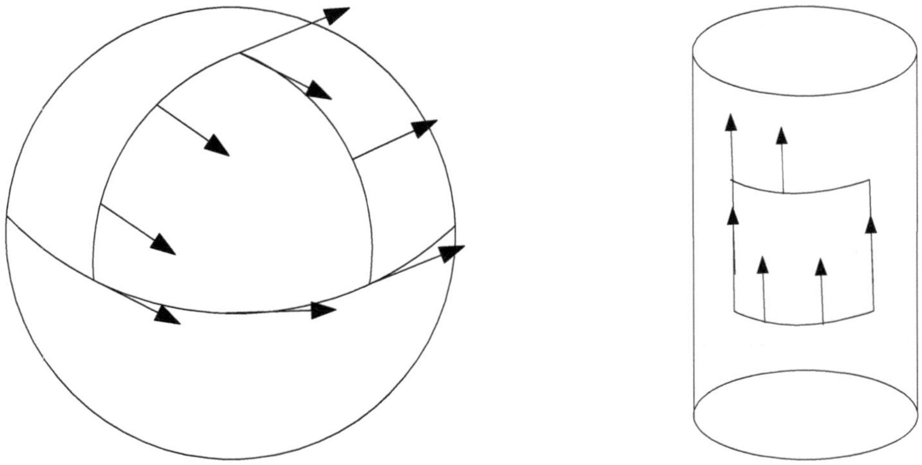

Figure 4. Parallel transport with respect to the metric connection: the curvature effect can be visualized as the angle defect along the parallel transport on smooth (infinitesimal) loops. For a sphere manifold, a vector parallel-transported along a loop does not coincide with itself (e.g., a sphere), while it always conside with itself for a (flat) manifold (e.g., a cylinder).

Historically, the Gaussian curvature at of point of a manifold has been defined as the product of the minimal and maximal sectional curvatures: $\kappa_G := \kappa_{\min}\kappa_{\max}$. For a cylinder, since $\kappa_{\min} = 0$, it follows that the Gaussian curvature of a cylinder is 0. Gauss's Theorema Egregium (meaning "remarkable theorem") proved that the Gaussian curvature is intrinsic and does not depend on how the surface is embedded into the ambient Euclidean space.

An affine connection is a torsion-free linear connection. Figure 5 summarizes the various concepts of differential geometry induced by an affine connection ∇ and a metric tensor g.

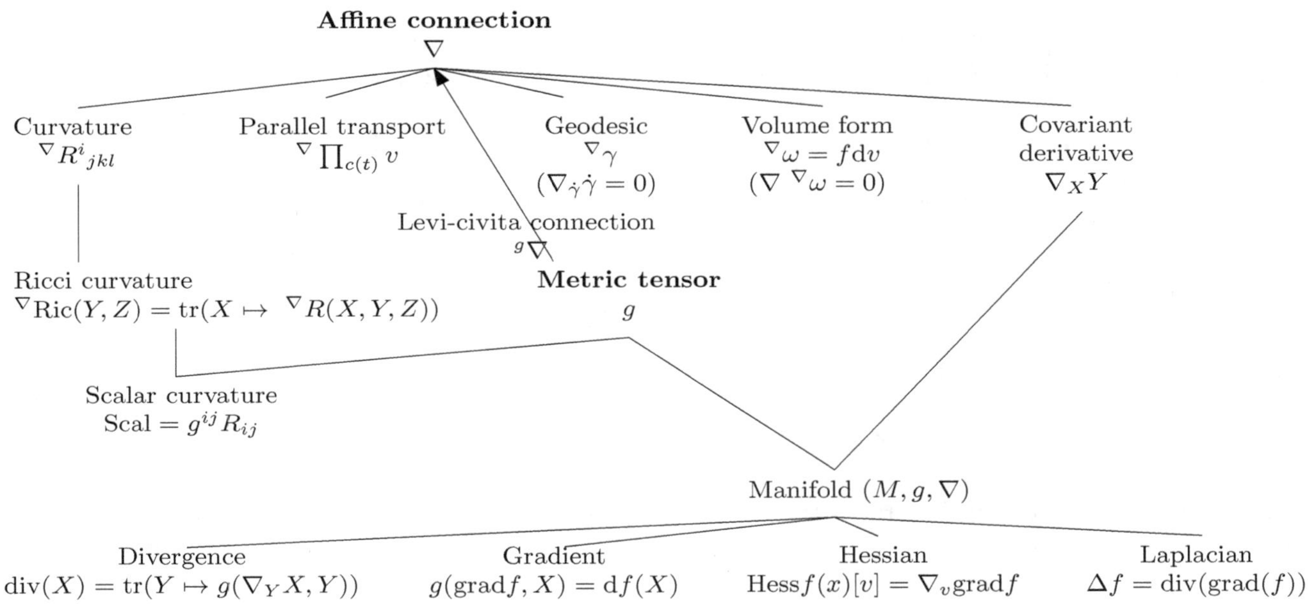

Figure 5. Differential-geometric concepts associated to an affine connection ∇ and a metric tensor g.

Curvature is a fundamental concept inherent to geometry [26]: there are several notions of curvatures in differential geometry: scalar curvature, sectional curvature, Gaussian curvature of surfaces to Riemannian–Christoffel 4-tensor, Ricci symmetric 2-tensor, synthetic Ricci curvature in Alexandrov geometry, etc.

For example, the real-valued Gaussian curvature $\sec(T_pM)$ on a 2D Riemannian manifold (M, g) with Riemann curvature $(1,3)$-tensor R at a point p (with local basis $\{\partial_1, \partial_2\}$ on its tangent plane T_pM) is defined by:

$$\sec(T_pM) = \frac{R_{2112}}{(g_{11}g_{22} - g_{12}^2)}, \quad (26)$$

$$= \frac{\langle (\nabla_{\partial_2}(\nabla_{\partial_1}\partial_1) - \nabla_{\partial_1}(\nabla_{\partial_2}\partial_1)), \partial_2 \rangle_p}{\det(g)}. \quad (27)$$

In general, the sectional curvatures are real values defined for 2-dimensional subspaces π_p of the tangent plane T_pM (called tangent 2-planes) as:

$$\sec_p(\pi) := \frac{\langle R(X,Y)Y, X \rangle_p}{Q_p(X,Y)}, \quad (28)$$

where X and Y are linearly independent vectors of T_p, and

$$Q_p(X,Y) := \langle X, X \rangle_p \langle Y, Y \rangle_p - \langle X, Y \rangle_p^2, \quad (29)$$

denotes the squared area of the parallelogram spanned by vectors X and Y of T_pM. It can be shown that $\sec_p(\pi)$ is independent of the chosen basis X and Y. In a local basis $\{\partial_i\}_i$ of D-dimensional tangent plane T_p, we thus get the sectional curvatures at point $p \in M$ as the following real values:

$$\kappa_{ij} := \sec_p(\mathrm{span}\{\partial_i, \partial_j\}), \quad i \neq j. \quad (30)$$

A Riemannian manifold (M, g) is said of constant curvature κ if and only if $\sec_p(\pi) = \kappa$ for all $p \in M$ and $\pi_p \subset T_pM$. In particular, the Riemannian manifold is said flat when it is of constant curvature 0. Notice that the definition of sectional curvatures relies on the metric tensor g but the Riemann–Christoffel curvature tensor is defined with respect to an affine connection (which can be taken as the default Levi–Civita metric connection induced by the metric g).

2.4. The Fundamental Theorem of Riemannian Geometry: The Levi–Civita Metric Connection

By definition, an affine connection ∇ is said metric compatible with g when it satisfies for any triple (X, Y, Z) of vector fields the following equation:

$$X\langle Y, Z \rangle = \langle \nabla_X Y, Z \rangle + \langle Y, \nabla_X Z \rangle, \quad (31)$$

which can be written equivalently as:

$$Xg(Y,Z) = g(\nabla_X Y, Z) + g(Y, \nabla_X Z) \quad (32)$$

Using local coordinates and natural basis $\{\partial_i\}$ for vector fields, the metric-compatibility property amounts to check that we have:

$$\partial_k g_{ij} = \langle \nabla_{\partial_k}\partial_i, \partial_j \rangle + \langle \partial_i, \nabla_{\partial_k}\partial_j \rangle \quad (33)$$

A property of using a metric-compatible connection is that the parallel transport $\prod^\nabla$ of vectors preserve the metric:

$$\langle u, v\rangle_{c(0)} = \left\langle \prod_{c(0)\to c(t)}^{\nabla} u, \prod_{c(0)\to c(t)}^{\nabla} v \right\rangle_{c(t)} \quad \forall t. \tag{34}$$

That is, the parallel transport preserves angles (and orthogonality) and lengths of vectors in tangent planes when transported along a smooth curve.

The fundamental theorem of Riemannian geometry states the existence of a unique torsion-free metric compatible connection:

Theorem 1 (Levi–Civita metric connection). *There exists a unique torsion-free affine connection compatible with the metric called the Levi–Civita connection* ${}^{\mathrm{LC}}\nabla$.

The Christoffel symbols of the Levi–Civita connection can be expressed from the metric tensor g as follows:

$$ {}^{\mathrm{LC}}\Gamma^k_{ij} \stackrel{\Sigma}{=} \frac{1}{2} g^{kl}\left(\partial_i g_{il} + \partial_j g_{il} - \partial_l g_{ij}\right), \tag{35}$$

where g^{ij} denote the matrix elements of the inverse matrix g^{-1}.

The Levi–Civita connection can also be defined coordinate-free with the Koszul formula:

$$2g(\nabla_X Y, Z) = X(g(Y,Z)) + Y(g(X,Z)) - Z(g(X,Y)) + g([X,Y],Z) - g([X,Z],Y) - g([Y,Z],X). \tag{36}$$

There exists metric-compatible connections with torsions studied in theoretical physics. See for example the flat Weitzenböck connection [27].

The metric tensor g induces the torsion-free metric-compatible Levi–Civita connection that determines the local structure of the manifold. However, the metric g does not fix the global topological structure: For example, although a cone and a cylinder have locally the same flat Euclidean metric, they exhibit different global structures.

2.5. Preview: Information Geometry versus Riemannian Geometry

In information geometry, we consider a pair of conjugate affine connections ∇ and ∇^* (often but not necessarily torsion-free) that are coupled to the metric g: the structure is conventionally written as (M, g, ∇, ∇^*). The key property is that those conjugate connections are metric compatible, and therefore the induced dual parallel transport preserves the metric:

$$\langle u, v\rangle_{c(0)} = \left\langle \prod_{c(0)\to c(t)}^{\nabla} u, \prod_{c(0)\to c(t)}^{\nabla^*} v \right\rangle_{c(t)}. \tag{37}$$

Thus the Riemannian manifold (M, g) can be interpreted as the self-dual information-geometric manifold obtained for $\nabla = \nabla^* = {}^{\mathrm{LC}}\nabla$ the unique torsion-free Levi–Civita metric connection: $(M, g) \equiv (M, g, {}^{\mathrm{LC}}\nabla, {}^{\mathrm{LC}}\nabla^* = {}^{\mathrm{LC}}\nabla)$. However, let us point out that for a pair of self-dual Levi–Civita conjugate connections, the information-geometric manifold does not induce a distance. This contrasts with the Riemannian modeling (M, g) which provides a Riemmanian metric distance $D_\rho(p, q)$ defined by the length of the geodesic γ connecting the two points $p = \gamma(0)$ and $q = \gamma(1)$:

$$D_\rho(p,q) := \int_0^1 \|\gamma'(t)\|_{\gamma(t)} \mathrm{d}t = \int_0^1 \sqrt{g_{\gamma(t)}(\dot\gamma(t), \dot\gamma(t))}\mathrm{d}t, \tag{38}$$

$$= \int_0^1 \sqrt{\dot\gamma(t)^\top g_{\gamma(t)} \dot\gamma(t)}\mathrm{d}t. \tag{39}$$

This geodesic length distance $D_\rho(p,q)$ can also be interpreted as the shortest path linking point p to point q: $D_\rho(p,q) = \inf_\gamma \int_0^1 \|\gamma'(t)\|_{\gamma(t)} \mathrm{d}t$ (with $p = \gamma(0)$ and $q = \gamma(1)$).

Usually, this Riemannian geodesic distance is not available in closed-form (and need to be approximated or bounded) because the geodesics cannot be explicitly parameterized (see geodesic shooting methods [28]).

We are now ready to introduce the key geometric structures of information geometry.

3. Information Manifolds

3.1. Overview

In this part, we explain the dualistic structures of manifolds in information geometry. In Section 3.2, we first present the core Conjugate Connection Manifolds (CCMs) (M, g, ∇, ∇^*), and show how to build Statistical Manifolds (SMs) (M, g, C) from a CCM in Section 3.3. From any statistical manifold, we can build a 1-parameter family $(M, g, \nabla^{-\alpha}, \nabla^{\alpha})$ of CCMs, the information α-manifolds. We state the fundamental theorem of information geometry in Section 3.5. These CCMs and SMs structures are not related to any distance a priori but require at first a pair (∇, ∇^*) of conjugate connections coupled to a metric tensor g. We show two methods to build an initial pair of conjugate connections. A first method consists of building a pair of conjugate connections $({}^D\nabla, {}^D\nabla^*)$ from any divergence D in Section 3.6. Thus we obtain self-conjugate connections when the divergence is symmetric: $D(\theta_1 : \theta_2) = D(\theta_2 : \theta_1)$. When the divergences are Bregman divergences (i.e., $D = B_F$ for a strictly convex and differentiable Bregman generator), we obtain Dually Flat Manifolds (DFMs) $(M, \nabla^2 F, {}^F\nabla, {}^F\nabla^*)$ in Section 3.7. DFMs nicely generalize the Euclidean geometry and exhibit Pythagorean theorems. We further characterize when orthogonal ${}^F\nabla$-projections and dual ${}^F\nabla^*$-projections of a point on submanifold a is unique. In Euclidean geometry, the orthogonal projection of a point p onto an affine subspace S is proved to be unique using the Pythagorean theorem. A second method to get a pair of conjugate connections $({}^e\nabla, {}^m\nabla)$ consists of defining these connections from a regular parametric family of probability distributions $\mathcal{P} = \{p_\theta(x)\}_\theta$. In that case, these 'e'xponential connection ${}^e\nabla$ and 'm'ixture connection ${}^m\nabla$ are coupled to the Fisher information metric ${}_\mathcal{P}g$. A statistical manifold $(\mathcal{P}, {}_\mathcal{P}g, {}_\mathcal{P}C)$ can be recovered by considering the skewness Amari–Chentsov cubic tensor ${}_\mathcal{P}C$, and it follows a 1-parameter family of CCMs, $(\mathcal{P}, {}_\mathcal{P}g, {}_\mathcal{P}\nabla^{-\alpha}, {}_\mathcal{P}\nabla^{+\alpha})$, the statistical expected α-manifolds. In this parametric statistical context, these information manifolds are called expected information manifolds because the various quantities are expressed from statistical expectations $E_\cdot[\cdot]$. Notice that these information manifolds can be used in information sciences in general, beyond the traditional fields of statistics. In statistics, we motivate the choice of the connections, metric tensors and divergences by studying statistical invariance criteria, in Section 3.10. We explain how to recover the expected α-connections from standard f-divergences that are the only separable divergences that satisfy the property of information monotonicity. Finally, in Section 3.11, the recall the Fisher–Rao expected Riemannian manifolds that are Riemannian manifolds $(\mathcal{P}, {}_\mathcal{P}g)$ equipped with a geodesic metric distance called the Fisher–Rao distance, or Rao distance for short.

3.2. Conjugate Connection Manifolds: (M, g, ∇, ∇^*)

We begin with a definition:

Definition 1 (Conjugate connections). *A connection ∇^* is said to be conjugate to a connection ∇ with respect to the metric tensor g if and only if we have for any triple (X, Y, Z) of smooth vector fields the following identity satisfied:*

$$X\langle Y, Z\rangle = \langle \nabla_X Y, Z\rangle + \langle Y, \nabla^*_X Z\rangle, \quad \forall X, Y, Z \in \mathfrak{X}(M). \tag{40}$$

We can notationally rewrite Equation (40) as:

$$Xg(Y,Z) = g(\nabla_X Y, Z) + g(Y, \nabla^*_X Z), \tag{41}$$

and further explicit that for each point $p \in M$, we have:

$$X_p g_p(Y_p, Z_p) = g_p((\nabla_X Y)_p, Z_p) + g_p(Y_p, (\nabla^*_X Z)_p). \tag{42}$$

We check that the right-hand-side is a scalar and that the left-hand-side is a directional derivative of a real-valued function, that is also a scalar.

Conjugation is an involution: $(\nabla^*)^* = \nabla$.

Definition 2 (Conjugate Connection Manifold). *The structure of the Conjugate Connection Manifold (CCM) is denoted by (M, g, ∇, ∇^*), where (∇, ∇^*) are conjugate connections with respect to the metric g.*

A remarkable property is that the dual parallel transport of vectors preserves the metric. That is, for any smooth curve $c(t)$, the inner product is conserved when we transport one of the vector u using the primal parallel transport $\prod_c^{\nabla}$ and the other vector v using the dual parallel transport $\prod_c^{\nabla^*}$.

$$\langle u, v\rangle_{c(0)} = \left\langle \prod_{c(0)\to c(t)}^{\nabla} u, \prod_{c(0)\to c(t)}^{\nabla^*} v \right\rangle_{c(t)}. \tag{43}$$

Property 1 (Dual parallel transport preserves the metric). *A pair (∇, ∇^*) of conjugate connections preserves the metric g if and only if:*

$$\forall t \in [0,1], \left\langle \prod_{c(0)\to c(t)}^{\nabla} u, \prod_{c(0)\to c(t)}^{\nabla^*} v \right\rangle_{c(t)} = \langle u, v\rangle_{c(0)}. \tag{44}$$

Property 2. *Given a connection ∇ on (M, g) (i.e., a structure (M, g, ∇)), there exists a unique conjugate connection ∇^* (i.e., a dual structure (M, g, ∇^*)).*

We consider a manifold M equipped with a pair of conjugate connections ∇ and ∇^* that are coupled with the metric tensor g so that the dual parallel transport preserves the metric. We define the mean connection $\bar{\nabla}$:

$$\bar{\nabla} = \frac{\nabla + \nabla^*}{2}, \tag{45}$$

with corresponding Christoffel coefficients denoted by $\bar{\Gamma}$. This mean connection coincides with the Levi–Civita metric connection:

$$\bar{\nabla} = {}^{LC}\nabla. \tag{46}$$

Property 3. *The mean connection $\bar{\nabla}$ is self-conjugate, and coincide with the Levi–Civita metric connection.*

3.3. Statistical Manifolds: (M, g, C)

Lauritzen introduced this corner structure [29] of information geometry in 1987. Beware that although it bears the name "statistical manifold", it is a purely geometric construction that may be used outside of the field of Statistics. However, as we shall mention later, we can always find a statistical model $\mathcal{P}$ corresponding to a statistical manifold [30]. We shall see how we can convert a conjugate connection manifold into such a statistical manifold, and how we can subsequently derive an infinite family of CCMs from a statistical manifold. In other words, once we have a pair of conjugate connections, we will be able to build a family of pairs of conjugate connections.

We define a cubic $(0,3)$-tensor (i.e., 3-covariant tensor) called the Amari–Chentsov tensor:

$$C_{ijk} := \Gamma^k_{ij} - \Gamma^{*k}_{ij}, \tag{47}$$

or in coordinate-free equation:

$$C(X,Y,Z) := \langle \nabla_X Y - \nabla^*_X Y, Z \rangle. \tag{48}$$

The cubic tensor is totally symmetric, meaning that $C_{ijk} = C_{\sigma(i)\sigma(j)\sigma(k)}$ for any permutation σ. The metric tensor is totally symmetric.

Using the local basis, this cubic tensor can be expressed as:

$$C_{ijk} = C(\partial_i, \partial_j, \partial_k) = \langle \nabla_{\partial_i}\partial_j - \nabla^*_{\partial_i}\partial_j, \partial_k \rangle \tag{49}$$

Definition 3 (Statistical manifold [29]). *A statistical manifold (M, g, C) is a manifold M equipped with a metric tensor g and a totally symmetric cubic tensor C.*

3.4. A Family $\{(M, g, \nabla^{-\alpha}, \nabla^\alpha = (\nabla^{-\alpha})^*)\}_{\alpha\in\mathbb{R}}$ of Conjugate Connection Manifolds

For any pair (∇, ∇^*) of conjugate connections, we can define a 1-parameter family of connections $\{\nabla^\alpha\}_{\alpha\in\mathbb{R}}$, called the α-connections such that $(\nabla^{-\alpha}, \nabla^\alpha)$ are dually coupled to the metric, with $\nabla^0 = \bar{\nabla} = {}^{\mathrm{LC}}\nabla$, $\nabla^1 = \nabla$ and $\nabla^{-1} = \nabla^*$. By observing that the scaled cubic tensor αC is also a totally symmetric cubic 3-covariant tensor, we can derive the α-connections from a statistical manifold (M, g, C) as:

$$\Gamma^\alpha_{ij,k} = \Gamma^0_{ij,k} - \frac{\alpha}{2} C_{ij,k}, \tag{50}$$

$$\Gamma^{-\alpha}_{ij,k} = \Gamma^0_{ij,k} + \frac{\alpha}{2} C_{ij,k}, \tag{51}$$

where $\Gamma^0_{ij,k}$ are the Levi–Civita Christoffel symbols, and $\Gamma_{ki,j} \stackrel{\Sigma}{=} \Gamma^l_{ij} g_{lk}$ (by index juggling).

The α-connection ∇^α can also be defined as follows:

$$g(\nabla^\alpha_X Y, Z) = g({}^{\mathrm{LC}}\nabla_X Y, Z) + \frac{\alpha}{2} C(X,Y,Z), \forall X,Y,Z \in \mathfrak{X}(M). \tag{52}$$

Theorem 2 (Family of information α-manifolds). *For any $\alpha \in \mathbb{R}$, $(M, g, \nabla^{-\alpha}, \nabla^\alpha = (\nabla^{-\alpha})^*)$ is a conjugate connection manifold.*

The α-connections ∇^α can also be constructed directly from a pair (∇, ∇^*) of conjugate connections by taking the following weighted combination:

$$\Gamma^\alpha_{ij,k} = \frac{1+\alpha}{2}\Gamma_{ij,k} + \frac{1-\alpha}{2}\Gamma^*_{ij,k}. \tag{53}$$

3.5. The Fundamental Theorem of Information Geometry: ∇ κ-Curved $\Leftrightarrow$ ∇^* κ-Curved

We now state the fundamental theorem of information geometry and its corollaries:

Theorem 3 (Dually constant curvature manifolds). *If a torsion-free affine connection ∇ has constant curvature κ then its conjugate torsion-free connection ∇^* has necessarily the same constant curvature κ.*

The proof is reported in [16] (Proposition 8.1.4, page 226).

A statistical manifold (M, g, C) is said α-flat if its induced α-connection is flat. It can be shown that $R^\alpha = -R^{-\alpha}$.

We get the following two corollaries:

Corollary 1 (Dually α-flat manifolds). *A manifold $(M, g, \nabla^{-\alpha}, \nabla^{\alpha})$ is ∇^{α}-flat if and only if it is $\nabla^{-\alpha}$-flat.*

Corollary 2 (Dually flat manifolds ($\alpha = \pm 1$)). *A manifold (M, g, ∇, ∇^*) is ∇-flat if and only if it is ∇^*-flat.*

Refer to Theorem 3.3 of [4] for a proof of this corollary.

Let us now define the notion of constant curvature of a statistical structure [31]:

Definition 4 (Constant curvature κ). *A statistical structure (M, g, ∇) is said of constant curvature κ when*

$$R^{\nabla}(X,Y)Z = \kappa\{g(Y,Z)X - g(X,Z)Y\}, \quad \forall\, X,Y,Z \in \Gamma(TM),$$

where $\Gamma(TM)$ denote the space of smooth vector fields.

It can be proved that the Riemann–Christoffel (RC) 4-tensors of conjugate α-connections [16] are related as follows:

$$g\left(R^{(\alpha)}(X,Y)Z, W\right) + g\left(Z, R^{(-\alpha)}(X,Y)W\right) = 0. \tag{54}$$

We have $g\left(R^{\nabla^*}(X,Y)Z, W\right) = -g\left(Z, R^{\nabla}(X,Y)W\right)$.

Thus once we are given a pair of conjugate connections, we can always build a 1-parametric family of manifolds. Manifolds with constant curvature κ are interesting from the computational viewpoint as dual geodesics have simple closed-form expressions.

3.6. Conjugate Connections from Divergences: $(M, D) \equiv (M,\, {}^{D}g,\, {}^{D}\nabla,\, {}^{D}\nabla^* = {}^{D^*}\nabla)$

Loosely speaking, a divergence $D(\cdot : \cdot)$ is a smooth distance [32], potentially asymmetric. In order to define precisely a divergence, let us first introduce the following handy notations: $\partial_{i,\cdot} f(x,y) = \frac{\partial}{\partial x^i} f(x,y)$, $\partial_{\cdot,j} f(x,y) = \frac{\partial}{\partial y^j} f(x,y)$, $\partial_{ij,k} f(x,y) = \frac{\partial^2}{\partial x^i \partial x^j} \frac{\partial}{\partial y^k} f(x,y)$ and $\partial_{i,jk} f(x,y) = \frac{\partial}{\partial x^i} \frac{\partial^2}{\partial y^j \partial y^k} f(x,y)$, etc.

Definition 5 (Divergence). *A divergence $D : M \times M \to [0, \infty)$ on a manifold M with respect to a local chart $\Theta \subset \mathbb{R}^D$ is a C^3-function satisfying the following properties:*

1. *$D(\theta : \theta') \geq 0$ for all $\theta, \theta' \in \Theta$ with equality holding iff $\theta = \theta'$ (law of the indiscernibles),*
2. *$\partial_{i,\cdot} D(\theta : \theta')|_{\theta=\theta'} = \partial_{\cdot,j} D(\theta : \theta')|_{\theta=\theta'} = 0$ for all $i,j \in [D]$,*
3. *$-\partial_{\cdot,i}\partial_{\cdot,j} D(\theta : \theta')|_{\theta=\theta'}$ is positive-definite.*

The dual divergence is defined by swapping the arguments:

$$D^*(\theta : \theta') := D(\theta' : \theta), \tag{55}$$

and is also called the reverse divergence (reference duality in information geometry). Reference duality of divergences is an involution: $(D^*)^* = D$.

The Euclidean distance is a metric distance but not a divergence. The squared Euclidean distance is a non-metric symmetric divergence. The metric tensor g yields Riemannian metric distance D_ρ but it is never a divergence.

From any given divergence D, we can define a conjugate connection manifold following the construction of Eguchi [33,34] (1983):

Theorem 4 (Manifold from divergence). *$(M, {}^{D}g, {}^{D}\nabla, {}^{D^*}\nabla)$ is an information manifold with:*

$$ {}^{D}g := -\partial_{i,j} D(\theta : \theta')|_{\theta=\theta'} = {}^{D^*}g, \tag{56}$$

$$ {}^{D}\Gamma_{ijk} := -\partial_{ij,k} D(\theta : \theta')|_{\theta=\theta'}, \tag{57}$$

$$ {}^{D^*}\Gamma_{ijk} := -\partial_{k,ij} D(\theta : \theta')|_{\theta=\theta'}. \tag{58}$$

The associated statistical manifold is $(M, {}^{D}g, {}^{D}C)$ with:

$$ {}^{D}C_{ijk} = {}^{D^*}\Gamma_{ijk} - {}^{D}\Gamma_{ijk}. \tag{59} $$

Since $\alpha^{D}C$ is a totally symmetric cubic tensor for any $\alpha \in \mathbb{R}$, we can derive a one-parameter family of conjugate connection manifolds:

$$ \left\{(M, {}^{D}g, {}^{D}C^{\alpha}) \equiv (M, {}^{D}g, {}^{D}\nabla^{-\alpha}, ({}^{D}\nabla^{-\alpha})^* = {}^{D}\nabla^{\alpha})\right\}_{\alpha\in\mathbb{R}}. \tag{60} $$

In the remainder, we use the shortcut (M, D) to denote the divergence-induced information manifold $(M, {}^{D}g, {}^{D}\nabla, {}^{D}\nabla^*)$. Notice that it follows from construction that:

$$ {}^{D}\nabla^* = {}^{D^*}\nabla. \tag{61} $$

3.7. *Dually Flat Manifolds (Bregman Geometry):* $(M, F) \equiv (M,\ {}^{B_F}g,\ {}^{B_F}\nabla,\ {}^{B_F}\nabla^* =\ {}^{B_{F^*}}\nabla)$

We consider dually flat manifolds that satisfy asymmetric Pythagorean theorems. These flat manifolds can be obtained from a canonical Bregman divergence.

Consider a strictly convex smooth function $F(\theta)$ called a potential function, with $\theta \in \Theta$ where Θ is an open convex domain. Notice that the function convexity does not change by an affine transformation. We associate to the potential function F a corresponding Bregman divergence (parameter divergence):

$$ B_F(\theta : \theta') := F(\theta) - F(\theta') - (\theta - \theta')^\top \nabla F(\theta'). \tag{62} $$

We write also the Bregman divergence between point P and point Q as $D(P : Q) := B_F(\theta(P) : \theta(Q))$, where $\theta(P)$ denotes the coordinates of a point P.

The information-geometric structure induced by a Bregman generator is $(M, {}^{F}g, {}^{F}C) := (M, {}^{B_F}g, {}^{B_F}C)$ with:

$$ {}^{F}g := {}^{B_F}g = -\left[\partial_i\partial_j B_F(\theta : \theta')|_{\theta'=\theta}\right] = \nabla^2 F(\theta), \tag{63} $$

$$ {}^{F}\Gamma := {}^{B_F}\Gamma_{ij,k}(\theta) = 0, \tag{64} $$

$$ {}^{F}C_{ijk} := {}^{B_F}C_{ijk} = \partial_i\partial_j\partial_k F(\theta). \tag{65} $$

Here, we define a Bregman generator as a proper, lower semi-continuous, and strictly convex and C^3 differentiable real-valued function.

Since all coefficients of the Christoffel symbols vanish (Equation (64)), the information manifold is ${}^{F}\nabla$-flat. The Levi–Civita connection ${}^{LC}\nabla$ is obtained from the metric tensor ${}^{F}g$ (usually not flat), and we get the conjugate connection $({}^{F}\nabla)^* = {}^{F}\nabla^1$ from $(M,\ {}^{F}g,\ {}^{F}C)$.

The Legendre–Fenchel transformation yields the convex conjugate F^* that is interpreted as the dual potential function:

$$ F^*(\eta) := \sup_{\theta\in\Theta}\{\theta^\top\eta - F(\theta)\}. \tag{66} $$

A function f is lower semicontinous (lsc) at x_0 iff $f(x_0) \leq \lim_{x\to x_0} \inf f(x)$. A function f is lsc if it is lsc at x for all x in the function domain. The following theorem states that the conjugation of lower semicontinuous and convex functions is an involution:

Theorem 5 (Fenchel–Moreau biconjugation [35])**.** *If F is a lower semicontinuous and convex function, then its Legendre–Fenchel transformation is involutive: $(F^*)^* = F$ (biconjugation).*

In a dually flat manifold, there exists two global dual affine coordinate systems $\eta = \nabla F(\theta)$ and $\theta = \nabla F^*(\eta)$, and therefore the manifold can be covered by a single chart. Thus if a probability family belongs to an exponential family then its natural parameters cannot belong to, say, a spherical space (that requires at least two charts).

We have the Crouzeix [36] identity relating the Hessians of the potential functions:

$$\nabla^2 F(\theta) \nabla^2 F^*(\eta) = I, \tag{67}$$

where I denote the $D \times D$ identity matrix. This Crouzeix identity reveals that $B = \{\partial_i\}_i$ and $B^* = \{\partial^j\}_j$ are the primal and reciprocal basis, respectively.

The Bregman divergence can be reinterpreted using Young–Fenchel (in)equality as the canonical divergence A_{F,F^*} [37]:

$$B_F(\theta : \theta') = A_{F,F^*}(\theta : \eta') = F(\theta) + F^*(\eta') - \theta^\top \eta' = A_{F^*,F}(\eta' : \theta). \tag{68}$$

The dual Bregman divergence $B_F{}^*(\theta : \theta') := B_F(\theta' : \theta) = B_{F^*}(\eta : \eta')$ yields

$$^F g^{ij}(\eta) = \partial^i \partial^j F^*(\eta), \quad \partial^l :=: \frac{\partial}{\partial \eta^l} \tag{69}$$

$$^F \Gamma^{*ijk}(\eta) = 0, \quad ^F C^{ijk} = \partial^i \partial^j \partial^k F^*(\eta) \tag{70}$$

Thus the information manifold is both $^F\nabla$-flat and $^F\nabla^*$-flat: This structure is called a dually flat manifold (DFM). In a DFM, we have two global affine coordinate systems $\theta(\cdot)$ and $\eta(\cdot)$ related by the Legendre–Fenchel transformation of a pair of potential functions F and F^*. That is, $(M, F) \equiv (M, F^*)$, and the dual atlases are $\mathcal{A} = \{(M, \theta)\}$ and $\mathcal{A}^* = \{(M, \eta)\}$.

In a dually flat manifold, any pair of points P and Q can either be linked using the ∇-geodesic (that is θ-straight) or the ∇^*-geodesic (that is η-straight). In general, there are $2^3 = 8$ types of geodesic triangles in a dually flat manifold.

On a Bregman manifold, the primal parallel transport of a vector does not change the contravariant vector components, and the dual parallel transport does not change the covariant vector components. Because the dual connections are flat, the dual parallel transports are path-independent.

Moreover, the dual Pythagorean theorems [38] illustrated in Figure 6 holds. Let $\gamma(P, Q) = \gamma_\nabla(P, Q)$ denote the ∇-geodesic passing through points P and Q, and $\gamma^*(P, Q) = \gamma_{\nabla^*}(P, Q)$ denote the ∇^*-geodesic passing through points P and Q. Two curves γ_1 and γ_2 are orthogonal at point $p = \gamma_1(t_1) = \gamma_2(t_2)$ with respect to the metric tensor g when $g(\dot\gamma_1(t_1), \dot\gamma_2(t_2)) = 0$.

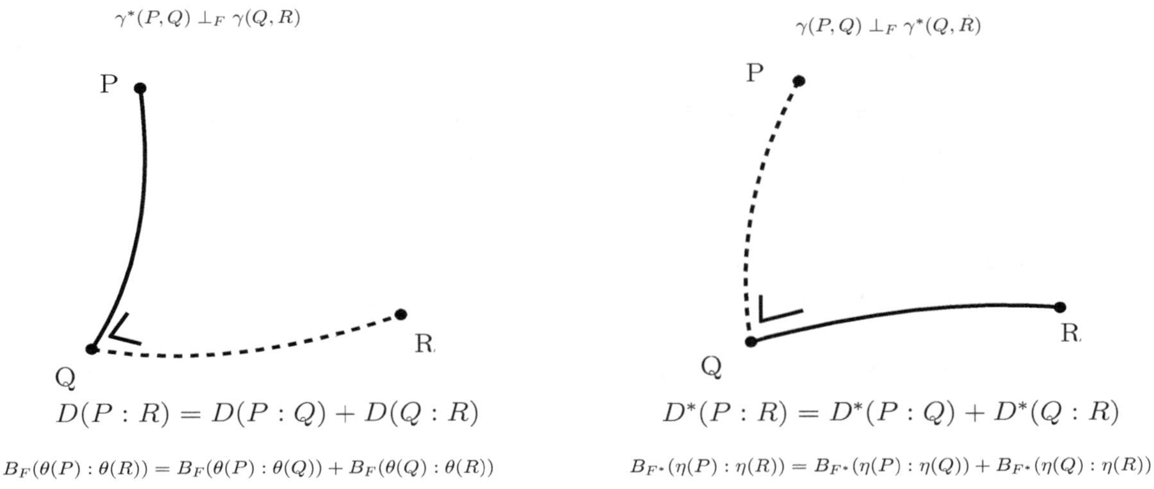

Figure 6. Dual Pythagorean theorems in a dually flat space.

Theorem 6 (Dual Pythagorean identities).

$$\gamma^*(P,Q) \perp \gamma(Q,R) \quad \Leftrightarrow \quad (\eta(P)-\eta(Q))^\top(\theta(Q)-\theta(R)) \stackrel{\Sigma}{=} (\eta_i(P)-\eta_i(Q))(\theta_i(Q)-\theta_i(R)) = 0,$$
$$\gamma(P,Q) \perp \gamma^*(Q,R) \quad \Leftrightarrow \quad (\theta(P)-\theta(Q))^\top(\eta(Q)-\eta(R)) \stackrel{\Sigma}{=} (\theta_i(P)-\theta_i(Q))^\top(\eta_i(Q)-\eta_i(R)) = 0.$$

We can define dual Bregman projections and characterize when these projections are unique: A submanifold $S \subset M$ is said ∇-flat (∇^*-flat) iff. it corresponds to an affine subspace in the θ-coordinate system (in the η-coordinate system, respectively).

Theorem 7 (Uniqueness of projections). *The ∇-projection P_S of P on S is unique if S is ∇^*-flat and minimizes the divergence $D(\theta(P):\theta(Q))$:*

$$\nabla\text{-projection:} \quad P_S = \arg\min_{Q\in S} D(\theta(P):\theta(Q)). \tag{71}$$

The dual ∇^*-projection P_S^* is unique if $M \subseteq S$ is ∇-flat and minimizes the divergence $D(\theta(Q):\theta(P))$:

$$\nabla^*\text{-projection:} \quad P_S^* = \arg\min_{Q\in S} D(\theta(Q):\theta(P)). \tag{72}$$

Let $S \subset M$ and $S' \subset M$, then we define the divergence between S and S' as

$$D(S:S') := \min_{s\in S, s'\in S'} D(s:s'). \tag{73}$$

When S is a ∇-flat submanifold and S' ∇^*-flat submanifold, the divergence $D(S:S')$ between submanifold S and submanifold S' can be calculated using the method of alternating projections [2]. Let us remark that Kurose [39] reported a Pythagorean theorem for dually constant curvature manifolds that generalizes the Pythagorean theorems of dually flat spaces.

We shall concisely explain the space of Bregman spheres explained in details in [40]. Let D denote the dimension of Θ. We define the lifting of primal coordinates θ to the primal potential function $\mathcal{F} = \{\hat{\theta} = (\theta, \theta_{D+1} = F(\theta)) \ : \ \theta \in \Theta\}$ using an extra dimension θ_{D+1}. A Bregman ball Σ

$$\mathrm{Ball}_F(C:r) := \{P \text{ such that } F(\theta(P)) + F^*(\eta(C)) - \langle\theta(P), \eta(C)\rangle \le r\} \tag{74}$$

can then be lifted to $\mathcal{F}$: $\hat{\Sigma} = \{\hat{\theta}(P) \ : \ P \in \sigma\}$. The boundary Bregman sphere $\sigma = \partial\Sigma$ is lifted to $\partial\hat{\Sigma} = \hat{\sigma}$, and the lifted points are all supported by a supporting $(D+1)$-dimensional hyperplane (of dimension D):

$$H_{\hat{\sigma}} : \theta_{D+1} = \langle\theta - \theta(C), \eta(C)\rangle + F(\theta(C)) + r. \tag{75}$$

Let $H_{\hat{\sigma}}^-$ denotes the halfspaces bounded by $H_{\hat{\sigma}}$ and containing $\hat{\theta}(C) = (\theta(C), F(\theta(C)))$. A point P belongs to a Bregman ball Σ iff $\hat{\theta(P)} \in H_{\hat{\sigma}}^-$, see [40]. Reciprocally, a $(D+1)$-dimensional hyperplane $H : \theta_{D+1} = \langle\theta, \eta_a\rangle + b$ cutting the potential function $\mathcal{F}$ yields a Bregman sphere σ_H of center C with $\theta(C) = \nabla F^*(\eta_a)$ and radius $r = \langle\nabla F^*(\eta_a), \eta_a\rangle - F(\theta_a) + b = F^*(\eta_a) + b$, where $\theta_a = \nabla F^*(\eta_a)$. It follows that the intersection of k Bregman balls is a $(D-k)$-dimensional Bregman ball, and that a Bregman sphere can be defined by $D+1$ points in general position since an hyperplane in the augmented space is defined by $D+1$ points. We can test whether a point P belongs to a Bregman ball with bounding Bregman sphere passing through $D+1$ points $P_1, \ldots, P_{D+1}$ or not by checking the sign of a $(D+2)\times(D+2)$ determinant:

$$\mathrm{InBregmanBall}_F(P_1, \ldots, P_{d+1}; P) := \mathrm{sign}\left(\begin{vmatrix} 1 & \ldots & 1 & 1 \\ \theta(P_1) & \ldots & \theta(P_{D+1}) & \theta(P) \\ F(\theta(P_1)) & \ldots & F(\theta(P_{D+1})) & F(\theta(P)) \end{vmatrix}\right). \tag{76}$$

We have:

$$\mathrm{InBregmanBall}_F(P_1,\ldots,P_{d+1};P):\begin{cases}=-1 & \Leftrightarrow \quad P\in\mathrm{InBregmanBall}_F^{\circ}(P_1,\ldots,P_{D+1};P)\\ =0 & \Leftrightarrow \quad P\in\partial\mathrm{InBregmanBall}_F(P_1,\ldots,P_{D+1};P)\\ =+1 & \Leftrightarrow \quad P\notin\mathrm{InBregmanBall}_F(P_1,\ldots,P_{D+1};P)\end{cases} \tag{77}$$

Similarly, a dual-type Bregman ball Σ^* can be defined by

$$\mathrm{Ball}_F^*(C:r):=\{P \text{ such that } F(\theta(C))+F^*(\eta(P))-\langle\theta(C),\eta(P)\rangle\le r\}, \tag{78}$$

and be lifted to the dual potential function $\mathcal{F}^*$. Notice that $\mathrm{Ball}_F^*(C:r)=\mathrm{Ball}_{F^*}(C:r)$. Figure 7 displays five concentric pairs of dual Itakura–Saito circles obtained for the separable Burg negentropy generator $F(x,y)=-\log(x)-\log(y)$ (with corresponding Bregman divergence the Itakura–Saito divergence).

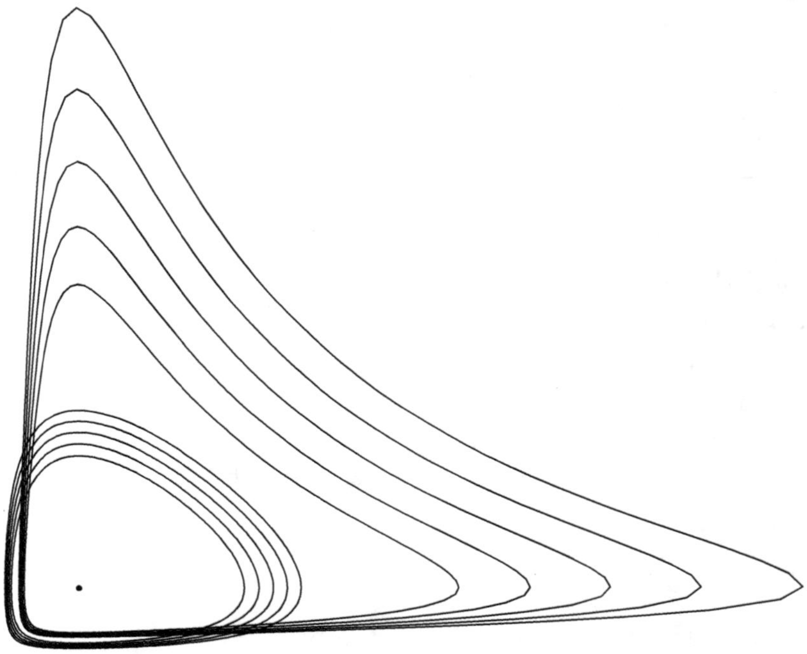

Figure 7. Five concentric pairs of dual Itakura–Saito circles.

Using the space of spheres, it is easy to design algorithms for calculating the union or intersection of Bregman spheres [40], or data-structures for proximity queries [41] (relying on the radical axis of two Bregman spheres). The Bregman spheres are considered for building Bregman Voronoi diagrams in [40,42].

The smallest enclosing Bregman ball [43,44] (SEBB) of a set of points $P_1,\ldots,P_n$ (with respective θ-coordinates $\theta_1,\ldots,\theta_n$) can also be modeled as a convex program; indeed, point P_i belongs to the lower halfspace H^- of equation $\theta_{D+1}\le\langle\eta_a,\theta\rangle+b$ (parameterized by vector $\eta_a\in\mathbb{R}^D$ and scalar $b\in\mathbb{R}$) iff $\langle\eta_a,\theta\rangle+b\ge F(\theta_i)$. Thus we seek to minimize $\min_{\eta_a,b} r=F^*(\eta_a)+b$ such that $\langle\theta_i,\eta_a\rangle+b-F(\theta_i)\ge 0$ for all $i\in\{1,\ldots,n\}$. This is a convex program since F^* is the convex conjugate of a convex generator F. When $F(\theta)=\frac{1}{2}\theta^\top\theta$ (i.e., Euclidean geometry), we recover the fact that the smallest enclosing ball of a point set in Euclidean geometry can be solved using quadratic programming [45]. Faster approximation algorithms for the smallest enclosing Bregman ball can be built based on core-sets [43].

In general, we have the following quadrilateral relation for Bregman divergences:

Property 4 (Bregman 4-parameter property [46]). *For any four points P_1, P_2, Q_1, Q_2, we have the following identity:*

$$B_F(\theta(P_1):\theta(Q_1)) + B_F(\theta(P_2):\theta(Q_2)) \quad -B_F(\theta(P_1):\theta(Q_2)) - B_F(\theta(P_2):\theta(Q_1)) \\ -(\theta(P_2)-\theta(P_1))^\top(\eta(Q_1)-\eta(Q_2)) = 0. \tag{79}$$

In summary, to define a dually flat space, we need a convex Bregman generator. When the α-geometries are neither dually flat (e.g., Cauchy manifolds [47], we may still build a dually flat structure on the manifold by considering some Bregman generator (e.g., Bregman- -Tsallis generator for the dually flat Cauchy manifold [47]). The dually flat geometry can be investigated under the wider scope of Hessian manifolds [48] which consider locally potential functions. In general, a dually flat space can be built from any smooth strictly convex generator F. For example, a dually flat geometry can be built on homogeneous cones with the characteristic function F of the cone [48]. Figure 8 illustrates several common constructions of dually flat spaces.

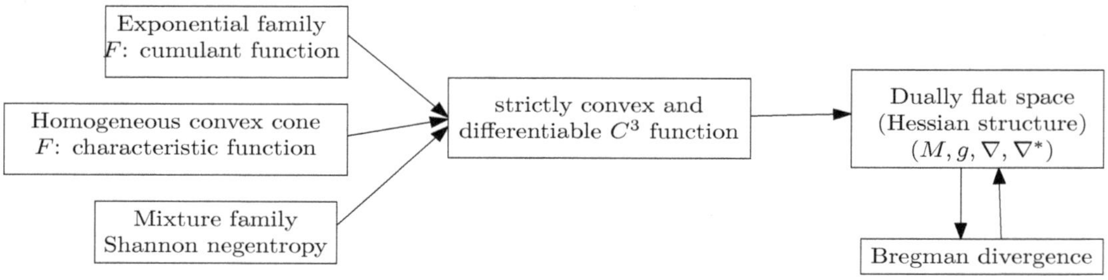

Figure 8. Common dually flat spaces associated to smooth and strictly convex generators.

3.8. Hessian α-Geometry: $(M, F, \alpha) \equiv (M, {}^Fg, {}^F\nabla^{-\alpha}, {}^F\nabla^{\alpha})$

The dually flat manifold is also called a manifold with a Hessian structure [48] induced by a convex potential function F. Since we built two dual affine connections ${}^{B_F}\nabla = {}^F\nabla$ and ${}^{B_F}\nabla^* = {}^F\nabla^* = {}^{F^*}\nabla$, we can build a family of α-geometry as follows:

$$^Fg_{ij}(\theta) = \partial_i\partial_j F(\theta), \quad {}^Fg^{ij}(\eta) = \partial^i\partial^j F(\eta), \tag{80}$$

and

$$^F\Gamma^\alpha_{ijk}(\theta) = \frac{1-\alpha}{2}\partial_i\partial_j\partial_k F(\theta), \quad {}^F\Gamma^{\alpha}_{ijk}{}^*(\eta) = {}^{F^*}\Gamma^\alpha_{ijk}(\eta) = \frac{1+\alpha}{2}\partial^i\partial^j\partial^k F^*(\eta). \tag{81}$$

Thus when $\alpha = \pm 1$, the Hessian α-geometry is dually flat since ${}^F\Gamma^1_{ijk}(\theta) = 0$ and ${}^F\Gamma^{-1}_{ijk}{}^*(\eta) = 0$.
We now consider information manifolds induced by parametric statistical models.

3.9. Expected α-Manifolds of a Family of Parametric Probability Distributions: $(\mathcal{P}, {}_{\mathcal{P}}g, {}_{\mathcal{P}}\nabla^{-\alpha}, {}_{\mathcal{P}}\nabla^{\alpha})$

Informally speaking, an expected manifold is an information manifold built on a regular parametric family of distributions. It is sometimes called "expected" manifold or "expected" geometry in the literature [49] because the components of the metric tensor g and the Amari–Chentsov cubic tensor C are expressed using statistical expectations.

Let $\mathcal{P}$ be a parametric family of probability distributions:

$$\mathcal{P} := \{p_\theta(x)\}_{\theta\in\Theta}, \tag{82}$$

with θ belonging to the open parameter space Θ. The order of the family is the dimension of its parameter space. We define the likelihood function $L(\theta; x) := p_\theta(x)$ as a function of θ, and its corresponding log-likelihood function:

$$l(\theta; x) := \log L(\theta; x) = \log p_\theta(x). \tag{83}$$

More precisely, the likelihood function is an equivalence class of functions defined modulo a positive scaling factor.

The score vector:

$$s_\theta = \nabla_\theta l = (\partial_i l)_i, \tag{84}$$

indicates the sensitivity of the likelihood $\partial_i l {:=:} \frac{\partial}{\partial \theta_i} l(\theta; x)$.

The Fisher information matrix (FIM) of $D \times D$ for $\dim(\Theta) = D$ is defined by:

$${}_{\mathcal{P}}I(\theta) := E_\theta \left[\partial_i l \partial_j l\right]_{ij} \succeq 0, \tag{85}$$

where $\succeq$ denotes the Löwner order. That is, for two symmetric positive-definite matrices A and B, $A \succeq B$ if and only if matrix $A - B$ is positive semidefinite. For regular models [16], the FIM is positive definite: ${}_{\mathcal{P}}I(\theta) \succ 0$, where $A \succ B$ if and only if matrix $A - B$ is positive-definite.

The FIM is invariant by reparameterization of the sample space $\mathcal{X}$, and covariant by reparameterization of the parameter space Θ, see [16]. That is, let $\bar{p}(x; \eta) = p(\theta(\eta); x)$. Then we have:

$$\bar{I}(\eta) = \left[\frac{\partial \theta_i}{\partial \eta_j}\right]_{ij}^\top \times I(\theta(\eta)) \times \left[\frac{\partial \theta_i}{\partial \eta_j}\right]_{ij}. \tag{86}$$

Matrix $J_{ij} = \left[\frac{\partial \theta_i}{\partial \eta_j}\right]_{ij}$ is the Jacobian matrix.

Let us give illustrate the covariance of the Fisher information matrix with the following example:

Example 1. *Consider the family*

$$\mathcal{N} = \left\{ p(x; \mu, \sigma) = \frac{1}{\sqrt{2\pi\sigma}} \exp(-\frac{(x-\mu)^2}{2\sigma^2})) \ : \ (\mu, \sigma) \in \mathbb{R} \times \mathbb{R}_{++} \right\} \tag{87}$$

of univariate normal distributions. The 2D parameter vector is $\lambda = (\mu, \sigma)$ with μ denoting the mean and σ the standard deviation. Another common parameterization of the normal family is $\lambda' = (\mu, \sigma^2)$. The λ' parameterization extends naturally to d-variance normal distributions with $\lambda' = (\mu, \Sigma)$, where Σ denotes the covariance matrix (with $\Sigma = \sigma^2$ when $d = 1$). For multivariate normal distributions, the λ-parameterization can be interpreted as $\lambda = (\mu, L^\top)$ where $L^\top$ is the upper triangular matrix in the Cholesky decomposition (when $d = 1$, $L^\top = \sigma$). We have the following Fisher information matrices in the λ-parameterization and λ'-parameterization:

$$I_\lambda(\lambda) = \begin{bmatrix} \frac{1}{\lambda_2^2} & 0 \\ 0 & \frac{2}{\lambda_2^2} \end{bmatrix} = \begin{bmatrix} \frac{1}{\sigma^2} & 0 \\ 0 & \frac{2}{\sigma^2} \end{bmatrix} \tag{88}$$

and

$$I_{\lambda'}(\lambda') = \begin{bmatrix} \frac{1}{\lambda_2} & 0 \\ 0 & \frac{1}{2\lambda_2^2} \end{bmatrix} = \begin{bmatrix} \frac{1}{\sigma^2} & 0 \\ 0 & \frac{1}{2\sigma^4} \end{bmatrix} \tag{89}$$

Since the FIM is covariant, we have the following the change of transformation:

$$I_{\lambda'}(\lambda') = J_{\lambda,\lambda'}^\top \times I_\lambda(\lambda(\lambda')) \times J_{\lambda,\lambda'}, \tag{90}$$

with

$$J_{\lambda',\lambda} = \begin{bmatrix} 1 & 0 \\ 0 & 2\sigma \end{bmatrix} \tag{91}$$

Thus we check that

$$I_\lambda(\lambda) = \begin{bmatrix} 1 & 0 \\ 0 & 2\sigma \end{bmatrix} \begin{bmatrix} \frac{1}{\sigma^2} & 0 \\ 0 & \frac{1}{2\sigma^4} \end{bmatrix} \begin{bmatrix} 1 & 0 \\ 0 & 2\sigma \end{bmatrix} = \begin{bmatrix} \frac{1}{\sigma^2} & 0 \\ 0 & \frac{2}{\sigma^2} \end{bmatrix} \tag{92}$$

Notice that the infinitesimal length elements are invariant: $\mathrm{d}s_\lambda = \mathrm{d}s_{\lambda'}$.

As a corollary, notice that we can recognize the Euclidean metric in any other coordinate system if the metric tensor g can be written $J_{\lambda,\lambda'}^{\top} J_{\lambda,\lambda'}$. For example, the Riemannian geometry induced by a dually flat space with a separable potential function is Euclidean [50].

In statistics, the FIM plays a role in the attainable precision of unbiased estimators. For any unbiased estimator, the Cramér–Rao lower bound [51] on the variance of the estimator is:

$$\mathrm{Var}_\theta[\hat{\theta}_n(X)] \succeq \frac{1}{n} {}_{\mathcal{P}}I^{-1}(\theta). \tag{93}$$

Figure 9 illustrates the Cramér–Rao lower bound (CRLB) for the univariate distributions: At regular grid locations (μ, σ) of the upper space of normal parameters, we repeat 200 runs (trials) of estimating the normal parameters $\widehat{(\mu, \sigma)}$ using the MLE on 100 iid samples $x_1, \ldots, x_n \sim N(\mu, \sigma)$. The sample mean and the sample covariance matrix are calculated for the number of trials and displayed as back ellipses. The Fisher information matrix is plotted as red ellipses at the grid locations: the red ellipses have semi-axes parallel to the coordinate system since the parameters μ and σ are orthogonal (diagonal FIM). This is not true anymore for the sample covariance matrix of the MLE estimator, and the centers of the sample covariance matrices deviate from the grid locations.

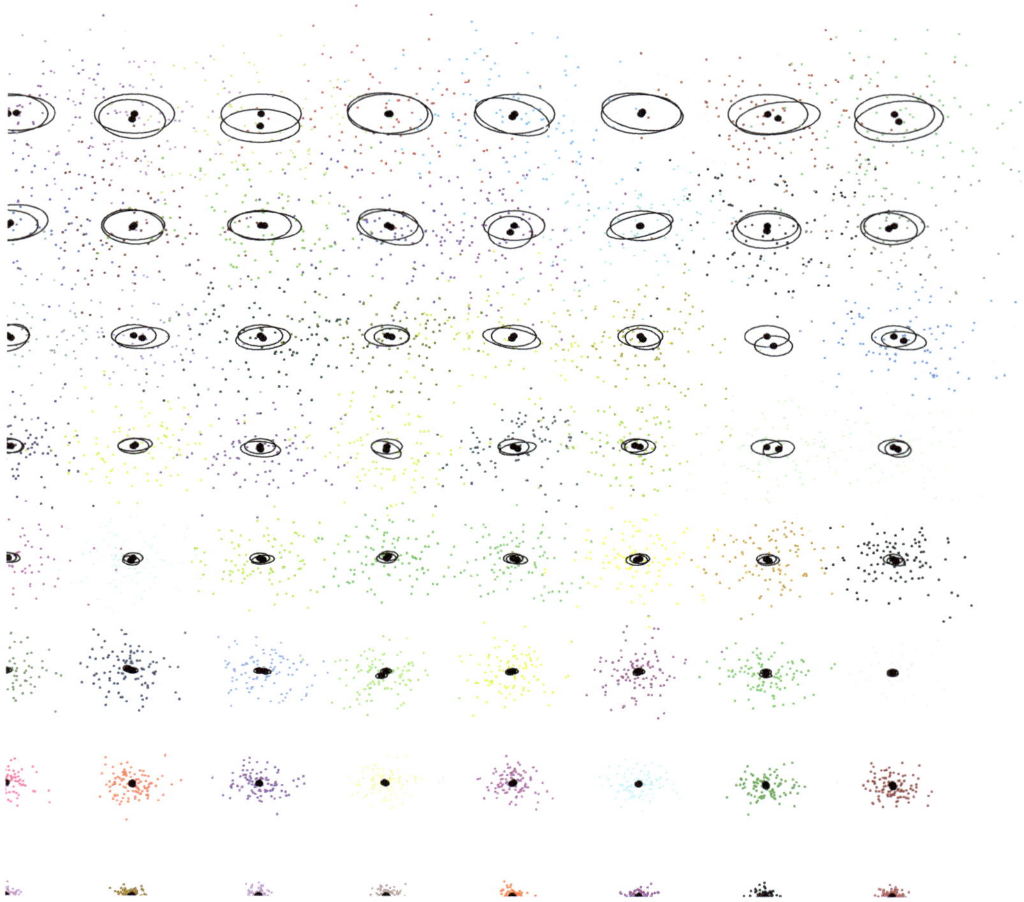

Figure 9. Visualizing the Cramér–Rao lower bound: the red ellipses display the Fisher information matrix of normal distributions $N(\mu, \sigma^2)$ at grid locations. The black ellipses are sample covariance matrices centered at the sample means calculated by repeating 200 runs of sampling 100 iid variates for the normal parameters of the grid.

We report the expression of the FIM for two important generic parametric family of probability distributions: (1) an exponential family (with its prominent multivariate normal family), and (2) a mixture family.

Example 2 (FIM of an exponential family $\mathcal{E}$). *An exponential family [52] $\mathcal{E}$ is defined for a sufficient statistic vector $t(x) = (t_1(x), \ldots, t_D(x))$, and an auxiliary carrier measure $k(x)$ by the following canonical density:*

$$\mathcal{E} = \left\{ p_\theta(x) = \exp\left(\sum_{i=1}^{D} t_i(x)\theta_i - F(\theta) + k(x) \right) \text{ such that } \theta \in \Theta \right\}, \tag{94}$$

where F is the strictly convex cumulant function (also called log-normalizer, and log partition function or free energy in statistical mechanics). Exponential families include the Gaussian family, the Gamma and Beta families, the probability simplex Δ, etc. The FIM of an exponential family is given by:

$$_{\mathcal{E}}I(\theta) = \mathrm{Cov}_{X\sim p_\theta(x)}[t(x)] = \nabla^2 F(\theta) = (\nabla^2 F^*(\eta))^{-1} \succ 0. \tag{95}$$

Indeed, under mild conditions [2], we have $I(\theta) = -E_{p_\theta}[\nabla^2 \log p_\theta(x)]$. Since $-\nabla^2 \log p_\theta(x) = \nabla^2 F(\theta)$, it follows that $_{\mathcal{E}}I(\theta) = \nabla^2 F(\theta)$. Natural parameters beyond vector types can also be used in the canonical decomposition of the density of an exponential family. For example, we may use a matrix type for defining the zero-centered multivariate Gaussian family or the Wishart family, a complex numbers for defining the complex-valued Gaussian distribution family, etc. We then replace the term $\sum_{i=1}^{D} t_i(x)\theta_i$ in Equation (94) by an inner product defined for the natural parameter type (e.g., dot product for vectors, matrix product trace for matrices, etc.). Furthermore, natural parameters can be of compound types: For example, the multivariate Gaussian distribution can be written using $\theta = (\theta_v, \theta_M)$ where θ_v is a vector part and θ_M a matrix part, see [52].

Let $\Sigma = [\sigma_{ij}]$ denote the covariance matrix and $\Sigma^{-1} = [\sigma^{ij}]$ the precision matrix of a multivariate normal distribution. The Fisher information matrix of the multivariate Gaussian [53,54] $N(\mu, \Sigma)$ is given by

$$I(\mu, \Sigma) = \begin{array}{c} \begin{array}{cc} \mu \;\;|\; & \Sigma = [\sigma_{ij}] \end{array} \\ \begin{bmatrix} \sigma^{ij} & 0 \\ 0 & \sigma^{il}\sigma^{jk} + \sigma^{ik}\sigma^{jl} \end{bmatrix} \end{array} \begin{array}{c} \mu \\ \Sigma = [\sigma_{kl}] \end{array} \tag{96}$$

Notice that the lower right block matrix is a 4D tensor of dimension $d \times d \times d \times d$. The zero subblock matrices in the FIM indicate that the parameters μ and Σ are orthogonal to each other. In particular, when $d = 1$, since $\sigma^{11} = \frac{1}{\sigma^2}$, we recover the Fisher information matrix of the univariate Gaussian:

$$I(\mu, \Sigma) = \begin{bmatrix} \frac{1}{\sigma^2} & 0 \\ 0 & \frac{1}{2\sigma^4} \end{bmatrix} \tag{97}$$

We refer to [55] for the FIM of a Gaussian distribution using other canonical parameterizations (natural/expectation parameters of exponential family).

Example 3 (FIM of a mixture family $\mathcal{M}$). *A mixture family is defined for $D+1$ functions $F_1, \ldots, F_D$ and C as:*

$$\mathcal{M} = \left\{ p_\theta(x) = \sum_{i=1}^{D} \theta_i F_i(x) + C(x) \text{ such that } \theta \in \Theta \right\}, \tag{98}$$

where the functions $\{F_i(x)\}_i$ are linearly independent on the common support $\mathcal{X}$ and satisfying $\int F_i(x)\mathrm{d}\mu(x) = 0$. Function C is such that $\int C(x)\mathrm{d}\mu(x) = 1$. Mixture families include statistical mixtures with prescribed component distributions and the probability simplex Δ. The FIM of a mixture family is given by:

$$_{\mathcal{M}}I(\theta) = E_{X\sim p_\theta(x)}\left[\frac{F_i(x)F_j(x)}{(p_\theta(x))^2} \right] = \int_{\mathcal{X}} \frac{F_i(x)F_j(x)}{p_\theta(x)} \mathrm{d}\mu(x) \succ 0. \tag{99}$$

The family of Gaussian mixture model (GMM) with prescribed component distributions (i.e., convex weight combinations of $D+1$ Gaussian densities) form a mixture family [56].

Notice that the probability simplex of discrete distributions can be both modeled as an exponential family or a mixture family [2].

The expected α-geometry is built from the expected dual $\pm\alpha$-connections. The Fisher "information metric" tensor is built from the FIM as follows:

$$ {}_{\mathcal{P}}g(u,v) := (u)_\theta^\top \, {}_{\mathcal{P}}I(\theta)\,(v)_\theta \tag{100}$$

The expected exponential connection and expected mixture connection are given by

$$ {}^{e}_{\mathcal{P}}\nabla := E_\theta\left[(\partial_i\partial_j l)(\partial_k l)\right], \tag{101}$$
$$ {}^{m}_{\mathcal{P}}\nabla := E_\theta\left[(\partial_i\partial_j l + \partial_i l\partial_j l)(\partial_k l)\right]. \tag{102}$$

The dualistic structure is denoted by $(\mathcal{P}, {}_{\mathcal{P}}g, {}^{m}_{\mathcal{P}}\nabla, {}^{e}_{\mathcal{P}}\nabla)$ with Amari–Chentsov cubic tensor called the skewness tensor:

$$ C_{ijk} := E_\theta\left[\partial_i l\partial_j l\partial_k l\right]. \tag{103}$$

It follows that we can build a one-family of expected information α-manifolds:

$$ \{(\mathcal{P}, {}_{\mathcal{P}}g, {}_{\mathcal{P}}\nabla^{-\alpha}, {}_{\mathcal{P}}\nabla^{+\alpha})\}_{\alpha\in\mathbb{R}}, \tag{104}$$

with

$$ {}_{\mathcal{P}}\Gamma^{\alpha}{}_{ij,k}(\theta) := E_\theta\left[\partial_i\partial_j l\partial_k l\right] + \frac{1-\alpha}{2}C_{ijk}(\theta), \tag{105}$$
$$ = E_\theta\left[\left(\partial_i\partial_j l + \frac{1-\alpha}{2}\partial_i l\partial_j l\right)(\partial_k l)\right]. \tag{106}$$

The Levi–Civita metric connection is recovered as follows:

$$ {}_{\mathcal{P}}\bar{\nabla} = \frac{{}_{\mathcal{P}}\nabla^{-\alpha} + {}_{\mathcal{P}}\nabla^{\alpha}}{2} = {}^{\mathrm{LC}}_{\mathcal{P}}\nabla := {}^{\mathrm{LC}}\nabla({}_{\mathcal{P}}g) \tag{107}$$

The α-Riemann–Christoffel curvature tensor is:

$$ {}_{\mathcal{P}}R_{ijkl} = \partial_i\Gamma^{\alpha}_{jk,l} - \partial_j\Gamma^{\alpha}_{ik,l} + g^{rs}\left(\Gamma^{\alpha}_{ik,r}\Gamma^{\alpha}_{js,l} - \Gamma^{\alpha}_{jk,r}\Gamma^{\alpha}_{is,l}\right), \tag{108}$$

with $R^{\alpha}_{ijkl} = -R^{-\alpha}_{ijlk}$. We check that the expected $\pm\alpha$-connections are coupled with the metric: $\partial_i g_{jk} = \Gamma^{\alpha}_{ij,k} + \Gamma^{-\alpha}_{ik,j}$.

In case of an exponential family $\mathcal{E}$ or a mixture family $\mathcal{M}$ equipped with the dual exponential/mixture connection, we get dually flat manifolds (Bregman geometry).

Indeed, for the exponential/mixture families, it is easy to check that the Christoffel symbols of ∇^e and ∇^m vanish:

$$ {}^{e}_{\mathcal{M}}\Gamma = {}^{m}_{\mathcal{M}}\Gamma = {}^{e}_{\mathcal{E}}\Gamma = {}^{m}_{\mathcal{E}}\Gamma = 0. \tag{109}$$

3.10. Criteria for Statistical Invariance

So far we have explained how to build an information manifold (or information α-manifold) from a pair of conjugate connections. Then we reported two ways to obtain such a pair of conjugate connections: (1) from a parametric divergence, or (2) by using the predefined expected exponential/mixture connections. We now ask the following question: which information manifold makes sense in Statistics? We can refine the question as follows:

- Which metric tensors g make sense in statistics?
- Which affine connections ∇ make sense in statistics?
- Which statistical divergences make sense in statistics (from which we can get the metric tensor and dual connections)?

By definition, an invariant metric tensor g shall preserve the inner product under important statistical mappings called Markov embeddings. Informally, we embed Δ_D into $\Delta_{D'}$ with $D' > D$ and the induced metric should be preserved (see [2], page 62).

Theorem 8 (Uniqueness of Fisher information metric [57,58]). *The Fisher information metric is the unique invariant metric tensor under Markov embeddings up to a scaling constant.*

A D-dimensional parameter (discrete) divergence satisfies the information monotonicity if and only if:

$$D(\theta_{\bar{\mathcal{A}}} : \theta'_{\bar{\mathcal{A}}}) \leq D(\theta : \theta') \tag{110}$$

for any coarse-grained partition $\mathcal{A} = \{\mathcal{A}_i\}_{i=1}^{E}$ of $[D] = \{1, \ldots, D\}$ ($\mathcal{A}$-lumping [59]) with $E \leq D$, where $\theta^i_{\bar{\mathcal{A}}} = \sum_{j \in \mathcal{A}_i} \theta^j$ for $i \in [E]$. This concept of coarse-graining is illustrated in Figure 10. This information monotonicity property could be renamed as the "distance coarse-binning inequality property."

p_1	p_2	p_3	p_4	p_5	p_6	p_7	p_8	p

coarse graining

$p_1 + p_2$	$p_3 + p_4 + p_5$	p_6	$p_7 + p_8$	$p_{\mathcal{A}}$

Figure 10. A divergence satisfies the property of information monotonicity iff $D(\theta_{\bar{\mathcal{A}}} : \theta'_{\bar{\mathcal{A}}}) \leq D(\theta : \theta')$. Here, parameter θ represents a discrete distribution.

A separable divergence $D(\theta_1 : \theta_2)$ is a divergence that can be expressed as the sum of elementary scalar divergences $d(x : y)$:

$$D(\theta_1 : \theta_2) := \sum_i d(\theta_1^i : \theta_2^j). \tag{111}$$

For example, the squared Euclidean distance $D(\theta_1 : \theta_2) = \sum_i (\theta_1^i - \theta_2^i)^2$ is a separable divergence for the scalar Euclidean divergence $d(x : y) = (x - y)^2$. The Euclidean distance $D_E(\theta_1, \theta_2) = \sqrt{\sum_i (\theta_1^i - \theta_2^i)^2}$ is not separable because of the square root operation.

The only invariant and decomposable divergences when $D > 1$ are f-divergences [60] defined for a convex functional generator f:

$$I_f(\theta : \theta') := \sum_{i=1}^{D} \theta_i f\left(\frac{\theta'_i}{\theta_i}\right) \geq f(1), \quad f(1) = 0. \tag{112}$$

The standard f-divergences are defined for f-generators satisfying $f'(1) = 0$ (choose $f_\lambda(u) := f(u) + \lambda(u - 1)$ since $I_{f_\lambda} = I_f$), and $f''(u) = 1$ (scale fixed).

Statistical f-divergences are invariant [61] under one-to-one/sufficient statistic transformations $y = t(x)$ of sample space: $p(x; \theta) = q(y(x); \theta)$:

$$\begin{aligned} I_f[p(x;\theta) : p(x;\theta')] &= \int_{\mathcal{X}} p(x;\theta) f\left(\frac{p(x;\theta')}{p(x;\theta)}\right) \mathrm{d}\mu(x), \\ &= \int_{\mathcal{Y}} q(y;\theta) f\left(\frac{q(y;\theta')}{q(y;\theta)}\right) \mathrm{d}\mu(y), \\ &= I_f[q(y;\theta) : q(y;\theta')]. \end{aligned}$$

The dual f-divergences for reference duality is

$$I_f{}^*[p(x;\theta) : p(x;\theta')] = I_f[p(x;\theta') : p(x;\theta)] = I_{f^\diamond}[p(x;\theta) : p(x;\theta')] \tag{113}$$

for the standard conjugate f-generator (diamond $f^\diamond$ generator) with:

$$f^\diamond(u) := uf\left(\frac{1}{u}\right). \tag{114}$$

One can check that $f^\diamond$ is a standard f-generator when f is standard.

Let us report some common examples of f-divergences:

- The family of α-divergences:

$$I_\alpha[p:q] := \frac{4}{1-\alpha^2}\left(1 - \int p^{\frac{1-\alpha}{2}}(x) q^{\frac{1+\alpha}{2}}(x) \mathrm{d}\mu(x)\right), \tag{115}$$

  obtained for $f(u) = \frac{4}{1-\alpha^2}(1 - u^{\frac{1+\alpha}{2}})$. The α-divergences include:

  - The Kullback–Leibler when $\alpha \to 1$:

$$\mathrm{KL}[p:q] = \int p(x) \log \frac{p(x)}{q(x)} \mathrm{d}\mu(x), \tag{116}$$

    for $f(u) = -\log u$.
  - The reverse Kullback–Leibler $\alpha \to -1$:

$$\mathrm{KL}^*[p:q] := \int q(x) \log \frac{q(x)}{p(x)} \mathrm{d}\mu(x) = \mathrm{KL}[q:p], \tag{117}$$

    for $f(u) = u \log u$.
  - The symmetric squared Hellinger divergence:

$$H^2[p:q] := \int \left(\sqrt{p(x)} - \sqrt{q(x)}\right)^2 \mathrm{d}\mu(x), \tag{118}$$

    for $f(u) = (\sqrt{u} - 1)^2$ (corresponding to $\alpha = 0$)
  - The Pearson and Neyman chi-squared divergences [62], etc.
- The Jensen–Shannon divergence:

$$\mathrm{JS}[p:q] := \frac{1}{2}\int \left(p(x) \log \frac{2p(x)}{p(x)+q(x)} + q(x) \log \frac{2q(x)}{p(x)+q(x)}\right) \mathrm{d}\mu(x), \tag{119}$$

  for $f(u) = -(u+1)\log\frac{1+u}{2} + u \log u$.
- The Total Variation

$$\mathrm{TV}[p:q] := \frac{1}{2}\int |p(x) - q(x)| \, \mathrm{d}\mu(x), \tag{120}$$

  for $f(u) = \frac{1}{2}|u-1|$. The total variation distance is the only metric f-divergence (up to a scaling factor).

The f-topology is the topology generated by open f-balls, open balls with respect to f-divergences. A topology T is said to be stronger than a topology T' if T contains all the open sets of T'. Csiszar's theorem [63] states that when $|\alpha| < 1$, the α-topology is equivalent to the topology induced by the total variation metric distance. Otherwise, the α-topology is stronger than the TV topology.

Let us state an important feature of f divergences:

Theorem 9. *The f-divergences are invariant by diffeomorphisms $m(x)$ of the sample space $\mathcal{X}$: Let $Y = m(X)$, and $X_i \sim p_i$ with $Y_i = m(X_i) \sim q_i$. Then we have $I_f[q_1 : q_2] = I_f[p_1 : p_2]$.*

Example 4. *Consider the exponential distributions and the Rayleigh distributions which are related by:*

$$X \sim \text{Exponential}(\lambda) \Leftrightarrow Y = m(X) = \sqrt{X} \sim \text{Rayleigh}\left(\sigma = \frac{1}{\sqrt{2\lambda}}\right).$$

The densities of the exponential distributions are defined by

$$p_\lambda(x) = \lambda \exp(-\lambda x) \text{ with support } \mathcal{X} = [0, \infty),$$

and the densities of the Rayleigh distributions are defined by

$$q_\sigma(x) = \frac{x}{\sigma^2} \exp\left(-\frac{x^2}{2\sigma^2}\right) \text{ with support } \mathcal{X} = [0, \infty).$$

We have

$$D_{\mathrm{KL}}\left[q_{\sigma_1} : q_{\sigma_2}\right] = \log\left(\frac{\lambda_2^2}{\lambda_1^2}\right) + \frac{\sigma_1^2 - \sigma_2^2}{\sigma_2^2}.$$

It follows that

$$\begin{aligned} D_{\mathrm{KL}}\left[p_{\lambda_1} : p_{\lambda_2}\right] &= D_{\mathrm{KL}}\left[q_{\frac{1}{\sqrt{2\lambda_1}}} : q_{\frac{1}{\sqrt{2\lambda_1}}}\right] \\ &= \log \frac{2\lambda_1}{2\lambda_2} + 2\lambda_2\left(\frac{1}{2\lambda_1} - \frac{1}{\lambda_2}\right) \\ &= \log\left(\frac{\lambda_1}{\lambda_2}\right) + \frac{\lambda_2}{\lambda_1} - 1. \end{aligned}$$

A remarkable property is that invariant standard f-divergences yield the Fisher information matrix and the α-connections. Indeed, the invariant standard f-divergences is related infinitesimally to the Fisher metric as follows:

$$I_f[p(x;\theta) : p(x;\theta + \mathrm{d}\theta)] = \int p(x;\theta) f\left(\frac{p(x;\theta+\mathrm{d}\theta)}{p(x;\theta)}\right) \mathrm{d}\mu(x) \quad (121)$$

$$\stackrel{\Sigma}{=} \frac{1}{2} {}_F g_{ij}(\theta) \mathrm{d}\theta^i \mathrm{d}\theta^j. \quad (122)$$

A statistical parameter divergence D on a parametric family of distributions $\mathcal{P}$ yields an equivalent parameter divergence ${}_\mathcal{P}D$:

$${}_\mathcal{P}D(\theta : \theta') := D[p(x;\theta) : p(x;\theta')]. \quad (123)$$

Thus we can build the information manifold induced by this parameter divergence ${}_\mathcal{P}D(\cdot : \cdot)$. For ${}_\mathcal{P}D(\cdot : \cdot) = I_f[\cdot : \cdot]$, the induced ± 1-divergence connections ${}_\mathcal{P}^{I_f}\nabla := {}^{\mathcal{P}I_f}\nabla$ and ${}_\mathcal{P}^{(I_f)^*}\nabla := {}^{\mathcal{P}I_f^*}\nabla$ are precisely the expected $\pm\alpha$-connections (derived from the exponential/mixture connections) with:

$$\alpha = 2f'''(1) + 3. \quad (124)$$

Thus the invariant connections which coincide with the connections induced by the invariant statistical divergences are the expected α-connections. Note that the curvature of an expected α-connection depends both on α and on the considered statistical model [64].

3.11. Fisher–Rao Expected Riemannian Manifolds: $(\mathcal{P}, {}_\mathcal{P}g)$

Historically, a first manifold modeling of a regular parametric family of distributions $\mathcal{P} = \{p_\theta(x)\}_\theta$ was to consider the Fisher Information Matrix (FIM) as the Riemannian metric tensor g (see [65,66]), with:

$$_{\mathcal{P}}I(\theta) := E_{p_\theta}\left[\partial_i l \partial_j l\right], \tag{125}$$

where $\partial_i l{:=:}\frac{\partial}{\partial\theta_i}\log p(x;\theta)$. Under some regularity conditions, we can rewrite the FIM:

$$_{\mathcal{P}}I(\theta) := -E_{p_\theta}\left[\partial_i\partial_j l\right]. \tag{126}$$

The Riemannian geodesic metric distance D_ρ is commonly called the Fisher–Rao distance:

$$D_\rho(p_{\theta_1}, p_{\theta_2}) = \int_0^1 \sqrt{\dot\gamma(t)^\top g_{\gamma(t)}\dot\gamma(t)}\mathrm{d}t, \tag{127}$$

where γ denotes the geodesic passing through $\gamma(0) = \theta_1$ and $\gamma(1) = \theta_2$. The Fisher–Rao distance can also be defined as the shortest path length: $D_\rho(p_{\theta_1}, p_{\theta_2}) = \inf_\gamma \int_0^1 \sqrt{\dot\gamma(t)^\top g_{\gamma(t)}\dot\gamma(t)}\mathrm{d}t$.

Definition 6 (Fisher–Rao distance). *The Fisher–Rao distance is the geodesic metric distance of the Fisher–Riemannian manifold $(\mathcal{P}, {}_{\mathcal{P}}g)$.*

Let us give some examples of Fisher–Riemannian manifolds:

- The Fisher–Riemannian manifold of the family of categorical distributions (also called finite discrete distributions in [2]) amount to the spherical geometry [14] (spherical manifold).
- The Fisher–Riemannian manifold of the family of bivariate location-scale families amount to hyperbolic geometry (hyperbolic manifold).
- The Fisher–Riemannian manifold of the family of location families amount to Euclidean geometry (Euclidean manifold).

The first fundamental form of the Riemannian geometry is $\mathrm{d}s^2 = \langle \mathrm{d}x, \mathrm{d}x\rangle \overset{\Sigma}{=} g_{ij}\mathrm{d}x^i\mathrm{d}x^j$ where ds denotes the line element. Let us give an example of Fisher–Rao geometry for location-scale families:

Example 5. *Consider the location-scale family induced by a symmetric probability density $f(x)$ with respect to 0 such that $\int_{\mathcal{X}} f(x)\mathrm{d}\mu(x) = 1$, $\int_{\mathcal{X}} xf(x)\mathrm{d}\mu(x) = 0$ and $\int_{\mathcal{X}} x^2 f(x)\mathrm{d}\mu(x) = 1$ (with support $\mathcal{X} = \mathbb{R}$):*

$$\mathcal{P} = \left\{p_\theta(x) = \frac{1}{\theta_2} f\left(\frac{x-\theta_1}{\theta_2}\right), \quad \theta = (\theta_1, \theta_2) \in \mathbb{R}\times\mathbb{R}_{++}\right\}. \tag{128}$$

The density $f(x)$ is called the standard density of the location-scale family, and corresponds to the parameter $\theta_0 = (0,1)$: $p_{(0,1)}(x) = f(x)$. The parameter space $\Theta = \mathbb{R}\times\mathbb{R}_{++}$ corresponds to the upper plane, and the Fisher information matrix can be structurally calculated [67] as the following diagonal matrix:

$$I(\theta) = \begin{bmatrix} a^2 & 0 \\ 0 & b^2 \end{bmatrix}, \tag{129}$$

with scalars:

$$a^2 := \int \left(\frac{f'(x)}{f(x)}\right)^2 f(x)\mathrm{d}\mu(x), \tag{130}$$

$$b^2 := \int \left(x\frac{f'(x)}{f(x)} + 1\right)^2 f(x)\mathrm{d}\mu(x). \tag{131}$$

By rescaling $\theta = (\theta_1, \theta_2)$ as $\theta' = (\theta'_1, \theta'_2)$ with $\theta'_1 = \frac{a}{b\sqrt{2}}\theta_1$ and $\theta'_2 = \theta_2$, we get the FIM with respect to θ' expressed as:

$$I(\theta') = \frac{b^2}{(\theta'_2)^2}\begin{bmatrix} 1 & 0 \\ 0 & 1 \end{bmatrix}. \tag{132}$$

We recognize that this metric is a constant time the metric of the Poincaré upper plane. Thus the Fisher–Rao manifold of a location-scale family (with symmetric standard probability density f) is isometric to the planar hyperbolic space of negative curvature $\kappa = -\frac{1}{b^2}$. In practice, the Klein non-conformal model of hyperbolic geometry is often used to implement computational geometric algorithms [20].

This Riemannian geometric structure applied to a family of parametric probability distributions was first proposed by Harold Hotelling [65] (in a handwritten note of 1929, reprinted typeset in [68]) and independently later by C. R. Rao [66] (1945, reprinted in [69]). In a similar vein, Jeffreys [70] proposed to use the volume element of a manifold as an invariant prior to 1946.

Notice that for a parametric family of probability distributions $\mathcal{P}$, the Riemannian structure $(\mathcal{P}, {}_{\mathcal{P}}g)$ coincides with the self-dual conjugate connection manifold $(\mathcal{P}, {}_{\mathcal{P}}g, {}^{I_f}_{\mathcal{P}}\nabla, {}^{I_f}_{\mathcal{P}}\nabla^*)$ induced by a symmetric f-divergence like the squared Hellinger divergence.

The exponential map $\exp_p$ at point $p \in M$ provides a way to map back a vector $v \in T_p$ to a point $\exp_p(v) \in M$ (when well-defined). The exponential map can be used to parameterize a geodesic γ with $\gamma(0) = p$ and unit tangent vector $\dot{\gamma}(0) = v$: $t \mapsto \exp_p(tv)$. For geodesically complete manifolds, the exponential map is defined everywhere.

3.12. The Monotone α-Embeddings and the Gauge Freedom of the Metric

Another common mathematically equivalent expression of the FIM [16] is given by:

$$I_{ij}(\theta) := 4\int \partial_i\sqrt{p(x;\theta)}\partial_j\sqrt{p(x;\theta)}\mathrm{d}\mu(x). \tag{133}$$

This form of the FIM is well-suited to prove that the FIM is always a positive semi-definite matrix [16] ($I(\theta) \succeq 0$). It turns out that we can define a family of equivalent representations of the FIM using the α-embedding [71] of the parametric family.

First, we define the α-representation of densities $l^\alpha(x;\theta) := k_\alpha(p(x;\theta))$ with:

$$k_\alpha(u) := \begin{cases} \frac{2}{1-\alpha}u^{\frac{1-\alpha}{2}}, & \text{if } \alpha \neq 1, \\ \log u, & \text{if } \alpha = 1. \end{cases} \tag{134}$$

The function $l^\alpha(x;\theta)$ is called the α-likelihood function. Then the α-representation of the FIM, the α-FIM for short, is expressed as:

$$I^\alpha_{ij}(\theta) := \int \partial_i l^\alpha(x;\theta)\partial_j l^{-\alpha}(x;\theta)\mathrm{d}\mu(x). \tag{135}$$

We can rewrite compactly the α-FIM, as $I^\alpha_{ij}(\theta) = \int \partial_i l^\alpha \partial_j l^{-\alpha}\mathrm{d}\mu(x)$. Expanding the α-FIM, we get:

$$I^\alpha_{ij}(\theta) = \begin{cases} \frac{1}{1-\alpha^2}\int \partial_i p(x;\theta)^{\frac{1-\alpha}{2}}\partial_j p(x;\theta)^{\frac{1+\alpha}{2}}\mathrm{d}\mu(x) & \text{for } \alpha \neq \pm 1 \\ \int \partial_i \log p(x;\theta)\partial_j p(x;\theta)\mathrm{d}\mu(x) & \text{for } \alpha \in \{-1, 1\}. \end{cases} \tag{136}$$

The 1-representation of the density is called the logarithmic representation (or e-representation), the -1-representation the mixture representation (or m-representation), and its 0-representation is called the square root representation. The set of α-scores vectors $B_\alpha := \{\partial_i l^\alpha\}_i$ are interpreted as the tangent basis vectors of the α-base B_α. Thus the FIM is α-independent.

Furthermore, the α-representation of the FIM can be rewritten under mild conditions [16] as:

$$I_{ij}^{\alpha}(\theta) = -\frac{2}{1+\alpha}\int p(x;\theta)^{\frac{1+\alpha}{2}}\partial_i\partial_j l^{\alpha}(x;\theta)\mathrm{d}\mu(x). \tag{137}$$

Since we have:

$$\partial_i\partial_j l^{\alpha}(x;\theta) = p^{\frac{1-\alpha}{2}}\left(\partial_i\partial_j l + \frac{1-\alpha}{2}\partial_i l\partial_j l\right), \tag{138}$$

it follows that:

$$I_{ij}^{\alpha}(\theta) = -\frac{2}{1+\alpha}\left(-I_{ij}(\theta) + \frac{1-\alpha}{2}I_{ij}\right) = I_{ij}(\theta). \tag{139}$$

Notice that when $\alpha = 1$, we recover the equivalent expression of the FIM (under mild conditions):

$$I_{ij}^{1}(\theta) = -E[\nabla^2 \log p(x;\theta)]. \tag{140}$$

In particular, when the family is an exponential family [52] with cumulant function $F(\theta)$ (satisfying the mild conditions), we have:

$$I(\theta) = \nabla^2 F(\theta). \tag{141}$$

Zhang [71,72] further discussed the representation/reference biduality which was confounded in the α-geometry.

Gauge freedom of the Riemannian metric tensor has been investigated under the framework of (ρ,τ)-monotone embeddings [71–73] in information geometry: let ρ and τ be two strictly increasing functions, and f a strictly convex function such that $f'(\rho(u)) = \tau(u)$ (with f^* denoting its convex conjugate). Observe that the set of strictly increasing real-valued univariate functions has a group structure for the group operation chosen as the functional composition $\circ$. Let us write $p_\theta(x) = p(x;\theta)$.

The (ρ,τ)-metric tensor ${}^{\rho,\tau}g(\theta) = [{}^{\rho,\tau}g_{ij}(\theta)]_{ij}$ can be derived from the (ρ,τ)-divergence:

$$D_{\rho,\tau}(p:q) = \int \left(f(\rho(p(x))) + f^*(\tau(q(x))) - \rho(p(x))\tau(q(x))\right)\mathrm{d}\nu(x). \tag{142}$$

We have:

$$\begin{aligned}
{}^{\rho,\tau}g_{ij}(\theta) &= \int \left(\partial_i\rho(p_\theta(x))\right)\left(\partial_j\tau(p_\theta(x))\right)\mathrm{d}\nu(x), && (143)\\
&= \int \rho'(p_\theta(x))\tau'(p_\theta(x))\left(\partial_i p_\theta(x)\right)\left(\partial_j p_\theta(x)\right)\mathrm{d}\nu(x), && (144)\\
&= \int f''(\rho(p_\theta(x)))\left(\partial_i\rho(p_\theta(x))\right)\left(\partial_j\rho(p_\theta(x))\right)\mathrm{d}\nu(x), && (145)\\
&= \int (f^*)''(\tau(p_\theta(x)))\left(\partial_i\tau(p_\theta(x))\right)\left(\partial_j\tau(p_\theta(x))\right)\mathrm{d}\nu(x). && (146)
\end{aligned}$$

3.13. Dually Flat Spaces and Canonical Bregman Divergences

We have described how to build a dually flat space from any strictly convex and smooth generator F: A Hessian structure is built from $F(\theta)$ with Riemannian Hessian metric $\nabla^2F(\theta)$, and the convex conjugate $F^*(\eta)$ (obtained by the Legendre–Fenchel duality) yields the dual Hessian structure with Riemannian Hessian metric $\nabla^2F^*(\eta)$. The dual connections ∇ and ∇^* are coupled with the metric. The connections are defined by their respective Christoffel symbols $\Gamma(\theta) = 0$ and $\Gamma^*(\eta) = 0$, showing that they are flat connections.

Conversely, it can be proved [2] that given two dually flat connections ∇ and ∇^*, we can reconstruct two dual canonical strictly convex potential functions $F(\theta)$ and $F^*(\eta)$ such that $\eta = \nabla F(\theta)$ and $\theta = \nabla F^*(\eta)$. The canonical divergence A_{F,F^*} yields the dual Bregman divergences B_F and B_{F^*}.

The only symmetric Bregman divergences are squared Mahalanobis distances M_Q^2 [40] with the Mahalanobis distance defined by:

$$M_Q(\theta,\theta') = \sqrt{(\theta'-\theta)^\top Q(\theta'-\theta)}. \tag{147}$$

Let $Q = LL^\top$ be the Cholesky decomposition of a positive-definite matrix $Q \succ 0$. It is well-known that the Mahalanobis distance M_Q amounts to the Euclidean distance on affinely transformed points:

$$\begin{aligned} M_Q^2(\theta,\theta') &= \Delta\theta^\top Q \Delta\theta, & (148)\\ &= \Delta\theta^\top LL^\top \Delta\theta, & (149)\\ &= M_I^2(L^\top\theta, L^\top\theta') = \|L^\top\theta - L^\top\theta'\|^2, & (150) \end{aligned}$$

where $\Delta\theta = \theta' - \theta$.

The squared Mahalanobis distance M_Q^2 does not satisfy the triangle inequality, but the Mahalanobis distance M_Q is a metric distance. We can convert a Mahalanobis distance M_{Q_1} into another Mahalanobis distance M_{Q_2}, and vice versa, as follows:

Proof. Let us write matrix $Q = L^\top L \succ 0$ using the Cholesky decomposition. Then we have

$$M_Q(\theta_1,\theta_2) = M_I(L^\top\theta_1, L^\top\theta_2) \Leftrightarrow M_I(\theta_1,\theta_2) = M_Q((L^\top)^{-1}\theta_1, ((L^\top)^{-1}\theta_2). \tag{151}$$

Then we have for two symmetric positive-definite matrices $Q_1 = L_1^\top L_1 \succ 0$ and $Q_2 = L_2^\top L_2 \succ 0$:

$$M_{Q_1}(\theta_1,\theta_2) = M_I(L_1^\top\theta_1, L_1^\top\theta_2) = M_{Q_2}((L_2^\top)^{-1}L_1^\top\theta_1, (L_2^\top)^{-1}L_1^\top\theta_2). \tag{152}$$

It follows that we have:

$$M_{Q_1}(\theta_1,\theta_2) = M_{Q_2}((L_2^\top)^{-1}L_1^\top\theta_1, (L_2^\top)^{-1}L_1^\top\theta_2). \tag{153}$$

□

We have $M_Q^2(\theta_1,\theta_2) = B_F(\theta_1,\theta_2)$ (Bregman divergence) with $F(\theta) = \frac{1}{2}\theta^\top Q\theta$ for a positive-definite matrix $Q \succ 0$. The convex conjugate $F^*(\eta) = \frac{1}{2}\eta^\top Q^{-1}\eta$ (with $Q^{-1} \succ 0$). We have $\eta = Q^{-1}\theta$ and $\eta = Q\theta$. We have the following identity between the dual Mahalanobis divergences M_Q^2 and $M_{Q^{-1}}^2$:

$$M_Q^2(\theta_1,\theta_2) = M_{Q^{-1}}^2(\eta_1,\eta_2). \tag{154}$$

When the Bregman generator is based on an integral, i.e., the log-normalizer $F(\theta) = \log\left(\int \exp(\langle t(x),\theta\rangle \mathrm{d}\mu(x)\right)$ for exponential families $\mathcal{E}$, or the negative Shannon entropy $F(\theta) = \int m_\theta(x)\log m(\eta)\mathrm{d}\mu(x)$ for mixture families $\mathcal{M}$, the associated Bregman divergences $B_{F,\mathcal{E}}$ or $B_{F,\mathcal{M}}$ can be relaxed and interpreted as a statistical distance. We explain how to obtain the reconstruction below:

- Consider an exponential family $\mathcal{E}$ of order D with densities defined according to a dominating measure μ:

  $$\mathcal{E} = \{p_\theta(x) = \exp(\theta^\top t(x) - F(\theta)) \;:\; \theta \in \Theta\}, \tag{155}$$

  where the natural parameter θ and the sufficient statistic vector $t(x)$ belong to $\mathbb{R}^D$. We have the integral-based Bregman generator:

  $$F(\theta) = F_{\mathcal{E}}(p_\theta) = \log\left(\int \exp(\theta^\top t(x))\mathrm{d}\mu(x)\right), \tag{156}$$

and the dual convex conjugate

$$F^*(\eta) = -h(p_\theta) = \int p(x) \log p(x) \mathrm{d}\mu(x), \tag{157}$$

where $h(p) = -\int p(x) \log p(x) \mathrm{d}\mu(x)$ denotes Shannon's entropy.

Let $\lambda(i)$ denotes the i-th coordinates of vector λ, and let us calculate the inner product $\theta_1^\top \eta_2 = \sum_i \theta_1(i)\eta_2(i)$ of the Legendre–Fenchel divergence. We have $\eta_2(i) = E_{p_{\theta_2}}[t_i(x)]$. Using the linear property of the expectation $E[\cdot]$, we find that $\sum_i \theta_1(i)\eta_2(i) = E_{p_{\theta_2}}[\sum_i \theta_1(i)t_i(x)]$. Moreover, we have $\sum_i \theta_1(i)t_i(x) = (\log p_{\theta_1}(x)) + F(\theta_1)$. Thus we have:

$$\theta_1^\top \eta_2 = E_{p_{\theta_2}}\left[\log p_{\theta_1} + F(\theta_1)\right] = F(\theta_1) + E_{p_{\theta_2}}\left[\log p_{\theta_1}\right]. \tag{158}$$

It follows that we get

$$\begin{aligned} B_{F,\mathcal{E}}[p_{\theta_1} : p_{\theta_2}] &= F(\theta_1) + F^*(\eta_2) - \theta_1^\top \eta_2, &(159)\\ &= F(\theta_1) - h(p_{\theta_2}) - E_{p_{\theta_2}}[\log p_{\theta_1}] - F(\theta_1), &(160)\\ &= E_{p_{\theta_2}}\left[\log \frac{p_{\theta_2}}{p_{\theta_1}}\right] =: D_{\mathrm{KL}^*}[p_{\theta_1} : p_{\theta_2}]. &(161) \end{aligned}$$

By relaxing the exponential family densities p_{θ_1} and p_{θ_2} to be arbitrary densities p_1 and p_2, we obtain the reverse KL divergence between p_1 and p_2 from the dually flat structure induced by the integral-based log-normalizer of an exponential family:

$$\begin{aligned} D_{\mathrm{KL}^*}[p_1 : p_2] &= E_{p_2}\left[\log \frac{p_2}{p_1}\right] = \int p_2(x) \log \frac{p_2(x)}{p_1(x)} \mathrm{d}\mu(x), &(162)\\ &= D_{\mathrm{KL}}[p_2 : p_1]. &(163) \end{aligned}$$

Thus we have recovered the reverse Kullback–Leibler divergence D_{KL^*} from $B_{F,\mathcal{E}}$.

The dual divergence $D^*[p_1 : p_2] := D[p_2 : p_1]$ is obtained by swapping the distribution parameter orders. We have:

$$D^*_{\mathrm{KL}^*}[p_1 : p_2] := D_{\mathrm{KL}^*}[p_2 : p_1] = E_{p_1}\left[\log \frac{p_1}{p_2}\right] =: D_{\mathrm{KL}}[p_1 : p_2], \tag{164}$$

and $D_{\mathrm{KL}^*}[p_1 : p_2] = D^*_{\mathrm{KL}^*}[p_2 : p_1] = D_{\mathrm{KL}}[p_2 : p_1]$.

To summarize, the canonical Legendre–Fenchel divergence associated with the log-normalizer of an exponential family amounts to the statistical reverse Kullback–Leibler divergence between p_{θ_1} and p_{θ_1} (or the KL divergence between the swapped corresponding densities): $D_{\mathrm{KL}}[p_{\theta_1} : p_{\theta_2}] = B_F(\theta_2 : \theta_1) = A_{F,F^*}(\theta_2 : \eta_1)$. Notice that it is easy to check that $D_{\mathrm{KL}}[p_{\theta_1} : p_{\theta_2}] = B_F(\theta_2 : \theta_1)$ [74,75]. Here, we took the opposite direction by constructing D_{KL} from B_F.

We may consider an auxiliary carrier term $k(x)$ so that the densities write $p_\theta(x) = \exp(\theta^\top t(x) - F(\theta) + k(x))$. Then the dual convex conjugate writes [76] as $F^*(\eta) = -h(p_\theta) + E_{p_\theta}[k(x)]$.

Notice that since the Bregman generator is defined up to an affine term, we may consider the equivalent generator $F(\theta) = -\log p_\theta(\omega)$ instead of the integral-based generator. This approach yields ways to build formula bypassing the explicit use of the log-normalizer for calculating various statistical distances [77].

- In this second example, we consider a mixture family

$$\mathcal{M} = \left\{ m_\theta = \sum_{i=1}^{D} \theta_i p_i(x) + (1 - \sum_{i=1}^{D} \theta_i) p_0(x) \right\}, \tag{165}$$

where $p_0, \ldots, p_D$ are $D+1$ linearly independent probability densities. The integral-based Bregman generator F is chosen as Shannon negentropy:

$$F(\theta) = F_{\mathcal{M}}(m_\theta) = -h(m_\theta) = \int m_\theta(x) \log m_\theta(x) \mathrm{d}\mu(x). \tag{166}$$

We have

$$\eta_i = [\nabla F(\theta)]_i = \int (p_i(x) - p_0(x)) \log m_\theta(x) \mathrm{d}\mu(x), \tag{167}$$

and the dual convex potential function is

$$F^*(\eta) = -\int p_0(x) \log m_\theta(x) \mathrm{d}\mu(x) = h^\times(p_0 : m_\theta), \tag{168}$$

i.e., the cross-entropy between the density p_0 and the mixture m_θ. Let us calculate the inner product $\theta_1^\top \eta_2$ of the Legendre–Fenchel divergence as follows:

$$\begin{aligned}\sum_i \theta_1(i) \int (p_i(x) - p_0(x)) \log m_{\theta_2}(x) \mathrm{d}\mu(x) &= \int \sum_i \theta_1(i) p_i(x) \log m_{\theta_2}(x) \mathrm{d}\mu(x) \\ &\quad - \sum_i \theta_1(i) p_0(x) \log m_{\theta_2}(x) \mathrm{d}\mu(x).\end{aligned} \tag{169}$$

That is

$$\theta_1^\top \eta_2 = \int \sum_i \theta_1(i) p_i \log m_{\theta_2} \mathrm{d}\mu - \sum_i \theta_1(i) p_0 \log m_{\theta_2} \mathrm{d}\mu. \tag{170}$$

Thus it follows that we have the following statistical distance:

$$\begin{aligned}
B_{F,\mathcal{M}}[m_{\theta_1} : m_{\theta_2}] &:= F(\theta_1) + F^*(\eta_2) - \theta_1^\top \eta_2, && (171)\\
&= -h(m_{\theta_1}) - \int p_0(x) \log m_{\theta_2}(x) \mathrm{d}\mu(x) - \int \sum_i \theta_1(i) p_i(x) \log m_{\theta_2}(x) \mathrm{d}\mu(x) \\
&\quad + \sum_i \theta_1(i) p_0(x) \log m_{\theta_2}(x) \mathrm{d}\mu(x), && (172)\\
&= -h(m_{\theta_1}) - \int ((1 - \sum_i \theta_1(i)) p_0(x) + \sum_i \theta_1(i) p_i(x)) \log m_{\theta_2}(x) \mathrm{d}\mu(x), && (173)\\
&= -h(m_{\theta_1}) - \int m_{\theta_1}(x) \log m_{\theta_2}(x) \mathrm{d}\mu(x), && (174)\\
&= \int m_{\theta_1}(x) \log \frac{m_{\theta_1}(x)}{m_{\theta_2}(x)} \mathrm{d}\mu(x), && (175)\\
&= D_{\mathrm{KL}}[m_{\theta_1} : m_{\theta_2}]. && (176)
\end{aligned}$$

Thus we have $D_{\mathrm{KL}}[m_{\theta_1} : m_{\theta_2}] = B_F(\theta_1 : \theta_2)$. By relaxing the mixture densities m_{θ_1} and m_{θ_2} to arbitrary densities m_1 and m_2, we find that the dually flat geometry induced by the negentropy of densities of a mixture family induces a statistical distance which corresponds to the (forward) KL divergence. That is, we have recovered the statistical distance D_{KL} from $B_{F,\mathcal{M}}$. Note that in general the entropy of a mixture is not available in closed-form (because of the log sum term), except when the component distributions have pairwise disjoint supports. This latter case includes the case of Dirac distributions whose mixtures represent the categorical distributions.

Let us consider the dually flat spaces induced by the family of discrete Poisson distributions and the family of continuous gamma distributions:

Example 6. *Consider the family $\mathcal{P}$ of Poisson distributions with rate parameter $\lambda > 0$:*

$$\mathcal{P} = \left\{ p_\lambda(x) = \frac{\lambda^x e^{-\lambda}}{x!}, \quad \lambda \in (0, \infty) \right\}. \tag{177}$$

This family is a univariate discrete exponential family of order one (i.e., $d = 1$ and $D = 1$) with the following canonical decomposition of its probability mass function $p_\lambda(x)$:

- *Base measure:* $\nu(x) = \frac{\mu(x)}{x!} = e^{k(x)}\mu(x)$ *where* μ *is the counting measure and* $k(x) = -\log(x!)$ *represents an auxiliary measure carrier term for defining the base measure* ν,
- *Sufficient statistics:* $t(x) = x$,
- *Natural parameter:* $\theta = \theta(\lambda) = \log(\lambda) \in \Theta = \mathbb{R}$,
- *Log-normalizer:* $F(\theta) = \exp(\theta)$ *since* $F(\theta(\lambda)) = \lambda$.

Thus we can rewrite the Poisson family as the following Discrete Exponential Family (DEF):

$$\mathcal{P} = \{p_\theta(x) = \exp(\theta t(x) - F(\theta))\mathrm{d}\nu(x) \,:\, \theta \in \Theta\}. \tag{178}$$

The expectation is $E_{p_\theta}[x] = F'(\theta) = \exp(\theta)$, *or equivalently* $E_{p_\lambda}[x] = \lambda$. *The variance* $\mathrm{Var}_{p_\theta}[x] = F''(\theta) = \exp(\theta)$, *or or equivalently* $\mathrm{Var}_{p_\lambda}[x] = \lambda$. *The Kullback–Leibler divergence between two Poisson distributions* p_{λ_1} *and* p_{λ_2} *is:*

$$\begin{aligned} D_{\mathrm{KL}}\left[p_{\lambda_1} : p_{\lambda_2}\right] &= B_F(\theta(\lambda_2) : \theta(\lambda_1)), &(179)\\ &= \lambda_1 \log\frac{\lambda_1}{\lambda_2} + \lambda_2 - \lambda_1. &(180)\end{aligned}$$

We recognize the expression of the univariate Kullback–Leibler divergence extended to the positive scalars.

We have $\eta = F'(\theta) = \lambda$ *and* $I_\eta(\eta) = (F^*)''(\eta) = \frac{1}{\eta}$ *where* $F^*(\eta) = \eta\log\eta - \eta$ *is the convex conjugate of* $F(\theta)$. *Since* $\eta = \lambda$, *we deduce that the Fisher information is* $I_\lambda(\lambda) = \frac{1}{\lambda}$. *Notice that* $I_\theta(\theta) = \exp(\theta) = \lambda$. *Thus we check the Crouzeix identity:* $F''(\theta)(F^*)''(\eta) = \lambda \times \frac{1}{\lambda} = 1$. *Beware that although* $I_\theta(\theta) = \lambda$, *this is not the FIM* I_λ. *Using the covariance equation of the FIM of Equation (86), we have:*

$$\begin{aligned} I_\lambda(\lambda) &= \frac{\mathrm{d}\theta}{\mathrm{d}\lambda} I_\theta(\theta(\lambda)) \frac{\mathrm{d}\theta}{\mathrm{d}\lambda}, &(181)\\ &= \frac{1}{\lambda}\exp(\log(\lambda))\frac{1}{\lambda} = \frac{1}{\lambda}. &(182)\end{aligned}$$

The Fisher–Rao distance [78] between two Poisson distributions p_{λ_1} *and* p_{λ_2} *is:*

$$D_\rho(\lambda_1, \lambda_2) = 2\left|\sqrt{\lambda_1} - \sqrt{\lambda_2}\right|. \tag{183}$$

In general, it is easy to get the Fisher–Rao distance of uniorder families because both the length elements and the geodesics are available in closed forms.

The following example demonstrates the computational intractability of the Fisher–Rao distance.

Example 7. *Consider the parametric family of Gamma distributions [79] with probability density:*

$$p_{\alpha,\beta}(x) = \frac{\beta^\alpha x^{\alpha-1}\exp(-\beta x)}{\Gamma(\alpha)}, \tag{184}$$

for shape parameter $\alpha > 0$, *rate parameter* $\beta > 0$ *and support* $x \in \mathcal{X} = (0, \infty)$. *Function* $\Gamma(z) = \int_0^\infty x^{z-1}e^{-x}\mathrm{d}x$ *is the Gamma function defined for* $z > 0$, *and satisfying* $\Gamma(n) = (n-1)!$ *for integers* n. *The Gamma distributions* $\{p_{\alpha,\beta},\ \alpha, \beta > 0\}$ *form an univariate exponential family of order 2 (i.e.,* $d = 1$ *and* $D = 2$*) with the following canonical decomposition:*

- *Natural parameters:* $\theta = (\theta_1, \theta_2)$ *with* $\theta(\lambda) = (-\beta, \alpha - 1)$ *with source parameter* $\lambda = (\alpha, \beta)$,
- *Sufficient statistics:* $t(x) = (x, \log(x))$,
- *Log-normalizer:* $F(\theta) = -(\theta_2 + 1)\log(-\theta_1) + \log\Gamma(\theta_2 + 1)$,

- *Dual parameterization:* $\eta = (\eta_1, \eta_2) = E_{p_\theta}[t(x)] = \nabla F(\theta) = \left(\frac{\theta_2+1}{-\theta_1}, -\log(-\theta_1) + \psi(\theta_2+1)\right)$, *where* $\psi(x) = \frac{d}{dx}\ln(\Gamma(x)) = \frac{\Gamma'(x)}{\Gamma(x)}$ *denotes the digamma function.*

It follows that the Kullback–Leibler divergence between two Gamma distributions p_{λ_1} *and* p_{λ_2} *with respective source parameters* $\lambda_1 = (\alpha_1, \beta_1)$ *and* $\lambda_2 = (\alpha_2, \beta_2)$ *is:*

$$D_{\mathrm{KL}}\left[p_{\alpha_1,\beta_1} : p_{\alpha_2,\beta_2}\right] = B_F(\theta(\lambda_2) : \theta(\lambda_1)), \tag{185}$$

$$= (\alpha_1 - \alpha_2)\,\psi(\alpha_1) - \log\Gamma(\alpha_1) + \log\Gamma(\alpha_2) \tag{186}$$

$$+\alpha_2(\log\beta_1 - \log\beta_2) + \alpha_1\frac{\beta_2 - \beta_1}{\beta_1}. \tag{187}$$

The Fisher information matrix is $I_\theta(\theta) = \nabla^2 F(\theta)$. *It can be expressed using the* λ*-parameterization [80] as:*

$$I_\lambda(\alpha, \beta) = \begin{bmatrix} \psi_1(\alpha) & -\frac{1}{\beta} \\ -\frac{1}{\beta} & \frac{\alpha}{\beta^2} \end{bmatrix}, \tag{188}$$

where $\psi_1(x)$ *is the trigamma function defined for* $x > 0$ *by:*

$$\psi_1(x) := \frac{d^2}{dx^2}\log\Gamma(x). \tag{189}$$

Because the Fisher information matrix is not diagonal using the λ*-parameterization, we deduce that the parameters* α *and* β *are correlated (non-orthogonal). In general, by mixing the natural parameters* θ *with the expectation parameters* η *of an exponential family, we obtain a block-diagonal Fisher information matrix [2,80]. That is, let* $\delta = (\theta_1, \ldots, \theta_l, \eta_1, \ldots, \eta_{D-l})$ *be a mixed primal/dual coordinate system for* $l \in \{1, \ldots, D-1\}$ *where D is the order of the family (or the dimension of the dually flat space). Then the Fisher information matrix* $I_\delta(\delta)$ *for the mixed parameterization is block diagonal. Thus we can always diagonalize the Fisher information matrix of an exponential family of order* 2*. For example, for the gamma manifold, let us choose the reparameterization* $\delta_1 = \frac{\alpha}{\beta}$ *and* $\delta_2 = \alpha$*. Then we have the Fisher information matrix that rewrites as:*

$$I_\delta(\delta) = \begin{bmatrix} \frac{\delta_2}{\delta_1^2} & 0 \\ 0 & \psi_1(\delta_2) - \frac{1}{\delta_2} \end{bmatrix}. \tag{190}$$

The parameters δ_1 *and* δ_2 *are not correlated and orthogonal since the FIM is diagonal.*

The numerical evaluation of the Fisher–Rao distance between two gamma distributions has been studied in [81]. Let $\omega = \log\frac{\alpha}{\beta}$*. The length element is shown to be in this* (α, ω) *parameterization:*

$$\mathrm{d}s^2 = \left(\psi_1(\alpha) - \frac{1}{\alpha}\right)(\mathrm{d}\alpha)^2 + \alpha(\mathrm{d}\omega)^2. \tag{191}$$

However, no closed-form expression is known for the Fisher–Rao distance between two gamma distributions because of the intractability of the geodesic equations on the gamma Fisher–Rao manifold [81]. This example highlights the fact that computing the Fisher–Rao distance for simple family of distributions can be challenging. In fact, we do not know the Fisher–Rao distance between any two multivariate Gaussian distibutions [82] (except in a few cases including the univariate case).

In general, dually flat spaces can be built from any strictly convex C^3 generator F. Vinberg and Koszul [48] showed how to obtain such a convex generator for homogeneous cones. A cone $\mathcal{C}$ in a vector space V yields a dual cone of positive linear functionals in the dual vector space V^*:

$$\mathcal{C}^* := \{\omega \in V^* : \forall v \in \mathcal{C}, \omega(v) \geq 0\}. \tag{192}$$

The characteristic function of the cone is defined by

$$\chi_C(\theta) := \int_{C^*} \exp(-\omega(\theta))\mathrm{d}\omega \geq 0, \tag{193}$$

and the function $\log \chi_C(\theta)$ defines a Bregman generator which induces a Hessian structure and a dually flat space.

Figure 11 displays the main types of information manifolds encountered in information geometry with their relationships.

4. Some Applications of Information Geometry

Information geometry [2] found broad applications in information sciences. For example, we can mention:

- Statistics: Asymptotic inference, Expectation-Maximization (EM and the novel information-geometric em), time series (AutoRegressive Moving Average model, ARMA) models,
- Pattern recognition [83] and machine learning: Restricted Boltzmann machines [2] (RBMs), neuromanifolds [84] and natural gradient [85],
- Signal processing: Principal Component Analysis (PCA), Independent Component Analysis (ICA), Non-negative Matrix Factorization (NMF),
- Mathematical programming: Barrier function of interior point methods,
- Game theory: Score functions.

Next, we shall describe a few applications, starting with the celebrated natural gradient descent.

4.1. *Natural Gradient in Riemannian Space*

The Natural Gradient [86] (NG) is an extension of the ordinary (Cartesian) gradient of Euclidean geometry to the gradient in a Riemannian space analyzed in an arbitrary coordinate system. We explain the natural gradient

4.1.1. The Vanilla Gradient Descent Method

Given a real-valued function $L_\theta(\theta)$ parameterized by a a D-dimensional vector θ on parameter space $\theta \in \Theta \subset \mathbb{R}^D$, we wish to minimize L_θ, i.e., solve $\min_{\theta\in\Theta} L_\theta(\theta)$. The gradient descent (GD) method, also called the steepest descent method, is a first-order local optimization procedure which starts by initializing the parameter to an arbitrary value (say, $\theta_0 \in \Theta$), and then iteratively updates at stage t the current location of θ_t to θ_{t+1} as follows:

$$\mathrm{GD}: \quad \theta_{t+1} = \theta_t - \alpha_t \nabla_\theta L_\theta(\theta_t). \tag{194}$$

The scalar $\alpha_t > 0$ is called the step size or learning rate in machine learning. The ordinary gradient (OG) $\nabla_\theta F_\theta(\theta)$ (vector of partial derivatives) represents the steepest vector at θ of the function graph $\mathcal{L}_\theta = \{(\theta, L_\theta(\theta)) : \theta \in \Theta\}$. The GD method was pioneered by Cauchy [87] (1847) and its convergence proof to a stationary point was first reported in Curry [88] (1944).

If we reparameterize the function L_θ using a one-to-one and onto differentiable mapping $\eta = \eta(\theta)$ (with reciprocal inverse mapping $\theta = \theta(\eta)$), the GD update rule transforms as:

$$\eta_{t+1} = \eta_t - \alpha_t \nabla_\eta L_\eta(\eta_t), \tag{195}$$

where

$$L_\eta(\eta) := L_\theta(\theta(\eta)). \tag{196}$$

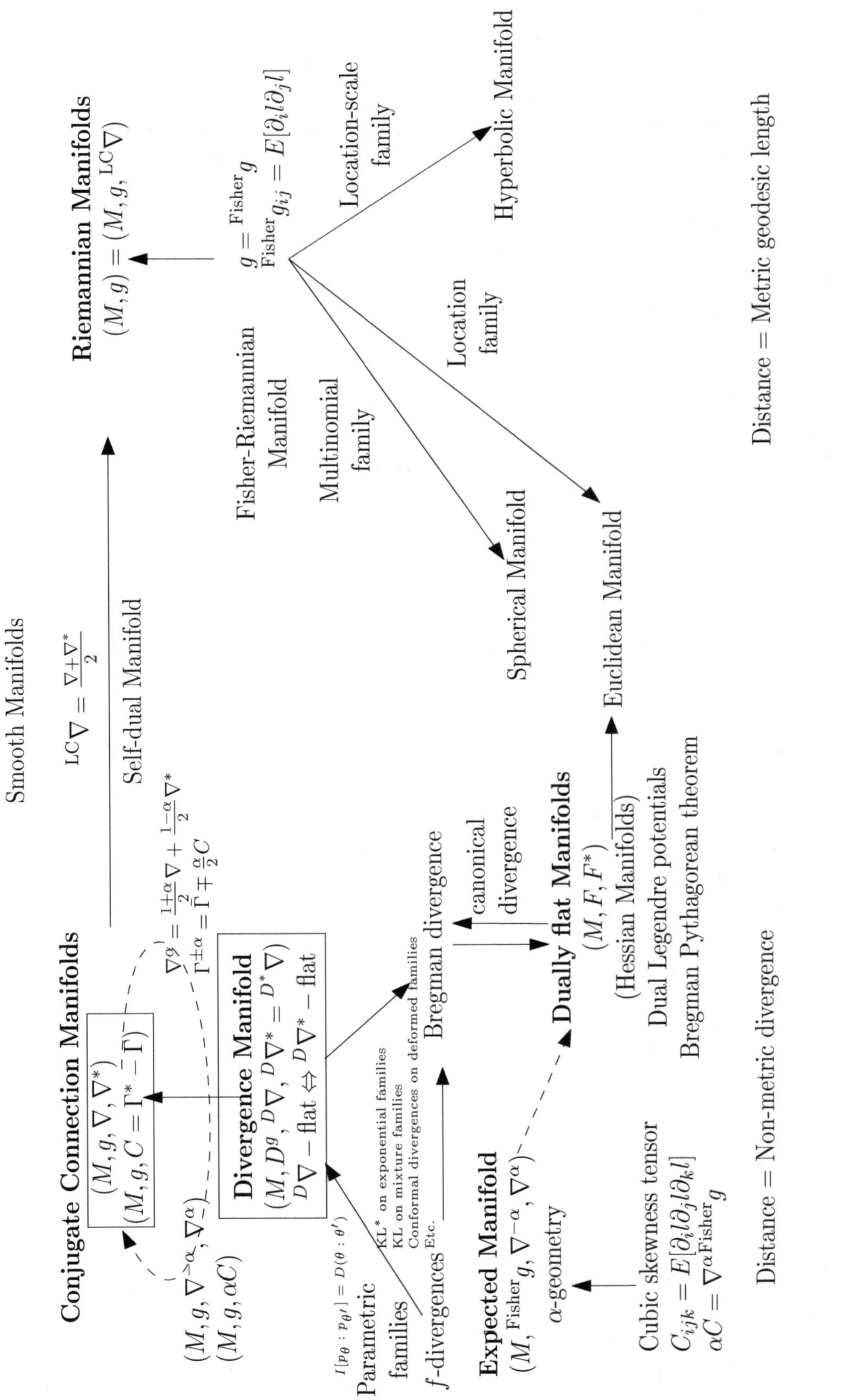

Figure 11. Overview of the main types of information manifolds with their relationships in information geometry.

Thus in general, the two gradient descent location sequences $\{\theta_t\}_t$ and $\{\eta_t\}_t$ (initialized at $\theta_0 = \theta(\eta_0)$ and $\eta_0 = \eta(\theta_0)$) are different (because usually $\eta(\theta) \neq \theta$), and the two GDs may potentially reach different stationary points. In other words, the GD local optimization depends on the choice of the parameterization of the function L (i.e., L_θ or L_η). For example, minimizing with the gradient descent a temperature function $L_\theta(\theta)$ with respect to Celsius degrees θ may yield a different result than minimizing the same temperature function $L_\eta(\eta) = L_\theta(\theta(\eta))$ expressed with respect to Fahrenheit degrees η. That is, the GD optimization is extrinsic since it depends on the choice of the parameterization of the function, and does not take into account the underlying geometry of the parameter space Θ.

The natural gradient precisely addresses this problem and solves it by choosing intrinsically the steepest direction with respect to a Riemannian metric tensor field on the parameter manifold. We shall explain the natural gradient descent method and highlight its connections with the Riemannian gradient descent, the mirror descent and even the ordinary gradient descent when the parameter space is dually flat.

4.1.2. Natural Gradient and Its Connection with the Riemannian Gradient

Let (M, g) be a D-dimensional Riemannian space [10] equipped with a metric tensor g, and $L \in C^\infty(M)$ a smooth function to minimize on the manifold M. The Riemannian gradient [89] uses the Riemannian exponential map $\exp_p : T_p \to M$ to update the sequence of points p_t's on the manifold as follows:

$$\text{RG}: \quad p_{t+1} = \exp_{p_t}(-\alpha_t \nabla_M L(p_t)), \tag{197}$$

where the Riemannian gradient ∇_M is defined according to a directional derivative ∇_v by:

$$\nabla_M L(p) := \nabla_v \left(L\left(\exp_p(v)\right)\right)\Big|_{v=0}, \tag{198}$$

with

$$\nabla_v L(p) := \lim_{h \to 0} \frac{L(p + hv) - L(p)}{h}. \tag{199}$$

However, the Riemannian exponential mapping $\exp_p(\cdot)$ is often computationally intractable since it requires to solve a system of second-order differential equations [10,22]. Thus instead of using $\exp_p$, we shall rather use a computable Euclidean retraction $R : T_p \to \mathbb{R}^D$ of the exponential map expressed in a local θ-coordinate system as:

$$\text{RetG}: \quad \theta_{t+1} = R_{\theta_t}\left(-\alpha_t \nabla_\theta L_\theta(\theta_t)\right). \tag{200}$$

Using the retraction [22] $R_p(v) = p + v$ which corresponds to a first-order Taylor approximation of the exponential map, we recover the natural gradient descent [86]:

$$\text{NG}: \theta_{t+1} = \theta_t - \alpha_t g_\theta^{-1}(\theta_t) \nabla_\theta L_\theta(\theta_t). \tag{201}$$

The natural gradient [86] (NG)

$$^{\text{NG}}\nabla L_\theta(\theta) := g_\theta^{-1}(\theta) \nabla_\theta L_\theta(\theta) \tag{202}$$

encodes the Riemannian steepest descent vector, and the natural gradient descent method yields the following update rule

$$\text{NG}: \theta_{t+1} = \theta_t - \alpha_t \, {}^{\text{NG}}\nabla L_\theta(\theta_t). \tag{203}$$

Notice that the natural gradient is a contravariant vector while the ordinary gradient is a covariant vector. Recall that a covariant vector $[v_i]$ is transformed into a contravariant vector $[v^i]$ by $v^i = \sum_j g^{ij} v_i$, that is by using the dual Riemannian metric $g_\eta^*(\eta) = g_\theta(\theta)^{-1}$. The natural gradient is invariant under an invertible smooth change of parameterization. However, the natural gradient descent does not guarantee that the locations θ_t's always stay on the manifold: Indeed, it may happen that for some t, $\theta_t \notin \Theta$ when $\Theta \neq \mathbb{R}^D$.

Property 5 ([89]). *The natural gradient descent approximates the intrinsic Riemannian gradient descent using a contravariant gradient vector induced by the Riemannian metric tensor g. The natural gradient is invariant to coordinate transformations.*

Next, we shall explain how the natural gradient descent is related to the mirror descent and the ordinary gradient when the Riemannian space Θ is dually flat.

4.1.3. Natural Gradient in Dually Flat Spaces: Connections to Bregman Mirror Descent and Ordinary Gradient

Recall that a dually flat space (M, g, ∇, ∇^*) is a manifold M equipped with a pair (∇, ∇^*) of dual torsion-free flat connections which are coupled to the Riemannian metric tensor g [2,90] in the sense that $\frac{\nabla+\nabla^*}{2} = {}^{LC}\nabla$, where ${}^{LC}\nabla$ denotes the unique metric torsion-free Levi–Civita connection.

On a dually flat space, there exists a pair of dual global Hessian structures [48] with dual canonical Bregman divergences [2,91]. The dual Riemannian metrics can be expressed as the Hessians of dual convex potential functions F and F^*. Examples of Hessian manifolds are the manifolds of exponential families or the manifolds of mixture families [92]. On a dually flat space induced by a strictly convex and C^3 function F (Bregman generator), we have two dual global coordinate system: $\theta(\eta) = \nabla F^*(\eta)$ and $\eta(\theta) = \nabla F(\theta)$, where F^* denotes the Legendre–Fenchel convex conjugate function [51,93]. The Hessian metric expressed in the primal θ-coordinate system is $g_\theta(\theta) = \nabla^2 F(\theta)$, and the dual Hessian metric expressed in the dual coordinate system is $g^*_\eta(\eta) = \nabla^2 F^*(\eta)$. Crouzeix's identity [36] shows that $g_\theta(\theta) g_\eta(\eta) = I$, where I denotes the $D \times D$ matrix identity.

The ordinary gradient descent method can be extended using a proximity function $\Phi(\cdot,\cdot)$ as follows:

$$\text{PGD}: \quad \theta_{t+1} = \arg\min_{\theta\in\Theta} \left\{ \langle \theta, \nabla L_\theta(\theta_t) \rangle + \frac{1}{\alpha_t} \Phi(\theta, \theta_t) \right\}. \tag{204}$$

When $\Phi(\theta, \theta_t) = \frac{1}{2}\|\theta - \theta_t\|^2$, the PGD update rule becomes the ordinary GD update rule.

Consider a Bregman divergence [91] B_F for the proximity function Φ: $\Phi(p, q) = B_F(p : q)$. Then the PGD yields the following mirror descent (MD):

$$\text{MD}: \quad \theta_{t+1} = \arg\min_{\theta\in\Theta} \left\{ \langle \theta, \nabla L(\theta_t) \rangle + \frac{1}{\alpha_t} B_F(\theta : \theta_t) \right\}. \tag{205}$$

This mirror descent can be interpreted as a natural gradient descent as follows:

Property 6 ([94]). *Bregman mirror descent on the Hessian manifold $(M, g = \nabla^2 F(\theta))$ is equivalent to natural gradient descent on the dual Hessian manifold $(M, g^* = \nabla^2 F(\eta))$, where F is a Bregman generator, $\eta = \nabla F(\theta)$ and $\theta = \nabla F^*(\eta)$.*

Indeed, the mirror descent rule yields the following natural gradient update rule:

$$\text{NG}^*: \eta_{t+1} = \eta_t - \alpha_t (g^*_\eta)^{-1}(\eta_t) \nabla_\eta L_\theta(\theta(\eta_t)), \tag{206}$$

$$= \eta_t - \alpha_t (g^*_\eta)^{-1}(\eta_t) \nabla_\eta L_\eta(\eta_t), \tag{207}$$

where $g^*_\eta(\eta) = \nabla^2 F^*(\eta) = (\nabla^2_\theta F(\theta))^{-1}$ and $\theta(\eta) = \nabla F^*(\theta)$.

The method is called mirror descent [95] because it performs that gradient step in the dual space (i.e., mirror space) $H = \{\eta = \nabla F(\theta) \ : \ \theta \in \Theta\}$, and thus solves the inconsistency contravariant/covariant type problem of subtracting a covariant vector from a contravariant vector of the ordinary GD (Equation (194)).

Let us prove now the following property of the natural gradient in a dually flat space or Bregman manifold [90]:

Property 7 ([96]). *In a dually flat space induced by potential convex function F, the natural gradient amounts to the ordinary gradient on the dually parameterized function:* ${}^{\mathrm{NG}}\nabla L_\theta(\theta) = \nabla_\eta L_\eta(\eta)$ *where* $\eta = \nabla_\theta F(\theta)$ *and* $L_\eta(\eta) = L_\theta(\theta(\eta))$.

Proof. Let (M, g, ∇, ∇^*) be a dually flat space. We have $g_\theta(\theta) = \nabla^2 F(\theta) = \nabla_\theta \nabla_\theta F(\theta) = \nabla_\theta \eta$ since $\eta = \nabla_\theta F(\theta)$. The function to minimize can be written either as $L_\theta(\theta) = L_\theta(\theta(\eta))$ or as $L_\eta(\eta) = L_\eta(\eta(\theta))$. Recall the chain rule in the calculus of differentiation:

$$\nabla_\theta L_\theta(\theta) = \nabla_\theta(L_\eta(\eta(\theta))) = (\nabla_\theta \eta)(\nabla_\eta L_\eta(\eta)). \tag{208}$$

Thus we have:

$$\begin{aligned} {}^{\mathrm{NG}}\nabla L_\theta(\theta) &:= g_\theta^{-1}(\theta)\nabla_\theta L_\theta(\theta), & (209)\\ &= (\nabla_\theta \eta)^{-1}(\nabla_\theta \eta)\nabla_\eta L_\eta(\eta), & (210)\\ &= \nabla_\eta L_\eta(\eta). & (211) \end{aligned}$$

□

It follows that the natural gradient descent on a loss function $L_\theta(\theta)$ amounts to an ordinary gradient descent on the dually parameterized loss function $L_\eta(\eta) := L_\theta(\theta(\eta))$. In short, ${}^{\mathrm{NG}}\nabla_\theta L_\theta = \nabla_\eta L_\eta$.

4.1.4. An Application of the Natural Gradient: Natural Evolution Strategies (NESs)

A nice family of applications of the natural gradient is the Natural Evolution Strategies (NESs) for black-box minimization [97]. Let $f(x)$ for $x \in \mathbb{X} \subset \mathbb{R}^d$ be a real-valued function to minimize. Berny [98] proposed to relax the optimization problem $\min_{x\in\mathbb{X}} f(x)$ by considering a parametric search distribution p_λ, and minimize instead:

$$\min_{\lambda\in\Lambda} E_{p_\lambda}[f(x)], \tag{212}$$

where $\lambda \in \Lambda \subset \mathbb{R}^D$ denotes the parameter space of the search distributions. Let $J(\lambda) = E_{p_\lambda}[f(x)]$. Minimizing $J(\lambda)$ instead of $f(x)$ is particularly useful when $\mathbb{X}$ is a discrete space: Indeed, the combinatorial optimization [98] $\min_{x\in\mathbb{X}} f(x)$ is replaced by a continuous optimization $\min_{\lambda\in\Lambda} J(\lambda)$ when Λ is a continuous parameter, and the ordinary or natural GD methods can be used. The gradient $\nabla J(\lambda)$ is called the search gradient, and it can be approximated stochastically using the log-likelihood trick [99] as

$$\tilde{\nabla} J(\lambda) := \frac{1}{n}\sum_{i=1}^n f(x_i)\nabla \log p_\lambda(x_i) \approx \nabla J(\lambda), \tag{213}$$

where $x_1, \ldots, x_n \sim p_\lambda$. Similarly, the Fisher information matrix (FIM) may be approximated by the following empirical FIM:

$$\tilde{I}(\lambda) = \frac{1}{n}\sum_{i=1}^n \nabla_\lambda l_\lambda(x_i)(\nabla_\lambda l_\lambda(x_i))^\top \approx I(\lambda), \tag{214}$$

where $l_\lambda(x) := \log p_\lambda(x)$ denote the log-likelihood function. Notice that the approximated FIM may potentially be degenerated and may not respect the structure of the true FIM. For example, we have $\nabla_\mu l(x;\mu,\sigma^2) = \frac{x-\mu}{\sigma^2}$ and $\nabla_{\sigma^2} = \frac{(x-\mu)^2}{2\sigma^4} - \frac{1}{2\sigma^2}$. The non-diagonal of the approximate FIM $\tilde{I}(\lambda)$ are close to but usually non-zero although the expected FIM is diagonal $I(\mu,\sigma^2) = \mathrm{diag}\left(\frac{1}{\sigma^2}, \frac{1}{2\sigma^4}\right)$. Thus we may estimate the FIM until the non-diagonal elements have absolute values less than a prescribed $\epsilon > 0$. For multivariate normals, we have $\nabla_\mu l(x;\mu,\Sigma) = \Sigma^{-1}(x-\mu)$ and $\nabla_\Sigma l(x;\mu,\Sigma) = \frac{1}{2}(\nabla_\mu l(x;\mu,\Sigma)\nabla_\mu l(x;\mu,\Sigma)^\top - \Sigma^{-1})$.

4.2. Some Illustrating Applications of Dually Flat Manifolds

In this part, we describe how to use the dually flat structures for handling an exponential family $\mathcal{E}$ (in a hypothesis testing problem detailed in Section 4.3) and the mixture family $\mathcal{M}$ (clustering statistical mixtures Section 4.4). Note that for a general divergence, neither $(\mathcal{E}, D)$ nor $(\mathcal{M}, D)$ is dually flat. However, when $D = \mathrm{KL}$, the Kullback–Leibler divergence, we get dually flat spaces that are computationally attractive since the primal/dual geodesics are straight lines in the corresponding global affine coordinate system.

4.3. Hypothesis Testing in the Dually Flat Exponential Family Manifold $(\mathcal{E}, \mathrm{KL}^*)$

Given two probability distributions $P_0 \sim p_0(x)$ and $P_1 \sim p_1(x)$, we ask to classify a set of iid. observations $X_{1:n} = \{x_1, \ldots, x_n\}$ as either sampled from P_0 or from P_1? This is a statistical decision problem [100]. For example, P_0 can represent the signal distribution and P_1 the noise distribution. Figure 12 displays the probability distributions and the unavoidable error that is made by any statistical decision rule (on observations x_1 and x_2).

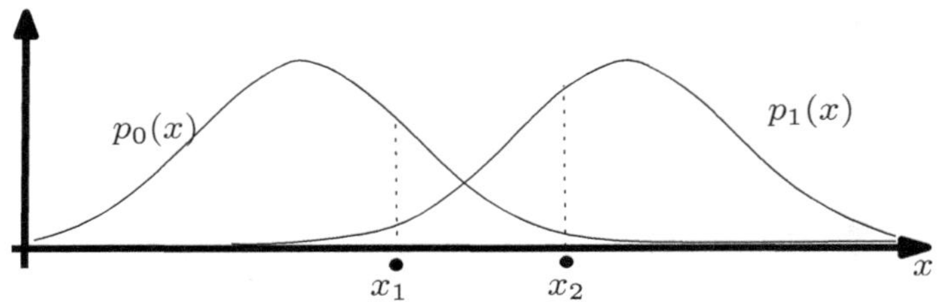

Figure 12. Statistical Bayesian hypothesis testing: the best Maximum A Posteriori (MAP) rule chooses to classify an observation from the class that yields the maximum likelihood.

Assume that both distributions $P_0 \sim P_{\theta_0}$ and $P_1 \sim P_{\theta_1}$ belong to the same exponential family $\mathcal{E} = \{P_\theta : \theta \in \Theta\}$, and consider the exponential family manifold with the dually flat structure $(\mathcal{E}, {}_\mathcal{E}g, {}_\mathcal{E}\nabla^e, {}_\mathcal{E}\nabla^m)$. That is, the manifold equipped with the Fisher information metric tensor field and the expected exponential connection and conjugate expected mixture connection. More generally, the expected α-geometry of an exponential family $\mathcal{E}$ with cumulant function F is given by:

$$g_{ij}(\theta) = \partial_i\partial_j F(\theta), \tag{215}$$

$$\Gamma^\alpha_{ij,k} = \frac{1-\alpha}{2}\partial_i\partial_j\partial_k F(\theta). \tag{216}$$

When $\alpha = 1$, $\Gamma^\alpha_{ij,k} = 0$ and ∇^1 is flat, and so is ∇^{-1} by using the fundamental theorem of information geometry.

The ± 1-structure can also be derived from a divergence manifold structure by choosing the reverse Kullback–Leibler divergence KL^*:

$$(\mathcal{E}, {}_\mathcal{E}g, {}_\mathcal{E}\nabla^e, {}_\mathcal{E}\nabla^m) \equiv (\mathcal{E}, \mathrm{KL}^*). \tag{217}$$

Therefore, the Kullback–Leibler divergence $\mathrm{KL}[P_\theta : P_{\theta'}]$ amounts to a Bregman divergence (for the cumulant function of the exponential family):

$$\mathrm{KL}^*[P_{\theta'} : P_\theta] = \mathrm{KL}[P_\theta : P_{\theta'}] = B_F(\theta' : \theta). \tag{218}$$

The best exponent error α^* of the best Maximum A Priori (MAP) decision rule is found by minimizing the Bhattacharyya distance to get the Chernoff information [101]:

$$C[P_1, P_2] = -\log \min_{\alpha \in (0,1)} \int_{x \in \mathcal{X}} p_1^{\alpha}(x) p_2^{1-\alpha}(x) \mathrm{d}\mu(x) \geq 0. \tag{219}$$

On the exponential family manifold $\mathcal{E}$, the Bhattacharyya distance:

$$B_\alpha[p_1 : p_2] = -\log \int_{x \in \mathcal{X}} p_1^{\alpha}(x) p_2^{1-\alpha}(x) \mathrm{d}\mu(x), \tag{220}$$

amounts to a skew Jensen parameter divergence [102] (also called Burbea-Rao divergence):

$$J_F^{\alpha}(\theta_1 : \theta_2) = \alpha F(\theta_1) + (1-\alpha) F(\theta_2) - F(\theta_1 + (1-\alpha)\theta_2). \tag{221}$$

It can be shown that the Chernoff information [100,103,104] (that minimizes α) is equivalent to a Bregman divergence: Namely, the Bregman divergence for exponential families at the optimal exponent value α^*.

Theorem 10 (Chernoff information [100]). *The Chernoff information between two distributions belonging to the same exponential family amount to a Bregman divergence:*

$$C[P_{\theta_1} : P_{\theta_2}] = B(\theta_1 : \theta_{12}^{\alpha^*}) = B(\theta_2 : \theta_{12}^{\alpha^*}), \tag{222}$$

where $\theta_{12}^{\alpha} = (1-\alpha)\theta_1 + \alpha\theta_2$, *and* α^* *denote the best exponent error.*

Let $\theta_{12}^* := \theta_{12}^{\alpha^*}$ denote the best exponent error. The geometry [100] of the best error exponent can be explained on the dually flat exponential family manifold as follows:

$$P^* = P_{\theta_{12}^*} = G_e(P_1, P_2) \cap \mathrm{Bi}_m(P_1, P_2), \tag{223}$$

where G_e denotes the exponential geodesic γ_{∇^e} and Bi_m the m-bisector:

$$\mathrm{Bi}_m(P_1, P_2) = \{P \ : \ F(\theta_1) - F(\theta_2) + \eta(P)^\top(\theta_2 - \theta_1) = 0\}. \tag{224}$$

Figure 13 illustrates how to retrieve the best error exponent from an exponential arc (θ-geodesic) intersecting the m-bisector.

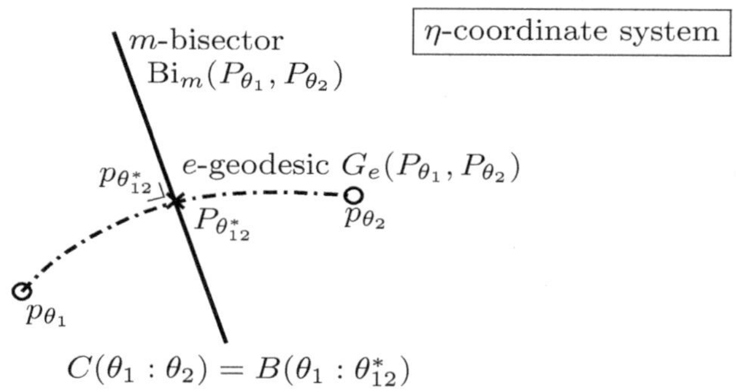

Figure 13. Exact geometric characterization (not necessarily i closed-form) of the best exponent error rate α^*.

Furthermore, instead of considering two distributions for this statistical binary decision problem, we may consider a set of n distributions of $P_1, \ldots, P_n \in \mathcal{E}$. The geometry of the error exponent in this multiple hypothesis testing setting has been investigated in [105]. On the dually flat exponential family manifold, it corresponds to check the exponential arcs between natural neighbors (sharing Voronoi subfaces) of a Bregman Voronoi diagram [40]. See Figure 14 for an illustration.

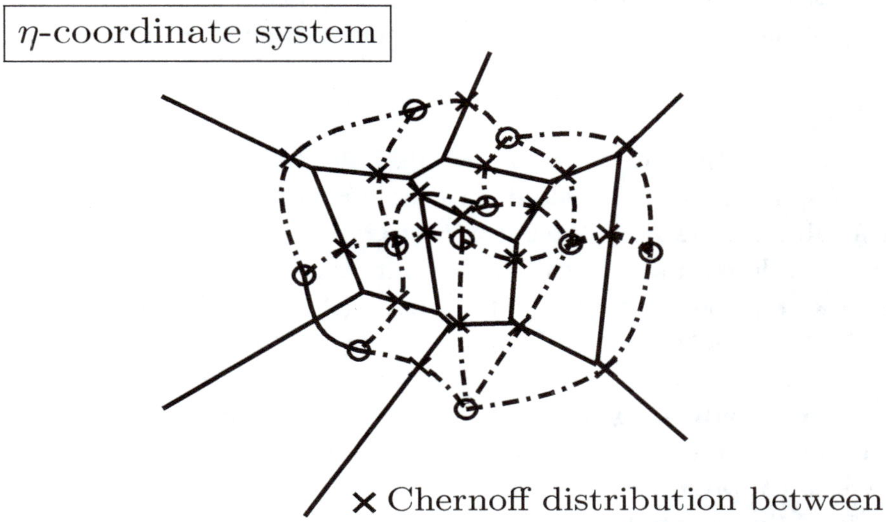

Figure 14. Geometric characterization of the best exponent error rate in the multiple hypothesis testing case.

4.4. *Clustering Mixtures in the Dually Flat Mixture Family Manifold* $(\mathcal{M}, \mathrm{KL})$

Given a set of k prescribed statistical distributions $p_0(x), \ldots, p_{k-1}(x)$, all sharing the same support $\mathcal{X}$ (say, $\mathbb{R}$), a mixture family $\mathcal{M}$ of order $D = k - 1$ consists of all strictly convex combinations of these component distributions [56]:

$$\mathcal{M} := \left\{ m(x;\theta) = \sum_{i=1}^{k-1} \theta_i p_i(x) + \left(1 - \sum_{i=1}^{k-1} \theta_i\right) p_0(x) \text{ such that } \theta_i > 0, \sum_{i=1}^{k-1} \theta_i < 1 \right\}. \quad (225)$$

Figure 15 displays two mixtures obtained as convex combinations of prescribed Laplacian, Gaussian and Cauchy component distributions ($D = 2$). When considering a set of prescribed Gaussian component distributions, we obtain a w-Gaussian Mixture Model, or w-GMM for short.

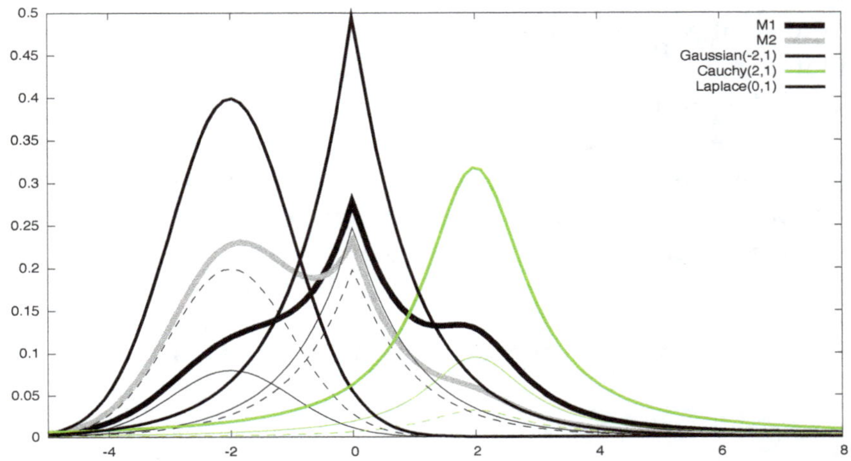

Figure 15. Example of a mixture family of order $D = 2$ (3 components: Laplacian, Gaussian and Cauchy prefixed distributions).

We consider the expected information manifold $(\mathcal{M}, {}^{\mathcal{M}}g, {}^{\mathcal{M}}\nabla^m, {}^{\mathcal{M}}\nabla^e)$ which is dually flat and equivalent to (M_Θ, KL). That is, the KL between two mixtures with prescribed components (w-mixtures, for short) is equivalent to a Bregman divergence for $F(\theta) = -h(m_\theta)$, where $h(p) = \int p(x) \log p(x) \mathrm{d}\mu(x)$ is the differential Shannon information (negative entropy) [56]:

$$\mathrm{KL}[m_{\theta_1} : m_{\theta_2}] = B_F(\theta_1 : \theta_2). \tag{226}$$

Consider a set $\{m_{\theta_1}, \ldots, m_{\theta_n}\}$ of n w-mixtures [56]. Because $F(\theta) = -h(m(x;\theta))$ is the negative differential entropy of a mixture (not available in closed form [106]), we approximate the untractable F by another close tractable generator $\tilde{F}$. We use Monte Carlo stochastic sampling to get Monte-Carlo convex $\tilde{F}_{\mathcal{S}}$ for an independent and identically distributed sample $\mathcal{S}$.

Thus we can build a nested sequence $(\mathcal{M}, \tilde{F}_{\mathcal{S}_1}), \ldots, (\mathcal{M}, \tilde{F}_{\mathcal{S}_m})$ of tractable dually flat manifolds for nested sample sets $\mathcal{S}_1 \subset \ldots \subset \mathcal{S}_m$ converging to the ideal mixture manifold $(\mathcal{M}, F)$: $\lim_{m\to\infty}(\mathcal{M}, \tilde{F}_{\mathcal{S}_m}) = (\mathcal{M}, F)$ (where convergence is defined with respect to the induced canonical Bregman divergence). A key advantage of this approach is that for a given sample $\mathcal{S}$, all computations carried inside the dually flat manifold $(\mathcal{M}, \tilde{F}_{\mathcal{S}})$ are consistent, see [56].

For example, we can apply Bregman k-means [107] on these Monte Carlo dually flat spaces [108] of w-GMMs (Gaussian Mixture Models) to cluster a set of w-GMMs. Figure 16 displays the result of such a clustering.

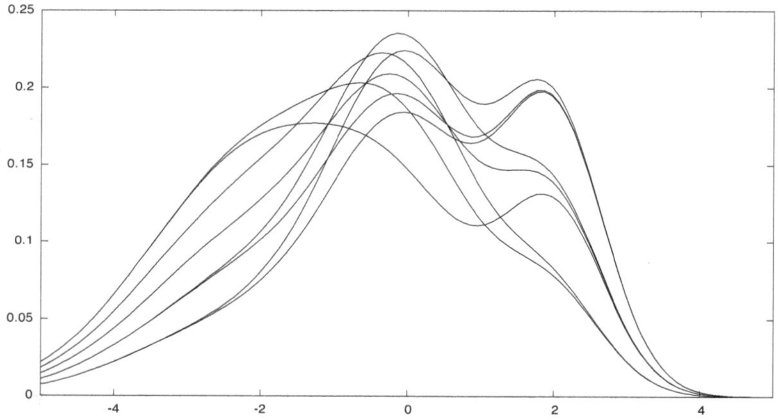

Figure 16. Example of w-GMM clustering into $k = 2$ clusters.

We have briefly described two applications using dually flat manifolds: (1) the dually flat exponential manifold induced by the statistical reverse Kullback–Leibler divergence on an exponential family (structure $(\mathcal{E}, \mathrm{KL}^*)$), and (2) the dually flat mixture manifold induced by the statistical Kullback–Leibler divergence on a mixture family (structure $(\mathcal{M}, \mathrm{KL})$). There are many other dually flat structures that can be met in a statistical context. For example, two other dually flat structures for the D-dimensional probability simplex Δ_D are reported in Amari's textbook [2]: (1) the conformally deforming of the α-geometry (page 88, Equation 4.95 of [2]), and (2) the χ-escort geometry (page 91, Equation 4.114 of [2]).

5. Conclusions: Summary, Historical Background, and Perspectives

5.1. Summary

We explained the dualistic nature of information manifolds (M, g, ∇, ∇^*) in information geometry. The dualistic structure is defined by a pair of conjugate connections coupled with the metric tensor that provides a dual parallel transport that preserves the metric. We showed how to extend this structure to a 1-parameter family of structures. From a pair of conjugate connections, the pipeline to build this 1-parameter family of structures can be informally summarized as:

$$(M, g, \nabla, \nabla^*) \Rightarrow (M, g, C) \Rightarrow (M, g, \alpha C) \Rightarrow (M, g, \nabla^{-\alpha}, \nabla^{\alpha}), \quad \forall \alpha \in \mathbb{R}. \tag{227}$$

We stated the fundamental theorem of information geometry on dual constant-curvature manifolds, including the special but important case of dually flat manifolds on which there exists two potential functions and global affine coordinate systems related by the Legendre–Fenchel transformation. Although, information geometry historically started with the Riemannian modeling $(\mathcal{P}, {}_{\mathcal{P}}g)$ of a parametric family of probability distributions $\mathcal{P}$ by letting the metric tensor be the Fisher information matrix, we have emphasized the dualistic view of information geometry which considers non-Riemannian manifolds that can be derived from any divergence, and not necessarily tied to a statistical context (e.g., information manifold can be used in mathematical programming [109]). Let us notice that for any symmetric divergence (e.g. any symmetrized f-divergence like the squared Hellinger divergence), the induced conjugate connections coincide with the Levi–Civita connection but the Fisher–Rao metric distance does not coincide with the squared Hellinger divergence.

On one hand, a Riemannian metric distance D_ρ is never a divergence because the rooted distance functions fail to be smooth at the extremities but a squared Riemmanian metric distance is always a divergence. On the other hand, taking the power δ of a divergence D (i.e., D^δ) for some $\delta > 0$ may yield a metric distance (e.g., the square root of the Jensen–Shannon divergence [110]), but this may not always be the case: the powered Jeffreys divergence J^δ is never a metric distance (see [111], page 889). Recently, the Optimal Transport (OT) theory [112] gained interest in statistics and machine learning. However, the optimal transport between two members of a same elliptically-contoured family has the same optimal transport formula distance (see [113] Eq. 16 and Eq. 17, although they have different Kullback–Leibler divergences). Another essential difference is that the Fisher–Rao manifold of location-scale families is hyperbolic but the Wasserstein manifold of location-scale families has positive curvature [113,114].

Notice that we may convert back and forth a similarity $S(p,q) \in (0,1]$ to a dissimilarity $D(p,q) \in [0,\infty)$ as follows:

$$S(p,q) = \exp(-D(p,q)) \in (0,1], \tag{228}$$
$$D(p,q) = -\log(S(p,q)) \in [0,\infty). \tag{229}$$

When the dissimilarity satisfies the (additive) triangle inequality (i.e., $D(p,q) + D(q,r) \geq D(p,r)$ for any triple (p,q,r)) then the corresponding similarity satisfies the multiplicative triangle inequality: $S(p,q) \times S(q,r) \leq S(p,r)$. A metric transform on a metric distance D is a transformation T such that $T(D(p,q))$ is a metric. The transformation $T(u) = \frac{1}{1+u}$ is a metric transform which bounds potentially unbounded metric distances; that is, if D is an unbounded metric, then $T(D(p,q)) = \frac{D(p,q)}{1+D(p,q)}$ is a bounded metric distance. The transformation $S(u) = u^2$ is not a metric transform since the squared of the Euclidean metric distance is not a metric distance.

5.2. A Brief Historical Review of Information Geometry

The field of Information Geometry (IG) was historically motivated by providing some differential-geometric structures to statistical models in order to reason geometrically about statistical problems with the endeavor goal of geometrizing mathematical statistics [4,12–14,115–118]: Professor Harold Hotelling [65] first considered in the late 1920s the Fisher Information Matrix (FIM) I as a Riemannian metric tensor g (i.e., the Fisher Information metric, FIm), and interpreted a parametric family of probability distributions M as a Riemannian manifold (M,g). Historically speaking, Hotelling attended the American Mathematical Society's Annual Meeting in Bethlehem (Bethlehem, PA, USA) on 26–29 December 1929, but left before his scheduled talk on December 27. His handwritten notes on the "Spaces of Statistical Parameters" was read by a colleague and are fully typeset in [68]. We warmly thank Professor Stigler for sending us the scanned handwritten notes and for discussing by emails some historical aspects of the birth of information geometry. In this pioneering work,

Hotelling mentioned that location-scale probability families yield Riemannian manifolds of constant non-positive curvatures. This Riemannian modeling of parametric family of densities was further independently studied by Calyampudi Radhakrishna Rao (C.R. Rao) in his celebrated paper [66] (1945) that also includes the Cramér–Rao lower bound [51] and the Rao–Blackwellization technique used in statistics. Nowadays the induced Riemannian metric distance is often called the Fisher–Rao distance [119] or Rao distance [81]. Yet another use of Riemannian geometry in statistics was pioneered by Harold Jeffreys [70] that proposed to use as an invariant prior the normalized volume element of the Fisher–Riemannian manifold. In those seminal papers, there was no theoretical justification of using the Fisher information matrix as a metric tensor (besides the fact that it is a well-defined positive-definite matrix for regular identifiable models). Nowadays, this Riemmanian metric tensor is called the information metric for short. Modern information geometry considers a generalization of this approach using a non-Riemannian dualistic modeling (M, g, ∇, ∇^*) which coincides with the Riemannian manifold when $\nabla = \nabla^* = {}^{LC}\nabla$, the Levi–Civita connection (the unique torsion-free affine connection compatible with the metric tensor). The Fisher–Rao geometry has also been explored in thermodynamics yielding the Ruppeiner geometry [120], and the geometry of thermodynamics is called nowadays called geometrothermodynamics [121].

In the 1960s, Nikolai Chentsov (also commonly written Čencov) studied the algebraic category of all statistical decision rules with its induced geometric structures: Namely, the α-geometries ("equivalent differential geometry") and the dually flat manifolds ("Nonsymmetric Pythagorean geometry" of the exponential families with respect to the Kullback–Leibler divergence). In the preface of the english translation of his 1972 russian monograph [115], the field of investigation is defined as "geometrical statistics." However in the original Russian monograph, Chentsov used the russian term geometrostatistics. According to Professor Alexander Holevo, the geometrostatistics term was coined by Andrey Kolmogorov to define the field of differential geometry of statistical models. In the monograph of Chentsov [115], the Fisher information metric is shown to be the unique metric tensor (up to a scaling factor) yielding statistical invariance under Markov morphisms (see [57] for a simpler proof that generalizes to positive measures).

The dual nature of the information geometry was thoroughly investigated by Professor Shun-ichi Amari [122]. In the preface of his 1985 monograph [116], Professor Amari coined the term information geometry as follows: "The differential-geometrical method developed in statistics is also applicable to other fields of sciences such as information theory and systems theory... They together will open a new field, which I would like to call information geometry." Professor Amari mentioned in [116] that he considered the Gaussian Riemannian manifold as a hyperbolic manifold in 1959, and was strongly influenced by Efron's paper on statistical curvature [123] (1975) to study the family of α-connections in the 1980s [122,124]. Professor Amari prepared his PhD under the supervision of Professor Kondo [125], an expert of differential geometry in touch with Professor Kawaguchi [126]. The role of differential geometry in statistics has been discussed in [127].

Note that the dual affine connections of information geometry have also been investigated independently in affine differential geometry [128] which considers invariance under volume-preserving affine transformations by defining a volume form instead of a metric form for Riemannian geometry. The notion of dual parallel transport compatible with the metric is due to Aleksandr Norden [129] and Rabindra Nath Sen [130–132] (see the Senian geometry in http://insaindia.res.in/detail/N54-0728).

We summarize the main fundamental structures of information manifolds below:

(M, g)	Riemannian manifold
$(\mathcal{P}, {}_{\mathcal{P}}g)$	Fisher–Riemannian (expected) Riemannian manifold
(M, g, ∇)	Riemannian manifold (M, g) with affine connection ∇
$(\mathcal{P}, {}_{\mathcal{P}}g, {}_{\mathcal{P}}{}^{e}\nabla^{\alpha})$	Chentsov's manifold with affine exponential α-connection
(M, g, ∇, ∇^*)	Amari's dualistic information manifold
$(\mathcal{P}, {}_{\mathcal{P}}g, {}_{\mathcal{P}}\nabla^{-\alpha}, {}_{\mathcal{P}}\nabla^{\alpha})$	Amari's (expected) information α-manifold, α-geometry
(M, g, C)	Lauritzen's statistical manifold [29]
$(M, {}^{D}g, {}^{D}\nabla, {}^{D^*}\nabla)$	Eguchi's conjugate connection manifold induced by divergence D
$(M, {}^{F}g, {}^{F}C)$	Chentsov/Amari's dually flat manifold induced by convex potential F

We use the $\equiv$ symbol to denote the equivalence of geometric structures. For example, we have $(M, g) \equiv (M, g, {}^{\mathrm{LC}}\nabla, {}^{\mathrm{LC}}\nabla^* = {}^{\mathrm{LC}}\nabla)$.

5.3. Perspectives

We recommend the two recent textbooks [2,16] for an indepth covering of (parametric) information geometry, and the book [133] for a thorough description of some infinite-dimensional statistical models. (Japanese readers may refer to [134,135]) We did not report the various coefficients of the metric tensors, Christoffel symbols and skewness tensors for the expected α-geometry of common parametric models like the multivariate Gaussian distributions, the Gamma/Beta distributions, etc. They can be found in [15,16] and in various articles dealing with less common family of distributions [15,64,136–140]. Although we have focused on the finite parametric setting, information geometry is also considering non-parametric families of distributions [141], and quantum information geometry [142].

We have shown that we can always create an information manifold (M, D) from any divergence function D. It is therefore important to consider generic classes of divergences in applications, that are ideally axiomatized and shown to have exhaustive characteristics. The α-skewed Jensen divergences [102] are defined by for a real-valued strictly convex function $F(\theta)$ by:

$$J_F^{\alpha}(\theta_1 : \theta_2) := (1-\alpha)F(\theta_1) + \alpha F(\theta_2) - F((1-\alpha)\theta_1 + \alpha\theta_2) > 0, \tag{230}$$

where both θ_1 and θ_2 belong to the parameter space Θ. Clearly, we have $J_F^{\alpha}(\theta_1 : \theta_2) = J_F^{1-\alpha}(\theta_2 : \theta_1)$. We have the following asymptotic properties of skewed Jensen divergences [102]:

$$\lim_{\alpha \to 0^+} \frac{1}{\alpha(1-\alpha)} J_F^{\alpha}(\theta_1 : \theta_2) = B_F(\theta_1 : \theta_2), \tag{231}$$

$$\lim_{\alpha \to 1^-} \frac{1}{\alpha(1-\alpha)} J_F^{\alpha}(\theta_1 : \theta_2) = B_F(\theta_2 : \theta_1), \tag{232}$$

where $B_F(\theta_1 : \theta_2)$ is the Bregman divergence [91] induced by a strictly convex and differentiable function $F(\theta)$:

$$B_F(\theta_1 : \theta_2) := F(\theta_1) - F(\theta_2) - (\theta_1 - \theta_2)F'(\theta_2). \tag{233}$$

Appendix C further reports how to interpret geometrically these Jensen/Bregman divergences from the chordal slope theorem. Beyond the three main Bregman/Csiszár/Jensen classes (theses classes overlap [143]), we may also mention the class of conformal divergences [73,144,145], the class of projective divergences [146,147], etc. Figure 17 illustrates the relationships between the principal classes of distances.

There are many perspectives on information geometry as attested by the new Springer journal (see online at https://www.springer.com/mathematics/geometry/journal/41884), and the biannual international conference "Geometric Sciences of Information" (GSI) [148–150] with its collective post-conference edited books [151,152]. We also mention the edited book [153] on the Occasion of Shun-ichi Amari's 80th birthday.

Additional materials are available online at https://FrankNielsen.github.io/SurveyIG/.

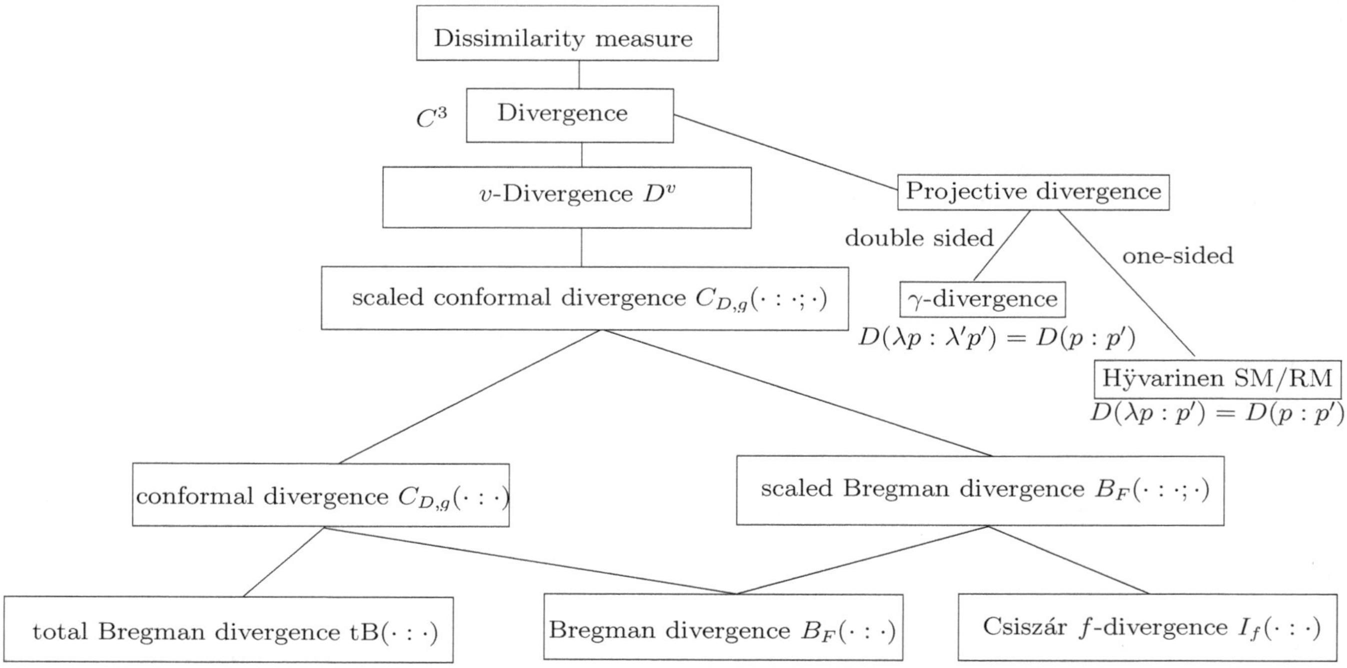

$D^v(P:Q) = D(v(P):v(Q))$ $\quad$ $\mathrm{tB}_F(P:Q) = \frac{B_F(P:Q)}{\sqrt{1+\|\nabla F(Q)\|^2}}$

$I_f(P:Q) = \int p(x) f\left(\left(\frac{q(x)}{p(x)}\right) \mathrm{d}\nu(x)\right.$ $\quad$ $C_{D,g}(P:Q) = g(Q)D(P:Q)$

$B_F(P:Q) = F(P) - F(Q) - \langle P-Q, \nabla F(Q)\rangle$ $\quad$ $B_{F,g}(P:Q;W) = WB_F\left(\frac{P}{Q}:\frac{Q}{W}\right)$

Figure 17. Principled classes of distances/divergences.

Funding: This research received no external funding.

Acknowledgments: FN would like to thank the organizers of the Geometry In Machine Learning workshop in 2018 (GiMLi, http://gimli.cc/2018/) for their kind keynote talk invitation, and specially Professor Søren Hauberg (Technical University of Denmark, DTU). This survey is based on the talk given at GiMLi. I am very thankful to Professor Stigler (University of Chicago, USA) and Professor Holevo (Steklov Mathematical Institute, Russia) for providing me feedback on some historical aspects of the field of information geometry. Finally, I would like to express my sincere thanks to Gaëtan Hadjeres (Sony Computer Science Laboratories Inc., Paris) for his careful proofreading and feedback.

Conflicts of Interest: The author declare no conflict of interest.

Appendix A. Monte Carlo Estimations of f-Divergences

Let (X, F, μ) be a probability space [154] with X denoting the sample space, F the σ-algebra, and μ a reference positive measure. The f-divergence [62,63] between two probability measures P and Q both absolutely continuous with respect to a positive measure μ for a convex generator $f:(0,\infty)\to\mathbb{R}$ strictly convex at 1 and satisfying $f(1)=0$ is

$$I_f(P:Q) = I_f(p:q) = \int p(x) f\left(\frac{q(x)}{p(x)}\right) \mathrm{d}\mu(x), \tag{A1}$$

where $P = p\mathrm{d}\mu$ and $Q = q\mathrm{d}\mu$ (i.e., p and q denote the Radon-Nikodym derivatives with respect to μ). We use the following conventions:

$$0f\left(\frac{0}{0}\right) = 0, \quad f(0) = \lim_{u\to 0^+} f(u), \quad \forall a>0, 0f\left(\frac{a}{0}\right) = \lim_{u\to 0^+} uf\left(\frac{a}{u}\right) = a\lim_{u\to\infty}\frac{f(u)}{u}. \tag{A2}$$

When $f(u) = -\log u$, we retrieve the Kullback–Leibler divergence (KLD):

$$D_{\mathrm{KL}}(p:q) = \int p(x) \log \frac{p(x)}{q(x)} \mathrm{d}\mu(x). \tag{A3}$$

The KLD is usually difficult to calculate in closed-form, say, for example, between statistical mixture models [155]. A common technique is to estimate the KLD using Monte Carlo sampling using a proposal distribution r:

$$\widehat{\mathrm{KL}}_n(p:q) = \frac{1}{n} \sum_{i=1}^{n} \frac{p(x_i)}{r(x_i)} \log \frac{p(x_i)}{q(x_i)}, \tag{A4}$$

where $x_1, \ldots, x_n \sim_{\mathrm{iid}} r$. When r is chosen as p, the KLD can be estimated as:

$$\widehat{\mathrm{KL}}_n(p:q) = \frac{1}{n} \sum_{i=1}^{n} \log \frac{p(x_i)}{q(x_i)}. \tag{A5}$$

Monte Carlo estimators are consistent under mild conditions: $\lim_{n\to\infty} \widehat{\mathrm{KL}}_n(p:q) = \mathrm{KL}(p:q)$.

In practice, one problem when implementing Equation (A5), is that we may end up potentially with $\widehat{\mathrm{KL}}_n(p:q) < 0$. This may have disastrous consequences as algorithms implemented by programs consider non-negative divergences to execute a correct workflow. The potential negative value problem of Equation (A5) comes from the fact that $\sum_i p(x_i) \neq 1$ and $\sum_i q(x_i) \neq 1$. One way to circumvent this problem is to consider the extended f-divergences:

Definition A1 (Extended f-divergence)**.** *The extended f-divergence for a convex generator f, strictly convex at 1 and satisfying $f(1) = 0$ is defined by*

$$I_f^e(p:q) = \int p(x) \left(f\left(\frac{q(x)}{p(x)}\right) - f'(1)\left(\frac{q(x)}{p(x)} - 1\right)\right) \mathrm{d}\mu(x). \tag{A6}$$

Indeed, for a strictly convex generator f, let us consider the scalar Bregman divergence [91]:

$$B_f(a:b) = f(a) - f(b) - (a-b)f'(b) \geq 0. \tag{A7}$$

Setting $a = \frac{q(x)}{p(x)}$ and $b = 1$ in Equation (A7), and using the fact that $f(1) = 0$, we get

$$f\left(\frac{q(x)}{p(x)}\right) - \left(\frac{q(x)}{p(x)} - 1\right) f'(1) \geq 0. \tag{A8}$$

Therefore we define the extended f-divergences as

$$I_f^e(p:q) = \int p(x) B_f\left(\frac{q(x)}{p(x)} : 1\right) \mathrm{d}\mu(x) \geq 0. \tag{A9}$$

That is, the formula for the extended f-divergences is

$$I_f^e(p:q) = \int p(x) \left(f\left(\frac{q(x)}{p(x)}\right) - f'(1)\left(\frac{q(x)}{p(x)} - 1\right)\right) \mathrm{d}\mu(x) \geq 0. \tag{A10}$$

Then we estimate the extended f-divergence using importance sampling of the integral with respect to distribution r, using n variates $x_1, \ldots, x_n \sim_{\mathrm{iid}} p$ as:

$$\hat{I}_{f,n}(p:q) = \frac{1}{n} \sum_{i=1}^{n} f\left(\frac{q(x_i)}{p(x_i)}\right) - f'(1)\left(\frac{q(x_i)}{p(x_i)} - 1\right) \geq 0. \tag{A11}$$

For example, for the KLD, we obtain the following Monte Carlo estimator:

$$\widehat{\mathrm{KL}}_n(p:q) = \frac{1}{n}\sum_{i=1}^{n}\left(\log\frac{p(x_i)}{q(x_i)} + \frac{q(x_i)}{p(x_i)} - 1\right) \geq 0, \tag{A12}$$

since the extended KLD is

$$D_{\mathrm{KL}^e}(p:q) = \int\left(p(x)\log\frac{p(x)}{q(x)} + q(x) - p(x)\right)\mathrm{d}\mu(x). \tag{A13}$$

Equation (A12) can be interpreted as a sum of scalar Itakura–Saito divergences since the Itakura–Saito divergence is scale-invariant: $\widehat{\mathrm{KL}}_n(p:q) = \frac{1}{n}\sum_{i=1}^{n} D_{\mathrm{IS}}(p(x_i):q(x_i))$ with the scalar Itakura–Saito divergence

$$D_{\mathrm{IS}}(a:b) = D_{\mathrm{IS}}\left(\frac{a}{b}:1\right) = \frac{a}{b} - \log\frac{a}{b} - 1 \geq 0, \tag{A14}$$

a Bregman divergence obtained for the generator $f(u) = -\log u$.

Notice that the extended f-divergence is a f-divergence for the generator

$$f_e(u) = f(u) - f'(1)(u-1). \tag{A15}$$

We check that the generator f_e satisfies both $f(1) = 0$ and $f'(1) = 0$, and we have $I_f^e(p:q) = I_{f_e}(p:q)$. Thus $D_{\mathrm{KL}^e}(p:q) = I_{f_{\mathrm{KL}}^e}(p:q)$ with $f_{\mathrm{KL}}^e(u) = -\log u + u - 1$.

Let us remark that we only need to have the scalar function strictly convex at 1 to ensure that $B_f\left(\frac{a}{b}:1\right) \geq 0$. Indeed, we may use the definition of Bregman divergences extended to strictly convex functions but not necessarily smooth functions [156,157]:

$$B_f(x:y) = \max_{g(y)\in\partial f(y)}\{f(x) - f(y) - (x-y)g(y)\}, \tag{A16}$$

where $\partial f(y)$ denotes the subderivative of f at y.

Furthermore, noticing that $I_{\lambda f}(p:q) = \lambda I_f(p:q)$ for $\lambda > 0$, we may enforce that $f''(1) = 1$, and obtain a standard f-divergence [2] which enjoys the property that $I_f(p_\theta(x):p_{\theta+\mathrm{d}\theta}(x)) = \mathrm{d}\theta^\top I(\theta)\mathrm{d}\theta$, where $I(\theta)$ denotes the Fisher information matrix of the parameteric family $\{p_\theta\}_\theta$ of densities.

Appendix B. The Multivariate Gaussian Family: An Exponential Family

We report the canonical decomposition of the multivariate Gaussian [158] family $\{N(\mu,\Sigma)$ such that $\mu\in\mathbb{R}^d, \Sigma\succ 0\}$ following [159]. The multivariate Gaussian family is also called the MultiVariate Normal family, or MVN family for short.

Let $\lambda := (\lambda_v, \lambda_M) = (\mu, \Sigma)$ denote the composite (vector,matrix) parameter of an MVN. The d-dimensional MVN density is given by

$$p_\lambda(x;\lambda) := \frac{1}{(2\pi)^{\frac{d}{2}}\sqrt{|\lambda_M|}}\exp\left(-\frac{1}{2}(x-\lambda_v)^\top\lambda_M^{-1}(x-\lambda_v)\right), \tag{A17}$$

where $|\cdot|$ denotes the matrix determinant. The natural parameters θ are also expressed using both a vector parameter θ_v and a matrix parameter θ_M in a compound object $\theta = (\theta_v, \theta_M)$. By defining the following compound inner product on a composite (vector,matrix) object

$$\langle\theta,\theta'\rangle := \theta_v^\top\theta'_v + \mathrm{tr}\left({\theta'_M}^\top\theta_M\right), \tag{A18}$$

where $\mathrm{tr}(\cdot)$ denotes the matrix trace, we rewrite the MVN density of Equation (A17) in the canonical form of an exponential family [52]:

$$p_\theta(x;\theta) \quad := \quad \exp\left(\langle t(x),\theta\rangle - F_\theta(\theta)\right) = p_\lambda(x;\lambda(\theta)), \tag{A19}$$

where

$$\theta = (\theta_v, \theta_M) = \left(\Sigma^{-1}\mu, -\frac{1}{2}\Sigma^{-1}\right) = \theta(\lambda) = \left(\lambda_M^{-1}\lambda_v, -\frac{1}{2}\lambda_M^{-1}\right), \tag{A20}$$

is the compound natural parameter and

$$t(x) = (x, -xx^\top) \tag{A21}$$

is the compound sufficient statistic. The function F_θ is the strictly convex and continuously differentiable log-normalizer defined by:

$$F_\theta(\theta) = \frac{1}{2}\left(d\log\pi - \log|\theta_M| + \frac{1}{2}\theta_v^\top\theta_M^{-1}\theta_v\right), \tag{A22}$$

The log-normalizer can be expressed using the ordinary parameters, $\lambda = (\mu, \Sigma)$, as:

$$\begin{aligned} F_\lambda(\lambda) &= \frac{1}{2}\left(\lambda_v^\top\lambda_M^{-1}\lambda_v + \log|\lambda_M| + d\log 2\pi\right), && \text{(A23)}\\ &= \frac{1}{2}\left(\mu^\top\Sigma^{-1}\mu + \log|\Sigma| + d\log 2\pi\right). && \text{(A24)}\end{aligned}$$

The moment/expectation parameters [2] are

$$\eta = (\eta_v, \eta_M) = E[t(x)] = \nabla F(\theta). \tag{A25}$$

We report the conversion formula between the three types of coordinate systems (namely the ordinary parameter λ, the natural parameter θ and the moment parameter η) as follows:

$$\left\{\begin{array}{l}\theta_v(\lambda) = \lambda_M^{-1}\lambda_v = \Sigma^{-1}\mu\\ \theta_M(\lambda) = \frac{1}{2}\lambda_M^{-1} = \frac{1}{2}\Sigma^{-1}\end{array}\right. \Leftrightarrow \left\{\begin{array}{l}\lambda_v(\theta) = \frac{1}{2}\theta_M^{-1}\theta_v = \mu\\ \lambda_M(\theta) = \frac{1}{2}\theta_M^{-1} = \Sigma\end{array}\right. \tag{A26}$$

$$\left\{\begin{array}{l}\eta_v(\theta) = \frac{1}{2}\theta_M^{-1}\theta_v\\ \eta_M(\theta) = -\frac{1}{2}\theta_M^{-1} - \frac{1}{4}(\theta_M^{-1}\theta_v)(\theta_M^{-1}\theta_v)^\top\end{array}\right. \Leftrightarrow \left\{\begin{array}{l}\theta_v(\eta) = -(\eta_M + \eta_v\eta_v^\top)^{-1}\eta_v\\ \theta_M(\eta) = -\frac{1}{2}(\eta_M + \eta_v\eta_v^\top)^{-1}\end{array}\right. \tag{A27}$$

$$\left\{\begin{array}{l}\lambda_v(\eta) = \eta_v = \mu\\ \lambda_M(\eta) = -\eta_M - \eta_v\eta_v^\top = \Sigma\end{array}\right. \Leftrightarrow \left\{\begin{array}{l}\eta_v(\lambda) = \lambda_v = \mu\\ \eta_M(\lambda) = -\lambda_M - \lambda_v\lambda_v^\top = -\Sigma - \mu\mu^\top\end{array}\right. \tag{A28}$$

The dual Legendre convex conjugate [2] is

$$F_\eta^*(\eta) = -\frac{1}{2}\left(\log(1 + \eta_v^\top\eta_M^{-1}\eta_v) + \log|-\eta_M| + d(1 + \log 2\pi)\right), \tag{A29}$$

and $\theta = \nabla_\eta F_\eta^*(\eta)$.

We check the Fenchel-Young equality when $\eta = \nabla F(\theta)$ and $\theta = \nabla F^*(\eta)$:

$$F_\theta(\theta) + F_\eta^*(\eta) - \langle\theta, \eta\rangle = 0. \tag{A30}$$

The Kullback-Leibler divergence between two d-dimensional Gaussians distributions $p_{(\mu_1,\Sigma_1)}$ and $p_{(\mu_2,\Sigma_2)}$ (with $\Delta_\mu = \mu_2 - \mu_1$) is

$$\mathrm{KL}(p_{(\mu_1,\Sigma_1)} : p_{(\mu_2,\Sigma_2)}) \quad = \quad \frac{1}{2}\left\{\mathrm{tr}(\Sigma_2^{-1}\Sigma_1) + \Delta_\mu^\top\Sigma_2^{-1}\Delta_\mu + \log\frac{|\Sigma_2|}{|\Sigma_1|} - d\right\} = \mathrm{KL}(p_{\lambda_1} : p_{\lambda_2}). \tag{A31}$$

We check that $\mathrm{KL}(p_{(\mu,\Sigma)} : p_{(\mu,\Sigma)}) = 0$ since $\Delta_\mu = 0$ and $\mathrm{tr}(\Sigma^{-1}\Sigma) = \mathrm{tr}(I) = d$. Notice that when $\Sigma_1 = \Sigma_2 = \Sigma$, we have

$$\mathrm{KL}(p_{(\mu_1,\Sigma)} : p_{(\mu_2,\Sigma)}) = \frac{1}{2}\Delta_\mu^\top \Sigma^{-1}\Delta_\mu = \frac{1}{2}D^2_{\Sigma^{-1}}(\mu_1, \mu_2), \tag{A32}$$

that is half the squared Mahalanobis distance for the precision matrix Σ^{-1} (a positive-definite matrix: $\Sigma^{-1} \succ 0$), where the Mahalanobis distance is defined for any positive matrix $Q \succ 0$ as follows:

$$D_Q(p_1 : p_2) = \sqrt{(p_1 - p_2)^\top Q(p_1 - p_2)}. \tag{A33}$$

The Kullback–Leibler divergence between two probability densities of the same exponential families amount to a Bregman divergence [2]:

$$\mathrm{KL}(p_{(\mu_1,\Sigma_1)} : p_{(\mu_2,\Sigma_2)}) = \mathrm{KL}(p_{\lambda_1} : p_{\lambda_2}) = B_F(\theta_2 : \theta_1) = B_{F^*}(\eta_1 : \eta_2), \tag{A34}$$

where the Bregman divergence is defined by

$$B_F(\theta : \theta') := F(\theta) - F(\theta') - \langle \theta - \theta', \nabla F(\theta')\rangle, \tag{A35}$$

with $\eta' = \nabla F(\theta')$. Define the canonical divergence [2]

$$A_F(\theta_1 : \eta_2) = F(\theta_1) + F^*(\eta_2) - \langle \theta_1, \eta_2\rangle = A_{F^*}(\eta_2 : \theta_1), \tag{A36}$$

since $F^{**} = F$. We have $B_F(\theta_1 : \theta_2) = A_F(\theta_1 : \eta_2)$.

Appendix C. Skew Jensen Divergences and Bregman Divergences

First, let us interpret geometrically the skewed Jensen divergences and the Bregman divergences using the chordal slope lemma of convexity theory [160] for univariate generators F:

Lemma A1 (Chordal slope lemma). *Let F be a strictly convex function on interval $I = (a, b)$ for $a < b$. For any $\theta_1 < \theta < \theta_2$ in I, we have:*

$$\frac{F(\theta) - F(\theta_1)}{\theta - \theta_1} < \frac{F(\theta_2) - F(\theta_1)}{\theta_2 - \theta_1} < \frac{F(\theta_2) - F(\theta)}{\theta_2 - \theta}. \tag{A37}$$

That is, define the points $P_1 = (\theta_1, F(\theta_1))$, $P = (\theta, F(\theta))$, and $P_2 = (\theta_2, F(\theta_2))$. Then the chordal slope lemma states that the slope of the chord $[P_1P]$ is less than the slope of the chord $[P_1P_2]$, which is less than the slope of $[PP_2]$ (Figure A1).

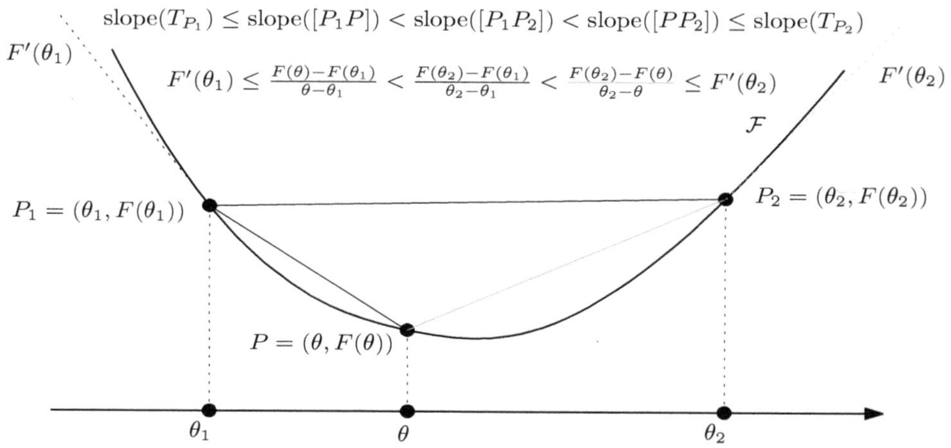

Figure A1. Illustration of the chordal slope lemma.

Let $\theta = (1-\alpha)\theta_1 + \alpha\theta_2$ for $\alpha \in (0,1)$. Then the inequalities of the chordal slope lemma can be rewritten as:

$$\frac{F(\theta) - F(\theta_1)}{\alpha(\theta_2 - \theta_1)} < \frac{F(\theta_2) - F(\theta_1)}{(\theta_2 - \theta_1)}, \tag{A38}$$

$$\frac{F(\theta_2) - F(\theta_1)}{(\theta_2 - \theta_1)} < \frac{F(\theta_2) - F(\theta)}{(1-\alpha)(\theta_2 - \theta_1)}, \tag{A39}$$

since $\theta - \theta_1 = \alpha(\theta_2 - \theta_1)$ and $\theta_2 - \theta = (1-\alpha)(\theta_2 - \theta_1)$. Thus we have:

$$F(\theta) - F(\theta_1) < \alpha(F(\theta_2) - F(\theta_1)), \tag{A40}$$

$$(1-\alpha)(F(\theta_2) - F(\theta_1)) < F(\theta_2) - F(\theta), \tag{A41}$$

or equivalently the following inequalities:

$$\alpha(F(\theta_2) - F(\theta_1)) - F(\theta) + F(\theta_1) > 0, \tag{A42}$$

$$0 < F(\theta_2) - F(\theta) + (1-\alpha)(F(\theta_1) - F(\theta_2)). \tag{A43}$$

That is, we recover for $\theta_1 < (1-\alpha)\theta_1 + \alpha\theta_2 < \theta_2$ (i.e., $\alpha \in (0,1)$ and $\theta_1 \neq \theta_2$) the α-skewed Jensen divergence [102]:

$$J_F^\alpha(\theta_1 : \theta_2) := (1-\alpha)F(\theta_1) + \alpha F(\theta_2) - F((1-\alpha)\theta_1 + \alpha\theta_2) > 0. \tag{A44}$$

Now, a consequence of the chordal slope lemma is that for a strictly convex and differentiable function F, we have:

$$F'(\theta_1) \leq \frac{F(\theta_2) - F(\theta_1)}{\theta_2 - \theta_1} \leq F'(\theta_2). \tag{A45}$$

This can be geometrically visualized in Figure A1.

That is,

$$F(\theta_2) - F(\theta_1) - (\theta_2 - \theta_1)F'(\theta_1) \geq 0, \tag{A46}$$

$$F(\theta_2) - F(\theta_1) - (\theta_2 - \theta_1)F'(\theta_2) \leq 0. \tag{A47}$$

We recognize the expressions of the Bregman divergences [91]:

$$B_F(\theta_1 : \theta_2) := F(\theta_1) - F(\theta_2) - (\theta_1 - \theta_2)F'(\theta_2), \tag{A48}$$

and get:

$$B_F(\theta_2 : \theta_1) \geq 0, \tag{A49}$$

$$B_F(\theta_1 : \theta_2) \geq 0. \tag{A50}$$

For multivariate strictly convex functions F, we observe that we can build a multivariate Bregman divergence from a family of 1D Bregman divergences [161] induced by the 1D strictly convex functions $F_{\theta_1,\theta_2}((1-\lambda)\theta_1 + \theta_2)$:

Theorem A1. *A multivariate Bregman divergence between two parameters θ_1 and θ_2 can be expressed as a univariate Bregman divergence for the generator F_{θ_1,θ_2} induced by the parameters:*

$$B_F(\theta_1 : \theta_2) = B_{F_{\theta_1,\theta_2}}(0 : 1), \quad \forall \theta_1 \neq \theta_2, \tag{A51}$$

where

$$F_{\theta_1,\theta_2}(u) := F(\theta_1 + u(\theta_2 - \theta_1)). \tag{A52}$$

Proof. The functions $\{F_{\theta_1,\theta_2}\}_{(\theta_1,\theta_2)\in\Theta^2}$ are 1D Bregman generators. Consider the directional derivative:

$$\nabla_{\theta_2-\theta_1} F_{\theta_1,\theta_2}(u) := \lim_{\epsilon\to 0} \frac{F(\theta_1+(\epsilon+u)(\theta_2-\theta_1)) - F(\theta_1+u(\theta_2-\theta_1))}{\epsilon}, \tag{A53}$$

$$= (\theta_2-\theta_1)^\top \nabla F(\theta_1+u(\theta_2-\theta_1)), \tag{A54}$$

Since $F_{\theta_1,\theta_2}(0) = F(\theta_1)$, $F_{\theta_1,\theta_2}(1) = F(\theta_2)$, and $F'_{\theta_1,\theta_2}(u) = \nabla_{\theta_2-\theta_1} F_{\theta_1,\theta_2}(u)$, it follows that

$$B_{F_{\theta_1,\theta_2}}(0:1) = F_{\theta_1,\theta_2}(0) - F_{\theta_1,\theta_2}(1) - (0-1)\nabla_{\theta_2-\theta_1} F_{\theta_1,\theta_2}(1), \tag{A55}$$

$$= F(\theta_1) - F(\theta_2) + (\theta_2-\theta_1)^\top \nabla F(\theta_2), \tag{A56}$$

$$= B_F(\theta_1:\theta_2). \tag{A57}$$

□

Notations

Below is a list of notations we used in this document:

$[D]$	$[D] := \{1,\ldots,D\}$
$\langle\cdot,\cdot\rangle$	inner product
$M_Q(u,v) = \|u-v\|_Q$	Mahalanobis distance $M_Q(u,v) = \sqrt{\sum_{i,j}(u^i-v^i)(u^j-v^j)Q_{ij}}$, $Q \succ 0$
$D(\theta:\theta')$	parameter divergence
$D[p(x):p'(x)]$	statistical divergence
D, D^*	Divergence and dual (reverse) divergence
Csiszár divergence I_f	$I_f(\theta:\theta') := \sum_{i=1}^D \theta_i f\left(\frac{\theta_i'}{\theta_i}\right)$ with $f(1)=0$
Bregman divergence B_F	$B_F(\theta:\theta') := F(\theta) - F(\theta') - (\theta-\theta')^\top \nabla F(\theta')$
Canonical divergence A_{F,F^*}	$A_{F,F^*}(\theta:\eta') = F(\theta) + F^*(\eta') - \theta^\top \eta'$
Bhattacharyya distance	$B_\alpha[p_1:p_2] = -\log \int_{x\in\mathcal{X}} p_1^\alpha(x) p_2^{1-\alpha}(x)\mathrm{d}\mu(x)$
Jensen/Burbea-Rao divergence	$J_F^{(\alpha)}(\theta_1:\theta_2) = \alpha F(\theta_1) + (1-\alpha)F(\theta_2) - F(\theta_1+(1-\alpha)\theta_2)$
Chernoff information	$C[P_1,P_2] = -\log \min_{\alpha\in(0,1)} \int_{x\in\mathcal{X}} p_1^\alpha(x) p_2^{1-\alpha}(x)\mathrm{d}\mu(x)$
F, F^*	Potential functions related by Legendre–Fenchel transformation
$D_\rho(p,q)$	Riemannian distance $D_\rho(p,q) := \int_0^1 \|\gamma'(t)\|_{\gamma(t)}\mathrm{d}t$
B, B^*	basis, reciprocal basis
$B = \{e_1 = \partial_1,\ldots,e_D = \partial_D\}$	natural basis
$\{\mathrm{d}x_i\}_i$	covector basis (one-forms)
$(v)_B := (v^i)$	contravariant components of vector v
$(v)_{B^*} := (v_i)$	covariant components of vector v
$u \perp v$	vector u is perpendicular to vector v ($\langle u,v\rangle = 0$)
$\|v\| = \sqrt{\langle v,v\rangle}$	induced norm, length of a vector v
M, S	Manifold, submanifold
T_p	tangent plane at p
TM	Tangent bundle $TM = \cup_p T_p = \{(p,v), p\in M, v\in T_p\}$
$\mathfrak{F}(M)$	space of smooth functions on M
$\mathfrak{X}(M) = \Gamma(TM)$	space of smooth vector fields on M
vf	direction derivative of f with respect to vector v
$X,Y,Z \in \mathfrak{X}(M)$	Vector fields
$g \overset{\Sigma}{=} g_{ij}\mathrm{d}x_i \otimes \mathrm{d}x_j$	metric tensor (field)
$(\mathcal{U},x)$	local coordinates x in a chat $\mathcal{U}$

$\partial_i{:=:}\frac{\partial}{\partial x_i}$	natural basis vector
$\partial^i{:=:}\frac{\partial}{\partial x^i}$	natural reciprocal basis vector
∇	affine connection
$\nabla_X Y$	covariant derivative
$\prod_c^\nabla$	parallel transport of vectors along a smooth curve c
$\prod_c^\nabla v$	Parallel transport of $v \in T_{c(0)}$ along a smooth curve c
γ, γ_∇	geodesic, geodesic with respect to connection ∇
$\Gamma_{ij,l}$	Christoffel symbols of the first kind (functions)
Γ_{ij}^k	Christoffel symbols of the second kind (functions)
R	Riemann–Christoffel curvature tensor
$[X,Y]$	Lie bracket $[X,Y](f) = X(Y(f)) - Y(X(f)), \forall f \in \mathfrak{F}(M)$
∇-projection	$P_S = \arg\min_{Q\in S} D(\theta(P) : \theta(Q))$
∇^*-projection	$P_S^* = \arg\min_{Q\in S} D(\theta(Q) : \theta(P))$
C	Amari–Chentsov totally symmetric cubic 3-covariant tensor
$\mathcal{P} = \{p_\theta(x)\}_{\theta_i n\Theta}$	parametric family of probability distributions
$\mathcal{E}, \mathcal{M}, \Delta_D$	exponential family, mixture family, probability simplex
${}_\mathcal{P}I(\theta)$	Fisher information matrix of family $\mathcal{P}$
${}_\mathcal{P}I(\theta)$	Fisher Information Matrix (FIM) for a parametric family $\mathcal{P}$
${}_\mathcal{P}g$	Fisher information metric tensor field
exponential connection ${}^e_\mathcal{P}\nabla$	${}^e_\mathcal{P}\nabla := E_\theta\left[(\partial_i\partial_j l)(\partial_k l)\right]$
mixture connection ${}^m_\mathcal{P}\nabla$	${}^m_\mathcal{P}\nabla := E_\theta\left[(\partial_i\partial_j l + \partial_i l \partial_j l)(\partial_k l)\right]$
expected skewness tensor C_{ijk}	$C_{ijk} := E_\theta\left[\partial_i l \partial_j l \partial_k l\right]$
expected α-connections	${}_\mathcal{P}\Gamma^{\alpha}{}_{ij}^{k} := -\frac{1+\alpha}{2} C_{ijk} = E_\theta\left[\left(\partial_i\partial_j l + \frac{1-\alpha}{2}\partial_i l \partial_j l\right)(\partial_k l)\right]$
$\equiv$	equivalence of geometric structures

References

1. Shannon, C.E. A mathematical theory of communication. *Bell Syst. Tech. J.* **1948**, *27*, 623–656. [CrossRef]
2. Amari, S. *Information Geometry and Its Applications*; Applied Mathematical Sciences; Springer: Tokyo, Japan, 2016.
3. Kakihara, S.; Ohara, A.; Tsuchiya, T. Information Geometry and Interior-Point Algorithms in Semidefinite Programs and Symmetric Cone Programs. *J. Optim. Theory Appl.* **2013**, *157*, 749–780. [CrossRef]
4. Amari, S.; Nagaoka, H. *Methods of Information Geometry*; American Mathematical Society: Providence, RL, USA 2007.
5. Peirce, C.S. *Chance, Love, and Logic: Philosophical Essays*; U of Nebraska Press: Lincoln, NE, USA, 1998.
6. Schurz, G. Patterns of abduction. *Synthese* **2008**, *164*, 201–234. [CrossRef]
7. Wald, A. Statistical decision functions. *Ann. Math. Stat.* **1949**, 165–205. [CrossRef]
8. Wald, A. *Statistical Decision Functions*; Wiley: Chichester, UK, 1950.
9. Dabak, A.G. A Geometry for Detection Theory. Ph.D. Thesis, Rice University, Houston, TX, USA, 1993.
10. Do Carmo, M.P. *Differential Geometry of Curves and Surfaces*; Courier Dover Publications: New York, NY, USA, 2016.
11. Amari, S.; Barndorff-Nielsen, O.E.; Kass, R.E.; Lauritzen, S.L.; Rao, C.R. *Differential Geometry in Statistical Inference*; Institute of Mathematical Statistics: Hayward, CA, USA, 1987.
12. Dodson, C.T.J. (Ed.) *Geometrization of Statistical Theory*; ULDM Publications; University of Lancaster, Department of Mathematics: Bailrigg, UK, 1987.
13. Murray, M.; Rice, J. *Differential Geometry and Statistics*; Number 48 in Monographs on Statistics and Applied Probability; Chapman and Hall: London, UK, 1993.
14. Kass, R.E.; Vos, P.W. *Geometrical Foundations of Asymptotic Inference*; Wiley-Interscience: New York, NY, USA, 1997.
15. Arwini, K.A.; Dodson, C.T.J. *Information Geometry: Near Randomness and Near Independance*; Springer: Berlin, Germany, 2008.
16. Calin, O.; Udriste, C. *Geometric Modeling in Probability and Statistics*; Mathematics and Statistics, Springer International Publishing: Cham, Switzerland, 2014.

17. Ay, N.; Jost, J.; Vân Lê, H.; Schwachhöfer, L. *Information Geometry*; Springer: Berlin, Germany, 2017; Volume 64.
18. Corcuera, J.; Giummolè, F. A characterization of monotone and regular divergences. *Ann. Inst. Stat. Math.* **1998**, *50*, 433–450. [CrossRef]
19. Mühlich, U. *Fundamentals of Tensor Calculus for Engineers with a Primer on Smooth Manifolds*; Springer: Berlin, Germany, 2017; Volume 230.
20. Nielsen, F.; Nock, R. Hyperbolic Voronoi diagrams made easy. In Proceedings of the IEEE International Conference on Computational Science and Its Applications (ICCSA), Fukuoka, Japan, 23–26 March 2010; pp. 74–80.
21. Whitney, H.; Eells, J.; Toledo, D. *Collected Papers of Hassler Whitney*; Nelson Thornes: London, UK, 1992.
22. Absil, P.A.; Mahony, R.; Sepulchre, R. *Optimization Algorithms on Matrix Manifolds*; Princeton University Press: Princeton, NJ, USA, 2009.
23. Cartan, E.J. *On Manifolds with an Affine Connection and the Theory of General Relativity*; Bibliopolis; Humanities Pr: London, UK, 1986.
24. Akivis, M.A.; Rosenfeld, B.A. *Élie Cartan (1869–1951)*; American Mathematical Society: Cambridge, MA, USA, 2011; Volume 123.
25. Wanas, M. Absolute parallelism geometry: Developments, applications and problems. *arXiv* **2002**, arXiv:gr-qc/0209050.
26. Bourguignon, J.P. Ricci curvature and measures. *Jpn. J. Math.* **2009**, 4, 27–45. [CrossRef]
27. Baez, J.C.; Wise, D.K. Teleparallel gravity as a higher gauge theory. *Commun. Math. Phys.* **2015**, *333*, 153–186. [CrossRef]
28. Ashburner, J.; Friston, K.J. Diffeomorphic registration using geodesic shooting and Gauss-Newton optimisation. *NeuroImage* **2011**, *55*, 954–967. [CrossRef]
29. Lauritzen, S.L. Statistical manifolds. *Differ. Geom. Stat. Inference* **1987**, *10*, 163–216.
30. Vân Lê, H. Statistical manifolds are statistical models. *J. Geom.* **2006**, *84*, 83–93.
31. Furuhata, H. Hypersurfaces in statistical manifolds. *Differ. Geom. Its Appl.* **2009**, *27*, 420–429. [CrossRef]
32. Zhang, J. Divergence functions and geometric structures they induce on a manifold. In *Geometric Theory of Information*; Nielsen, F., Ed.; Springer: Berlin, Germany, 2014; pp. 1–30.
33. Eguchi, S. Second order efficiency of minimum contrast estimators in a curved exponential family. *Ann. Stat.* **1983**, *11*, 793–803. [CrossRef]
34. Eguchi, S. A differential geometric approach to statistical inference on the basis of contrast functionals. *Hiroshima Math. J.* **1985**, *15*, 341–391. [CrossRef]
35. Hiriart-Urruty, J.B.; Lemaréchal, C. *Fundamentals of Convex Analysis*; Springer Science & Business Media: Berlin, Germany, 2012.
36. Crouzeix, J.P. A relationship between the second derivatives of a convex function and of its conjugate. *Math. Program.* **1977**, *13*, 364–365. [CrossRef]
37. Ay, N.; Amari, S. A novel approach to canonical divergences within information geometry. *Entropy* **2015**, *17*, 8111–8129. [CrossRef]
38. Nielsen, F. What is ... an information projection? *Not. AMS* **2018**, *65*, 321–324. [CrossRef]
39. Kurose, T. On the divergences of 1-conformally flat statistical manifolds. *Tohoku Math. J. Second Ser.* **1994**, *46*, 427–433. [CrossRef]
40. Boissonnat, J.D.; Nielsen, F.; Nock, R. Bregman Voronoi diagrams. *Discret. Comput. Geom.* **2010**, *44*, 281–307. [CrossRef]
41. Nielsen, F.; Piro, P.; Barlaud, M. Bregman vantage point trees for efficient nearest neighbor queries. In Proceedings of the 2009 IEEE International Conference on Multimedia and Expo, New York, NY, USA, 28 June–3 July 2009; pp. 878–881.
42. Nielsen, F.; Boissonnat, J.D.; Nock, R. Visualizing Bregman Voronoi diagrams. In Proceedings of the Twenty-Third Annual Symposium on Computational Geometry, Gyeongju, Korea, 6–8 June 2007; pp. 121–122.
43. Nock, R.; Nielsen, F. Fitting the smallest enclosing Bregman ball. In Proceedings of the European Conference on Machine Learning, Porto, Portugal, 3–7 October 2005; Springer: Berlin, Germany, 2005; pp. 649–656.
44. Nielsen, F.; Nock, R. On the smallest enclosing information disk. *Inf. Process. Lett.* **2008**, *105*, 93–97. [CrossRef]

45. Fischer, K.; Gärtner, B.; Kutz, M. Fast smallest-enclosing-ball computation in high dimensions. In Proceedings of the European Symposium on Algorithms, Budapest, Hungary, 16–19 September 2003; Springer: Berlin, Germany, 2003; pp. 630–641.
46. Della Pietra, S.; Della Pietra, V.; Lafferty, J. Inducing features of random fields. *IEEE Trans. Pattern Anal. Mach. Intell.* **1997**, *19*, 380–393. [CrossRef]
47. Nielsen, F. On Voronoi Diagrams on the Information-Geometric Cauchy Manifolds. *Entropy* **2020**, *22*, 713. [CrossRef]
48. Shima, H. *The Geometry of Hessian Structures*; World Scientific: New Jersey, NJ, USA, 2007.
49. Zhang, J. Reference duality and representation duality in information geometry. *AIP Conf. Proc.* **2015**, *1641*, 130–146.
50. Gomes-Gonçalves, E.; Gzyl, H.; Nielsen, F. Geometry and Fixed-Rate Quantization in Riemannian Metric Spaces Induced by Separable Bregman Divergences. In Proceedings of the 4th International Conference on Geometric Science of Information (GSI), Toulouse, France, 27–29 August 2019; Nielsen, F., Barbaresco, F., Eds.; Lecture Notes in Computer Science; Springer: Berlin, Germany, 2019; Volume 11712, pp. 351–358. [CrossRef]
51. Nielsen, F. Cramér-Rao lower bound and information geometry. In *Connected at Infinity II*; Springer: Berlin, Germany, 2013; pp. 18–37.
52. Nielsen, F.; Garcia, V. Statistical exponential families: A digest with flash cards. *arXiv* **2009**, arXiv:0911.4863.
53. Sato, Y.; Sugawa, K.; Kawaguchi, M. The geometrical structure of the parameter space of the two-dimensional normal distribution. *Rep. Math. Phys.* **1979**, *16*, 111–119. [CrossRef]
54. Skovgaard, L.T. A Riemannian geometry of the multivariate normal model. *Scand. J. Stat.* **1984**, *11*, 211–223.
55. Malagò, L.; Pistone, G. Information geometry of the Gaussian distribution in view of stochastic optimization. In Proceedings of the 2015 ACM Conference on Foundations of Genetic Algorithms XIII, Aberystwyth, UK, 17–20 January 2015; pp. 150–162.
56. Nielsen, F.; Nock, R. On the geometry of mixtures of prescribed distributions. In Proceedings of the 2018 IEEE International Conference on Acoustics, Speech and Signal Processing (ICASSP), Calgary, AB, Canada, 15–20 April 2018; pp. 2861–2865.
57. Campbell, L.L. An extended Čencov characterization of the information metric. *Proc. Am. Math. Soc.* **1986**, *98*, 135–141.
58. Vân Lê, H. The uniqueness of the Fisher metric as information metric. *Ann. Inst. Stat. Math.* **2017**, *69*, 879–896.
59. Csiszár, I.; Shields, P.C. *Information Theory and Statistics: A Tutorial*; Foundations and Trends® in Communications and Information Theory; Now Publishers Inc.: Hanover, MA, USA, 2004; Volume 1, pp. 417–528.
60. Jiao, J.; Courtade, T.A.; No, A.; Venkat, K.; Weissman, T. Information measures: The curious case of the binary alphabet. *IEEE Trans. Inf. Theory* **2014**, *60*, 7616–7626. [CrossRef]
61. Qiao, Y.; Minematsu, N. A Study on Invariance of f-Divergence and Its Application to Speech Recognition. *IEEE Trans. Signal Process.* **2010**, *58*, 3884–3890. [CrossRef]
62. Nielsen, F.; Nock, R. On the chi square and higher-order chi distances for approximating f-divergences. *IEEE Signal Process. Lett.* **2013**, *21*, 10–13. [CrossRef]
63. Csiszár, I. Information-type measures of difference of probability distributions and indirect observation. *Stud. Sci. Math. Hung.* **1967**, *2*, 229–318.
64. Mitchell, A.F.S. Statistical manifolds of univariate elliptic distributions. *Int. Stat. Rev.* **1988**, *56*, 1–16. [CrossRef]
65. Hotelling, H. Spaces of statistical parameters. *Bull. Am. Math. Soc. (AMS)* **1930**, *36*, 191.
66. Rao, R.C. Information and the accuracy attainable in the estimation of statistical parameters. *Bull. Calcutta Math. Soc.* **1945**, *37*, 81–91.
67. Komaki, F. Bayesian prediction based on a class of shrinkage priors for location-scale models. *Ann. Inst. Stat. Math.* **2007**, *59*, 135–146. [CrossRef]
68. Stigler, S.M. The epic story of maximum likelihood. *Stat. Sci.* **2007**, *22*, 598–620. [CrossRef]
69. Rao, C.R. Information and the accuracy attainable in the estimation of statistical parameters. In *Breakthroughs in Statistics*; Springer: Berlin, Germany, 1992; pp. 235–247.

70. Jeffreys, H. An invariant form for the prior probability in estimation problems. *Proc. R. Soc. Lond. A* **1946**, *186*, 453–461.
71. Zhang, J. On monotone embedding in information geometry. *Entropy* **2015**, *17*, 4485–4499. [CrossRef]
72. Naudts, J.; Zhang, J. Rho–tau embedding and gauge freedom in information geometry. *Inf. Geom.* **2018**. [CrossRef]
73. Nock, R.; Nielsen, F.; Amari, S. On Conformal Divergences and Their Population Minimizers. *IEEE TIT* **2016**, *62*, 527–538. [CrossRef]
74. Azoury, K.S.; Warmuth, M.K. Relative loss bounds for on-line density estimation with the exponential family of distributions. *Mach. Learn.* **2001**, *43*, 211–246. [CrossRef]
75. Banerjee, A.; Merugu, S.; Dhillon, I.S.; Ghosh, J. Clustering with Bregman divergences. *J. Mach. Learn. Res.* **2005**, *6*, 1705–1749.
76. Nielsen, F.; Nock, R. Entropies and cross-entropies of exponential families. In Proceedings of the 2010 IEEE International Conference on Image Processing, Hong Kong, China, 26–29 September 2010; pp. 3621–3624.
77. Nielsen, F.; Nock, R. Cumulant-free closed-form formulas for some common (dis)similarities between densities of an exponential family. *Tech. Rep.* **2020**. [CrossRef]
78. Amari, S. Differential geometry of a parametric family of invertible linear systems: Riemannian metric, dual affine connections, and divergence. *Math. Syst. Theory* **1987**, *20*, 53–82. [CrossRef]
79. Schwander, O.; Nielsen, F. Fast learning of Gamma mixture models with *k*-MLE. In Proceedings of the International Workshop on Similarity-Based Pattern Recognition, York, UK, 3–5 July 2013; Springer: Berlin, Germany, 2013; pp. 235–249.
80. Miura, K. An introduction to maximum likelihood estimation and information geometry. *Interdiscip. Inf. Sci.* **2011**, *17*, 155–174. [CrossRef]
81. Reverter, F.; Oller, J.M. Computing the Rao distance for Gamma distributions. *J. Comput. Appl. Math.* **2003**, *157*, 155–167. [CrossRef]
82. Pinele, J.; Strapasson, J.E.; Costa, S.I. The Fisher-Rao Distance between Multivariate Normal Distributions: Special Cases, Bounds and Applications. *Entropy* **2020**, *22*, 404. [CrossRef]
83. Nielsen, F. Pattern learning and recognition on statistical manifolds: An information-geometric review. In Proceedings of the International Workshop on Similarity-Based Pattern Recognition, York, UK, 3–5 July 2013; Springer: Berlin, Germany, 2013; pp. 1–25.
84. Sun, K.; Nielsen, F. Lightlike Neuromanifolds, Occam's Razor and Deep Learning. *arXiv* **2019**, arXiv:1905.11027.
85. Sun, K.; Nielsen, F. Relative Fisher Information and Natural Gradient for Learning Large Modular Models. In Proceedings of the 34th International Conference on Machine Learning (ICML), Sydney, Australia, 6–11 August 2017; pp. 3289–3298.
86. Amari, S. Natural gradient works efficiently in learning. *Neural Comput.* **1998**, *10*, 251–276. [CrossRef]
87. Cauchy, A. Methode générale pour la résolution des systèmes d'équations simultanées. *C. R. l'Académie Sci.* **1847**, *25*, 536–538.
88. Curry, H.B. The method of steepest descent for non-linear minimization problems. *Q. Appl. Math.* **1944**, *2*, 258–261. [CrossRef]
89. Bonnabel, S. Stochastic gradient descent on Riemannian manifolds. *IEEE Trans. Autom. Control* **2013**, *58*, 2217–2229. [CrossRef]
90. Nielsen, F. On geodesic triangles with right angles in a dually flat space. *arXiv* **2019**, arXiv:1910.03935.
91. Bregman, L.M. The relaxation method of finding the common point of convex sets and its application to the solution of problems in convex programming. *USSR Comput. Math. Math. Phys.* **1967**, *7*, 200–217. [CrossRef]
92. Nielsen, F.; Hadjeres, G. Monte Carlo information-geometric structures. In *Geometric Structures of Information*; Springer, Berlin, Germany, 2019; pp. 69–103.
93. Nielsen, F. *Legendre Transformation and Information Geometry*; Springer, Berlin/Heidelberg, Germany, 2010.
94. Raskutti, G.; Mukherjee, S. The information geometry of mirror descent. *IEEE Trans. Inf. Theory* **2015**, *61*, 1451–1457. [CrossRef]
95. Bubeck, S. *Convex Optimization: Algorithms and Complexity*; Foundations and Trends® in Machine Learning: Hanover, MA, USA, 2015; Volume 8, pp. 231–357.
96. Zhang, G.; Sun, S.; Duvenaud, D.; Grosse, R. Noisy natural gradient as variational inference. In Proceedings of the International Conference on Machine Learning, Stockholm, Sweden, 10–15 July 2018; pp. 5852–5861.

97. Beyer, H.G.; Schwefel, H.P. Evolution strategies–A comprehensive introduction. *Nat. Comput.* **2002**, *1*, 3–52. [CrossRef]
98. Berny, A. Selection and reinforcement learning for combinatorial optimization. In Proceedings of the International Conference on Parallel Problem Solving from Nature, Paris, France, 18–20 September 2000; Springer: Berlin, Germany, 2000; pp. 601–610.
99. Wierstra, D.; Schaul, T.; Glasmachers, T.; Sun, Y.; Peters, J.; Schmidhuber, J. Natural evolution strategies. *J. Mach. Learn. Res.* **2014**, *15*, 949–980.
100. Nielsen, F. An Information-Geometric Characterization of Chernoff Information. *IEEE Sig. Proc. Lett.* **2013**, *20*, 269–272. [CrossRef]
101. Pham, G.; Boyer, R.; Nielsen, F. Computational Information Geometry for Binary Classification of High-Dimensional Random Tensors. *Entropy* **2018**, *20*, 203. [CrossRef]
102. Nielsen, F.; Boltz, S. The Burbea-Rao and Bhattacharyya centroids. *IEEE Trans. Inf. Theory* **2011**, *57*, 5455–5466. [CrossRef]
103. Nielsen, F. Chernoff Information of Exponential Families. *arXiv* **2011** arXiv:1102.2684.
104. Nielsen, F. Generalized Bhattacharyya and Chernoff upper bounds on Bayes error using quasi-arithmetic means. *Pattern Recognit. Lett.* **2014**, *42*, 25–34. [CrossRef]
105. Nielsen, F. Hypothesis Testing, Information Divergence and Computational Geometry. In Proceedings of the International Conference on Geometric Science of Information Geometric Science of Information (GSI), Paris, France, 28–30 August 2013; pp. 241–248.
106. Nielsen, F.; Sun, K. Guaranteed Bounds on Information-Theoretic Measures of Univariate Mixtures Using Piecewise Log-Sum-Exp Inequalities. *Entropy* **2016**, *18*, 442. [CrossRef]
107. Nielsen, F.; Nock, R. Sided and symmetrized Bregman centroids. *IEEE Trans. Inf. Theory* **2009**, 55, 2882–2904. [CrossRef]
108. Nielsen, F.; Hadjeres, G. Monte Carlo Information Geometry: The dually flat case. *arXiv* **2018**, arXiv:1803.07225.
109. Ohara, A.; Tsuchiya, T. *An Information Geometric Approach to Polynomial-Time Interior-Point Algorithms: Complexity Bound via Curvature Integral*; Research Memorandum; The Institute of Statistical Mathematics: Tokyo, Japan, 2007; Volume 1055.
110. Fuglede, B.; Topsøe, F. Jensen-Shannon divergence and Hilbert space embedding. In Proceedings of the IEEE International Symposium on Information Theory (ISIT), Chicago, IL, USA, 27 June–2 July 2004; p. 31.
111. Vajda, I. On metric divergences of probability measures. *Kybernetika* **2009**, *45*, 885–900.
112. Villani, C. *Optimal Transport: Old and New*; Springer Science & Business Media: Berlin, Germany, 2008; Volume 338.
113. Dowson, D.C.; Landau, B.V. The Fréchet distance between multivariate normal distributions. *J. Multivar. Anal.* **1982**, *12*, 450–455. [CrossRef]
114. Takatsu, A. Wasserstein geometry of Gaussian measures. *Osaka J. Math.* **2011**, *48*, 1005–1026.
115. Chentsov, N.N. *Statistical Decision Rules and Optimal Inference*; Monographs; American Mathematical Society: Providence, RI, USA, 1982.
116. Amari, S. *Differential-Geometrical Methods in Statistics*; Lecture Notes on Statistics; Second Edition in 1990; Springer: New York, NY, USA, 1985; Volume 28.
117. Amari, S.; Nagaoka, H. *Methods of Information Geometry*; Jouhou kika no Houhou; Iwanami Shoten: Tokyo, Japan, 1993. (In Japanese)
118. Gibilisco, P.; Riccomagno, E.; Rogantin, M.P.; Wynn, H.P. (Eds.) *Algebraic and Geometric Methods in Statistics*; Cambridge University Press: Cambridge, UK, 2009.
119. Srivastava, A.; Wu, W.; Kurtek, S.; Klassen, E.; Marron, J.S. Registration of Functional Data Using Fisher-Rao Metric. *arXiv* **2011**, arXiv:1103.3817.
120. Wei, S.W.; Liu, Y.X.; Mann, R.B. Ruppeiner geometry, phase transitions, and the microstructure of charged AdS black holes. *Phys. Rev. D* **2019**, *100*, 124033. [CrossRef]
121. Quevedo, H. Geometrothermodynamics. *J. Math. Phys.* **2007**, *48*, 013506. [CrossRef]
122. Amari, S. *Theory of Information Spaces: A Differential Geometrical Foundation of Statistics*; Post RAAG Reports; Tokyo, Japan, 1980. Available online: Available online: https://bsi-ni.brain.riken.jp/database/item/92 (accessed on 8 May 2012). (accessed on 29 September 2020).

123. Efron, B. Defining the curvature of a statistical problem (with applications to second order efficiency). *Ann. Stat.* **1975**, *3*, 1189–1242. [CrossRef]
124. Nagaoka, H.; Amari, S. *Differential Geometry of Smooth Families of Probability Distributions*; Technical Report; METR 82-7; University of Tokyo: Tokyo, Japan, 1982.
125. Croll, G.J. The Natural Philosophy of Kazuo Kondo. *arXiv* **2007**, arXiv:0712.0641.
126. Kawaguchi, M. An introduction to the theory of higher order spaces I. The theory of Kawaguchi spaces. *RAAG Memoirs* **1960**, *3*, 718–734.
127. Barndorff-Nielsen, O.E.; Cox, D.R.; Reid, N. The role of differential geometry in statistical theory. *Int. Stat. Rev.* **1986**, *54*, 83–96. [CrossRef]
128. Nomizu, K.; Katsumi, N.; Sasaki, T. *Affine Differential Geometry: Geometry of Affine Immersions*; Cambridge University Press: Cambridge, UK, 1994.
129. Norden, A.P. On Pairs of Conjugate Parallel Displacements in Multidimensional Spaces. In *Doklady Akademii nauk SSSR*; Kazan State University, Comptes rendus de l'Académie des Sciences de l'URSS: Kazan, Russia, 1945; Volume 49, pp. 1345–1347.
130. Sen, R.N. On parallelism in Riemannian space I. *Bull. Calcutta Math. Soc.* **1944**, *36*, 102–107.
131. Sen, R.N. On parallelism in Riemannian space II. *Bull. Calcutta Math. Soc.* **1944**, *37*, 153–159.
132. Sen, R.N. On parallelism in Riemannian space III. *Bull. Calcutta Math. Soc.* **1946**, *38*, 161–167.
133. Giné, E.; Nickl, R. *Mathematical Foundations of Infinite-Dimensional Statistical Models*; Cambridge University Press: Cambridge, UK, 2015; Volume 40.
134. Amari, S. *New Developments of Information Geometry*; Jouhou Kikagaku no Shintenkai; Saiensu'sha: Tokyo, Japan, 2014. (In Japanese)
135. Fujiwara, A. *Foundations of Information Geometry*; Jouhou Kikagaku no Kisou; Makino Shoten: Tokyo, Japan, 2015; p. 223. (In Japanese)
136. Mitchell, A.F.S. The information matrix, skewness tensor and α-connections for the general multivariate elliptic distribution. *Ann. Inst. Stat. Math.* **1989**, *41*, 289–304. [CrossRef]
137. Zhang, Z.; Sun, H.; Zhong, F. Information geometry of the power inverse Gaussian distribution. *Appl. Sci.* **2007**, *9*, 194–203.
138. Peng, T.L.L.; Sun, H. The geometric structure of the inverse gamma distribution. *Contrib. Algebra Geom.* **2008**, *49*, 217–225.
139. Zhong, F.; Sun, H.; Zhang, Z. The geometry of the Dirichlet manifold. *J. Korean Math. Soc.* **2008**, *45*, 859–870. [CrossRef]
140. Peng, L.; Sun, H.; Jiu, L. The geometric structure of the Pareto distribution. *Bol. Asoc. Mat. Venez.* **2007**, *14*, 5–13.
141. Pistone, G. Nonparametric information geometry. In *Geometric Science of Information*; Springer: Berlin, Germany, 2013; pp. 5–36.
142. Hayashi, M. *Quantum Information*; Springer: Berlin, Germany, 2006.
143. Pardo, M.d.C.; Vajda, I. About distances of discrete distributions satisfying the data processing theorem of information theory. *IEEE Trans. Inf. Theory* **1997**, *43*, 1288–1293. [CrossRef]
144. Nielsen, F.; Nock, R. Total Jensen divergences: Definition, properties and k-means++ clustering. *arXiv* **2013**, arXiv:1309.7109.
145. Nielsen, F.; Nock, R. Total Jensen divergences: Definition, properties and clustering. In Proceedings of the 2015 IEEE International Conference on Acoustics, Speech and Signal Processing (ICASSP), Brisbane, Australia, 19–24 April 2015; pp. 2016–2020.
146. Nielsen, F.; Nock, R. Patch Matching with Polynomial Exponential Families and Projective Divergences. In Proceedings of the International Conference on Similarity Search and Applications (SISAP), Tokyo, Japan, 24–26 October 2016, pp. 109–116.
147. Nielsen, F.; Sun, K.; Marchand-Maillet, S. On Hölder Projective Divergences. *Entropy* **2017**, *19*, 122. [CrossRef]
148. Nielsen, F.; Barbaresco, F. (Eds.) *Geometric Science of Information*; Lecture Notes in Computer Science; Springer: Berlin, Germany, 2013; Volume 8085. [CrossRef]
149. Nielsen, F.; Barbaresco, F. (Eds.) *Geometric Science of Information*; Lecture Notes in Computer Science; Springer: Berlin, Germany, 2015; Volume 9389. [CrossRef]
150. Nielsen, F.; Barbaresco, F. (Eds.) *Geometric Science of Information*; Lecture Notes in Computer Science; Springer: Berline, Germany, 2017; Volume 10589. [CrossRef]

151. Nielsen, F. *Geometric Structures of Information*; Springer: Berlin, Germany, 2018.
152. Nielsen, F. *Geometric Theory of Information*; Springer: Berlin, Germany, 2014.
153. Ay, N.; Gibilisco, P.; Matús, F. *Information Geometry and its Applications: On the Occasion of Shun-ichi Amari's 80th Birthday, IGAIA IV Liblice, Czech Republic, 12–17 June 2016*; Springer Proceedings in Mathematics & Statistics; Springer: Berlin, Germany, 2018; Volume 252.
154. Keener, R.W. *Theoretical Statistics: Topics for a Core Course*; Springer: Berlin, Germany, 2011.
155. Nielsen, F.; Sun, K. Guaranteed bounds on the Kullback–Leibler divergence of univariate mixtures. *IEEE Signal Process. Lett.* **2016**, *23*, 1543–1546. [CrossRef]
156. Gordon, G.J. Approximate Solutions to Markov Decision Processes. Ph.D. Thesis, Carnegie Mellon University, Pittsburgh, PA, USA, 1999.
157. Telgarsky, M.; Dasgupta, S. Agglomerative Bregman clustering. In Proceedings of the 29th International Conference on Machine Learning, Edinburgh, UK, 26 June–1 July 2012; Omnipress: Madison, WI, USA, pp. 1011–1018.
158. Yoshizawa, S.; Tanabe, K. Dual differential geometry associated with Kullback–Leibler information on the Gaussian distributions and its 2-parameter deformations. *SUT J. Math.* **1999**, *35*, 113–137.
159. Nielsen, F. On the Jensen–Shannon symmetrization of distances relying on abstract means. *Entropy* **2019**, *21*, 485. [CrossRef]
160. Niculescu, C.; Persson, L.E. *Convex Functions and Their Applications*; Springer: Berlin, Germany, 2006.
161. Nielsen, F.; Nock, R. The Bregman chord divergence. In Proceedings of the International Conference on Geometric Science of Information, Toulouse, France, 27–29 August 2019; Springer: Berlin, Germany, pp. 299–308.

2

Information Geometry of Randomized Quantum State Tomography

Akio Fujiwara [1,*] and Koichi Yamagata [2]

1 Department of Mathematics, Osaka University, Toyonaka, Osaka 560-0043, Japan
2 Graduate School of Informatics and Engineering, The University of Electro-Communications, Chofu, Tokyo 182-8585, Japan; koichi.yamagata@uec.ac.jp
* Correspondence: fujiwara@math.sci.osaka-u.ac.jp

Abstract: Suppose that a d-dimensional Hilbert space $\mathcal{H} \simeq \mathbb{C}^d$ admits a full set of mutually unbiased bases $\left\{|1^{(a)}\rangle, \ldots, |d^{(a)}\rangle\right\}$, where $a = 1, \ldots, d+1$. A randomized quantum state tomography is a scheme for estimating an unknown quantum state on $\mathcal{H}$ through iterative applications of measurements $M^{(a)} = \left\{|1^{(a)}\rangle\langle 1^{(a)}|, \ldots, |d^{(a)}\rangle\langle d^{(a)}|\right\}$ for $a = 1, \ldots, d+1$, where the numbers of applications of these measurements are random variables. We show that the space of the resulting probability distributions enjoys a mutually orthogonal dualistic foliation structure, which provides us with a simple geometrical insight into the maximum likelihood method for the quantum state tomography.

Keywords: quantum state tomography; mutually unbiased bases; information geometry; dualistic foliation; mixed coordinate system

---

1. Introduction

Quantum state tomography is a method of estimating an unknown quantum state represented on some Hilbert space $\mathcal{H}$, consisting of a fixed set of measurements that provides sufficient information about the unknown quantum state, as well as a data processing that maps each measurement outcome into the quantum state space $\mathcal{S}(\mathcal{H})$ on $\mathcal{H}$ [1]. A set of measurements that fulfils this requirement is sometimes called a measurement basis. For mathematical simplicity, we restrict ourselves to Hilbert spaces of finite dimensions.

To elucidate our motivation, let us treat the simplest case when $\mathcal{H} \simeq \mathbb{C}^2$. It is well known that there is a one-to-one affine correspondence between the qubit state space

$$\mathcal{S}(\mathbb{C}^2) := \{\rho \in \mathbb{C}^{2\times 2} \mid \rho \geq 0,\ \mathrm{Tr}\,\rho = 1\}$$

and the unit ball (called the Bloch ball)

$$B := \left\{ x = (x_1, x_2, x_3) \in \mathbb{R}^3 \;\middle|\; \|x\|^2 := (x_1)^2 + (x_2)^2 + (x_3)^2 \leq 1 \right\}.$$

In fact, the correspondence is explicitly given by the Stokes parametrization

$$x \longmapsto \rho_x = \frac{1}{2}(I + x_1\sigma_1 + x_2\sigma_2 + x_3\sigma_3),$$

where σ_1, σ_2, and σ_3 are the standard Pauli matrices. Since $E_{\rho_x}[\sigma_i] := \mathrm{Tr}\,\rho_x\sigma_i = x_i$ for $i \in \{1,2,3\}$, the set $\sigma = (\sigma_1, \sigma_2, \sigma_3)$ of observables is regarded as an unbiased estimator [2–4] for the Stokes parameter $x = (x_1, x_2, x_3)$. This is the basic idea behind the standard qubit state tomography, which runs as follows: suppose that, among N independent experiments, the ith Pauli matrix σ_i was measured $N/3$ times, and outcomes $+1$ (spin-up) and -1 (spin-down) were obtained n_i^+ and n_i^- times, respectively. Then a naive estimate for the true value of the parameter $x = (x_1, x_2, x_3)$ is

$$\hat{x} = (\hat{x}_1, \hat{x}_2, \hat{x}_3) := \left(\frac{n_1^+ - n_1^-}{N/3}, \frac{n_2^+ - n_2^-}{N/3}, \frac{n_3^+ - n_3^-}{N/3} \right).$$

When the estimate $\hat{x} \in [-1, 1]^3$ falls outside the Bloch ball B, it needs to be corrected so that the new estimate lies in the Bloch ball B. The maximum likelihood method is a canonical one to obtain a corrected estimate [2,5–10]. From the point of view of information geometry [11–13], the maximum likelihood estimate (MLE) is the orthogonal projection from the temporary estimate $\hat{x}$ onto the Bloch ball B with respect to the standard Fisher metric along the $\nabla^{(m)}$-geodesic [14], (cf., Appendix A).

Now let us deal with a slightly generalized situation: suppose that the ith Pauli matrix σ_i was measured N_i times and outcomes $+1$ and -1 were obtained n_i^+ and n_i^- times, respectively, where $\{N_i\}_{i=1,2,3}$ were random variables. Such a situation arises in an actual experiment due to unexpected particle loss [15]. We shall call such a generalized estimation scheme a *randomized state tomography*. A naive estimate in this case is the following:

$$\hat{x} = (\hat{x}_1, \hat{x}_2, \hat{x}_3) := \left(\frac{n_1^+ - n_1^-}{N_1}, \frac{n_2^+ - n_2^-}{N_2}, \frac{n_3^+ - n_3^-}{N_3} \right).$$

One may invoke the maximum likelihood method when $\hat{x}$ falls outside the Bloch ball. It is then interesting to ask if there is also a useful geometrical picture for the MLE even when the numbers N_i of measurements are random variables.

The above mentioned problem is naturally extended to quantum state tomography on an arbitrary Hilbert space that admits a full set of mutually unbiased bases [16,17]. In a d-dimensional Hilbert space $\mathcal{H} \simeq \mathbb{C}^d$, k orthonormal bases

$$\left\{ |\alpha^{(1)}\rangle \right\}_{\alpha \in \{1,\ldots,d\}}, \left\{ |\beta^{(2)}\rangle \right\}_{\beta \in \{1,\ldots,d\}}, \ldots, \left\{ |\gamma^{(k)}\rangle \right\}_{\gamma \in \{1,\ldots,d\}}$$

are called *mutually unbiased* if they satisfy

$$\left| \left\langle \alpha^{(a)} \,\middle|\, \beta^{(b)} \right\rangle \right|^2 = \frac{1}{d}$$

for all $a, b \in \{1, \ldots, k\}$ with $a \neq b$, and $\alpha, \beta \in \{1, \ldots, d\}$. It is known that the number k of mutually unbiased bases (MUBs) is at most $d+1$ [18]. If there are $d+1$ MUBs, the Hilbert space $\mathcal{H}$ is said to admit a full set of MUBs. For example, when the dimension d of $\mathcal{H}$ is a power of a prime, $\mathcal{H}$ admits a full set of MUBs [19]. Whether or not any Hilbert space admits a full set of MUBs is an open question [16].

In what follows, unless otherwise stated, we assume that the Hilbert space $\mathcal{H} \simeq \mathbb{C}^d$ under consideration admits a full set of MUBs. As demonstrated in Appendix B (cf., [17,20]), each density operator $\rho \in \mathcal{S}(\mathcal{H})$ can be uniquely represented as

$$\rho = \rho(\xi) := \sum_{a=1}^{d+1} \left\{ \sum_{\alpha=1}^{d-1} \xi_\alpha^{(a)} M_\alpha^{(a)} + \left(1 - \sum_{\beta=1}^{d-1} \xi_\beta^{(a)} \right) M_d^{(a)} \right\} - I, \tag{1}$$

where

$$M^{(a)} := \left\{ M_1^{(a)}, \ldots, M_d^{(a)} \right\} = \left\{ |1^{(a)}\rangle \langle 1^{(a)}|, \ldots, |d^{(a)}\rangle \langle d^{(a)}| \right\}$$

is the projection-valued measure (PVM) associated with the ath orthogonal basis in the MUBs, and

$$\xi := \left(\xi_\alpha^{(a)}\right)_{(a,\alpha)\in\{1,\dots,d+1\}\times\{1,\dots,d-1\}}$$

is a (d^2-1)-dimensional real parameter that is chosen so that $\rho(\xi) \geq 0$. A simple calculation shows that, if the ath measurement $M^{(a)}$ is applied to the state $\rho(\xi)$, one obtains each outcome $\alpha \in \{1,\dots,d\}$ with probability

$$p_\alpha^{(a)} = \operatorname{Tr}\rho(\xi)M_\alpha^{(a)} = \begin{cases} \xi_\alpha^{(a)}, & \text{for } \alpha = 1,\dots,d-1, \\ 1 - \sum_{\beta=1}^{d-1} \xi_\beta^{(a)}, & \text{for } \alpha = d. \end{cases} \tag{2}$$

This implies that the parametrization $\xi \mapsto \rho(\xi)$ establishes an affine isomorphism between the quantum state space

$$\mathcal{S}(\mathbb{C}^d) := \left\{\rho \in \mathbb{C}^{d\times d} \,\middle|\, \rho \geq 0,\ \operatorname{Tr}\rho = 1\right\}$$

and the convex set

$$B := \left\{\xi \in \mathbb{R}^{d^2-1} \,\middle|\, \rho(\xi) \geq 0\right\}.$$

Incidentally, the Stokes parametrization $x \mapsto \rho_x$ for the qubit state space $\mathcal{S}(\mathbb{C}^2)$ is regarded as a special case of the above parametrization $\xi \mapsto \rho(\xi)$ for $\mathcal{S}(\mathbb{C}^d)$. In fact, the eigenvectors of the Pauli matrices $\sigma_1, \sigma_2, \sigma_3$ form a full set of MUBs on $\mathbb{C}^2$, and the Stokes parametrization $x = (x_1, x_2, x_3)$ is related to the above parametrization $\xi = (\xi_1^{(1)}, \xi_1^{(2)}, \xi_1^{(3)})$ as

$$\xi_1^{(a)} = \frac{x_a + 1}{2}, \qquad (a = 1,2,3).$$

Now that a standard affine parametrization $\xi \mapsto \rho(\xi)$ has been established on an arbitrary Hilbert space $\mathcal{H} \simeq \mathbb{C}^d$ that admits a full set of MUBs, the scheme of randomized state tomography is naturally extended to $\mathcal{H}$ as follows. Suppose that the ath measurement $M^{(a)}$ was applied $N^{(a)}$ times and the outcome $\alpha \in \{1,\dots,d\}$ was obtained $n_\alpha^{(a)}$ times, where $\left\{N^{(a)}\right\}_{a=1,\dots,d+1}$ were random variables. Then, due to (2), a naive estimate for the parameter $\xi_\alpha^{(a)}$ is

$$\hat{\xi}_\alpha^{(a)} = \frac{n_\alpha^{(a)}}{N^{(a)}}.$$

When the estimate $\hat{\xi} := \left(\hat{\xi}_\alpha^{(a)}\right) \in [0,1]^{d^2-1}$ falls outside the parameter space B, one may invoke the maximum likelihood method to obtain a corrected estimate.

The objective of the present paper is to clarify that the $\nabla^{(m)}$-projection interpretation for the MLE is still valid for the randomized state tomography by changing the standard Fisher metric into a deformed one depending on the realization of the random variables $N^{(a)}$, which might as well be called a randomized Fisher metric. Such a novel geometrical picture will provide important insights into the quantum metrology.

The paper is organized as follows. In Section 2, we first introduce a statistical model on an extended sample space Ω that represents the randomized state tomography. We then clarify that the probability simplex $\mathcal{P}(\Omega)$ is decomposed into mutually orthogonal dualistic foliation by means of certain $\nabla^{(m)}$- and $\nabla^{(e)}$-autoparallel submanifolds. In Section 3, we give a statistical interpretation of the above-mentioned dualistic foliation structure. In particular, we point out that the MLE is the $\nabla^{(m)}$-projection with respect to a deformed Fisher metric that depends on the realization of the random variables $N^{(a)}$. These results are demonstrated by several illustrative examples in Section 4. Finally, some concluding remarks are presented in Section 5. For the reader's convenience, some background information is provided in Appendices A and B, including information geometry of the MLE and affine parametrization of a quantum state space $\mathcal{S}(\mathcal{H})$.

2. Geometry of Randomized State Tomography

We identify the randomized state tomography on $\mathcal{H} \simeq \mathbb{C}^d$ with the following scheme [21]: at each step of the measurement, one chooses a PVM $M^{(a)}$ at random with probability $s^{(a)}$, $(a = 1, \ldots, d+1)$, and applies the chosen PVM to yield an outcome $\alpha \in \{1, \ldots, d\}$. The sample space Ω for this statistical picture is

$$\Omega = \{(a, \alpha) \,|\, a \in \{1, \ldots, d+1\},\ \alpha \in \{1, \ldots, d\}\}.$$

Suppose that the unknown state ρ is specified by the coordinate $\xi \in B$ as (1). Then the corresponding probability distribution on Ω is represented by the $d(d+1)$-dimensional probability vector

$$\begin{aligned} p_{(s,\xi)} := \Bigg(& s^{(1)} \left(\xi_1^{(1)}, \ldots, \xi_{d-1}^{(1)}, 1 - \sum_{\alpha=1}^{d-1} \xi_\alpha^{(1)} \right), \ldots, s^{(d)} \left(\xi_1^{(d)}, \ldots, \xi_{d-1}^{(d)}, 1 - \sum_{\alpha=1}^{d-1} \xi_\alpha^{(d)} \right), \\ & \left(1 - \sum_{a=1}^{d} s^{(a)} \right) \left(\xi_1^{(d+1)}, \ldots, \xi_{d-1}^{(d+1)}, 1 - \sum_{\alpha=1}^{d-1} \xi_\alpha^{(d+1)} \right) \Bigg) \end{aligned}$$

where the parameter $s := (s^{(1)}, \ldots, s^{(d)})$ belongs to the domain

$$D := \left\{ s \in \mathbb{R}^d \;\middle|\; s^{(a)} > 0 \text{ for } a \in \{1, \ldots, d\}, \text{ and } \sum_{a=1}^{d} s^{(a)} < 1 \right\}.$$

Note that the family

$$\left\{ p_{(s,\xi)} \middle| s \in D, \xi \in \Xi \right\}$$

with

$$\begin{aligned} \Xi := \Bigg\{ \xi \in \mathbb{R}^{d^2-1} \;\Bigg|\; & \xi_\alpha^{(a)} > 0 \text{ for } (a, \alpha) \in \{1, \ldots, d+1\} \times \{1, \ldots, d-1\}, \\ & \text{and } \sum_{\alpha=1}^{d-1} \xi_\alpha^{(a)} < 1 \text{ for } a \in \{1, \ldots, d+1\} \Bigg\} \end{aligned}$$

forms a $(d^2 + d - 1)$-dimensional open probability simplex $\mathcal{P}(\Omega)$, and the parameters (s, ξ) form a coordinate system of $\mathcal{P}(\Omega)$. Since we are only interested in estimating the parameter $\xi \in \Xi$, the remaining parameter $s \in D$ is understood as a set of nuisance parameters [2,12]. In what follows, we regard $\mathcal{P}(\Omega)$ as a statistical manifold endowed with the standard dualistic structure $\left(g, \nabla^{(e)}, \nabla^{(m)}\right)$, where g is the Fisher metric, and $\nabla^{(e)}$ and $\nabla^{(m)}$ are the exponential and mixture connections [12].

Let us consider the following submanifolds of $\mathcal{P}(\Omega)$:

$$M(s) := \left\{ p_{(s,\xi)} \mid \xi \in \Xi \right\}$$

for each $s \in D$, and

$$E(\xi) := \left\{ p_{(s,\xi)} \mid s \in D \right\}$$

for each $\xi \in \Xi$. Since $M(s)$ and $E(\xi)$ are convex subsets of $\mathcal{P}(\Omega)$, they are both $\nabla^{(m)}$-autoparallel. In addition, we have the following.

Proposition 1. *For each $\xi \in \Xi$, the submanifold $E(\xi)$ is $\nabla^{(e)}$-autoparallel. Furthermore, for each $s \in D$ and $\xi \in \Xi$, the submanifolds $M(s)$ and $E(\xi)$ are mutually orthogonal with respect to the Fisher metric g.*

Proof. Let us change the coordinate system (s,ξ) into $(\eta_{\langle a\rangle},\eta_{\langle b,\alpha\rangle})$, where

$$\eta_{\langle a\rangle} := s^{(a)}$$

for $a\in\{1,\dots,d\}$, and

$$\eta_{\langle b,\alpha\rangle} := s^{(b)}\xi_\alpha^{(b)}$$

for $(b,\alpha)\in\{1,\dots,d+1\}\times\{1,\dots,d-1\}$. With this coordinate transformation, the probability vector $p_{(s,\xi)}$ is rewritten as

$$p_\eta = \bigoplus_{a=1}^{d+1}\left(\eta_{\langle a,1\rangle},\dots,\eta_{\langle a,d-1\rangle},\eta_{\langle a\rangle}-\sum_{\alpha=1}^{d-1}\eta_{\langle a,\alpha\rangle}\right). \tag{3}$$

Here, $\eta_{\langle d+1\rangle}$ is a function of $\{\eta_{\langle a\rangle}\}_{a\in\{1,\dots,d\}}$ defined by

$$\eta_{\langle d+1\rangle} := 1-\sum_{a=1}^{d}\eta_{\langle a\rangle},$$

and is not a component of the coordinate system $\eta := (\eta_{\langle a\rangle},\eta_{\langle b,\alpha\rangle})$. We see from the representation (3) that the coordinate system η is $\nabla^{(m)}$-affine. The potential function for η is given by the negative entropy

$$\begin{aligned}\varphi(\eta) &:= \sum_{\omega\in\Omega} p_\eta(\omega)\log p_\eta(\omega)\\ &= \sum_{a=1}^{d+1}\left\{\sum_{\alpha=1}^{d-1}\eta_{\langle a,\alpha\rangle}\log\eta_{\langle a,\alpha\rangle}+\left(\eta_{\langle a\rangle}-\sum_{\beta=1}^{d-1}\eta_{\langle a,\beta\rangle}\right)\log\left(\eta_{\langle a\rangle}-\sum_{\beta=1}^{d-1}\eta_{\langle a,\beta\rangle}\right)\right\}\end{aligned}$$

and the dual $\nabla^{(e)}$-affine coordinate system θ is given by

$$\theta^{\langle a\rangle} = \frac{\partial\varphi}{\partial\eta_{\langle a\rangle}} = \log\frac{s^{(a)}}{\left(1-\sum_{b=1}^{d}s^{(b)}\right)}+\log\frac{\left(1-\sum_{\beta=1}^{d-1}\xi_\beta^{(a)}\right)}{\left(1-\sum_{\beta=1}^{d-1}\xi_\beta^{(d+1)}\right)}$$

for $a\in\{1,\dots,d\}$, and

$$\theta^{\langle b,\alpha\rangle} = \frac{\partial\varphi}{\partial\eta_{\langle b,\alpha\rangle}} = \log\frac{\xi_\alpha^{(b)}}{\left(1-\sum_{\beta=1}^{d-1}\xi_\beta^{(b)}\right)}$$

for $(b,\alpha)\in\{1,\dots,d+1\}\times\{1,\dots,d-1\}$. Thus, fixing ξ is equivalent to fixing the coordinates $\left(\theta^{\langle b,\alpha\rangle}\right)_{(b,\alpha)\in\{1,\dots,d+1\}\times\{1,\dots,d-1\}}$, and the submanifold $E(\xi)$ is generated by changing the remaining parameters $\left(\theta^{\langle a\rangle}\right)_{a\in\{1,\dots,d\}}$. This implies that $E(\xi)$ is $\nabla^{(e)}$-autoparallel, proving the first part of the claim.

To prove the second part, let us introduce a mixed coordinate system [11]

$$\left(\eta_{\langle a\rangle};\theta^{\langle b,\alpha\rangle}\right)_{a\in\{1,\dots,d\},\,(b,\alpha)\in\{1,\dots,d+1\}\times\{1,\dots,d-1\}}$$

of $\mathcal{P}(\Omega)$. Since $\eta_{\langle a\rangle}=s^{(a)}$, the submanifold $M(s)$ is rewritten as

$$M(s) = \left\{p_{(s,\xi)}\,\middle|\,\left(\eta_{\langle a\rangle}\right)_{a\in\{1,\dots,d\}}\text{ are fixed and }\left(\theta^{\langle b,\alpha\rangle}\right)_{(b,\alpha)\in\{1,\dots,d+1\}\times\{1,\dots,d-1\}}\text{ are arbitrary}\right\}.$$

On the other hand, the submanifold $E(\xi)$ is rewritten as

$$E(\xi) = \left\{ p_{(s,\xi)} \,\middle|\, \left(\theta^{\langle b,\alpha\rangle}\right)_{(b,\alpha)\in\{1,\ldots,d+1\}\times\{1,\ldots,d-1\}} \text{ are fixed and } \left(\eta_{\langle a\rangle}\right)_{a\in\{1,\ldots,d\}} \text{ are arbitrary} \right\}.$$

Thus, the orthogonality of $M(s)$ and $E(\xi)$ is an immediate consequence of the orthogonality of the dual affine coordinate systems θ and η with respect to the Fisher metric g. □

Proposition 1 implies that the manifold $\mathcal{P}(\Omega)$ is decomposed into mutually orthogonal dualistic foliation based on the submanifolds $M(s)$ and $E(\xi)$, as illustrated in Figure 1. We shall exploit this geometrical structure in the next section.

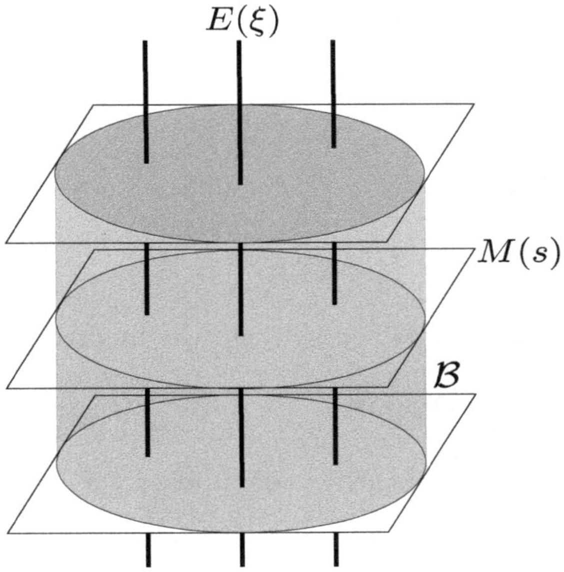

Figure 1. Mutually orthogonal dualistic foliation of $\mathcal{P}(\Omega)$ based on $M(s)$ and $E(\xi)$. Each section $M(s)$ is affinely isomorphic to the parameter space Ξ. The greyish cylindrical area indicates the subset $\mathcal{B} = \{p_{(s,\xi)} | s \in D, \xi \in B\}$ of $\mathcal{P}(\Omega)$. In particular, for each $s \in D$, the intersection $M(s) \cap \mathcal{B}$ is affinely isomorphic to the physical domain B that corresponds to the state space $\mathcal{S}(\mathcal{H})$.

3. Estimation of the Parameter ξ

Let us proceed to the problem of estimating the unknown parameter ξ using the randomized tomography. Suppose that, among N independent repetitions of experiments, the ath measurement $M^{(a)}$ was applied $N^{(a)}$ times and outcomes $\alpha \in \{1,\ldots,d\}$ were obtained $n_\alpha^{(a)}$ times. Then temporary estimates $(\hat{s},\hat{\xi})$ for the parameters (s,ξ) are given by

$$\hat{s}^{(a)} := \frac{N^{(a)}}{N}$$

for $a \in \{1,\ldots,d\}$, and

$$\hat{\xi}_\beta^{(b)} := \frac{n_\beta^{(b)}}{N^{(b)}}$$

for $(b,\beta) \in \{1,\ldots,d+1\} \times \{1,\ldots,d-1\}$. If $\hat{\xi}$ has fallen outside the physical domain B, one may seek a corrected estimate by the maximum likelihood method. Observe that, due to (2), the empirical distribution $\hat{q}_N \in \mathcal{P}(\Omega)$ is represented as

$$\hat{q}_N = p_{(\hat{s},\hat{\xi})}. \tag{4}$$

On the other hand, the physical domain B in the parameter space Ξ corresponds to the subset

$$\mathcal{B} := \{p_{(s,\xi)} \mid s \in D,\ \xi \in B\}$$

of $\mathcal{P}(\Omega)$, (see Figure 1). The MLE p^* in $\mathcal{P}(\Omega)$ is then given by

$$p^* = \arg\min_{p \in \mathcal{B}} D(\hat{q}_N \| p), \tag{5}$$

where $D(\,\cdot\,\|\,\cdot\,)$ is the Kullback-Leibler divergence (cf., Appendix A). A crucial observation is the following.

Proposition 2. *The minimum in (5) is achieved on $M(\hat{s}) \cap \mathcal{B}$.*

Proof. Let us take a point $p_{(s,\xi)} \in \mathcal{B}$ arbitrarily. It then follows from the mutually orthogonal dualistic foliation of $\mathcal{P}(\Omega)$ established in Proposition 1 that

$$\begin{aligned} D(\hat{q}_N \| p_{(s,\xi)}) &= D(p_{(\hat{s},\hat{\xi})} \| p_{(s,\xi)}) \\ &= D(p_{(\hat{s},\hat{\xi})} \| p_{(\hat{s},\xi)}) + D(p_{(\hat{s},\xi)} \| p_{(s,\xi)}) \\ &\geq D(p_{(\hat{s},\hat{\xi})} \| p_{(\hat{s},\xi)}). \end{aligned}$$

In the second equality, the generalized Pythagorean theorem was used. Consequently,

$$\min_{\xi \in B} D(p_{(\hat{s},\hat{\xi})} \| p_{(s,\xi)}) \geq \min_{\xi \in B} D(p_{(\hat{s},\hat{\xi})} \| p_{(\hat{s},\xi)})$$

for all $s \in D$, and the right-hand side is achieved if and only if $s = \hat{s}$. $\square$

The geometrical implication of Proposition 2 is illustrated in Figure 2. The MLE $p^* = p_{(\hat{s},\xi^*)}$ is the $\nabla^{(m)}$-projection from the empirical distribution $p_{(\hat{s},\hat{\xi})}$ to $\mathcal{B}$, and is on the section $M(\hat{s})$ specified by the temporary estimate $\hat{s}$.

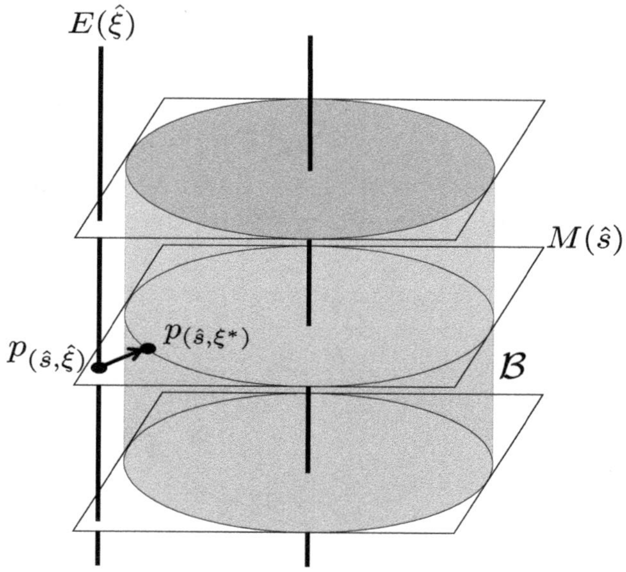

Figure 2. The maximum likelihood method in the framework of randomized tomography. Given a temporary estimate $(\hat{s}, \hat{\xi})$ with $\hat{\xi} \notin B$, we can restrict ourselves to the section $M(\hat{s})$ as the search space for the MLE p^*, and $p^* = p_{(\hat{s},\xi^*)}$ is the $\nabla^{(m)}$-projection from the empirical distribution $p_{(\hat{s},\hat{\xi})}$ to $\mathcal{B}$ on the section $M(\hat{s})$.

Now we arrive at a geometrical picture behind the parameter estimation based on randomized state tomography. Suppose we are given a temporary estimate $(\hat{s}, \hat{\xi})$ with $\hat{\xi} \notin B$. Due to Proposition 2, we can restrict ourselves to section $M(\hat{s})$ as the search space for the MLE p^*. Since each section $M(\hat{s})$ is affinely isomorphic to the parameter space Ξ, we can introduce a dualistic structure $(\tilde{g}, \tilde{\nabla}^{(e)}, \tilde{\nabla}^{(m)})$ on Ξ in the following way. Firstly, we identify the metric $\tilde{g}$ with the Fisher metric g restricted on $M(\hat{s})$, that is,

$$\begin{aligned}
\tilde{g}_{(\hat{s},\xi)}\left(\frac{\partial}{\partial \xi_\alpha^{(a)}}, \frac{\partial}{\partial \xi_\beta^{(b)}}\right) &= \frac{\partial \eta_{\langle a',\alpha'\rangle}}{\partial \xi_\alpha^{(a)}} \frac{\partial \eta_{\langle b',\beta'\rangle}}{\partial \xi_\beta^{(b)}} g_{(s,\xi)}\left(\frac{\partial}{\partial \eta_{\langle a',\alpha'\rangle}}, \frac{\partial}{\partial \eta_{\langle b',\beta'\rangle}}\right)\Bigg|_{s=\hat{s}} \\
&= s^{(a)} s^{(b)} \frac{\partial^2 \varphi(\eta)}{\partial \eta_{\langle a,\alpha\rangle} \partial \eta_{\langle b,\beta\rangle}}\Bigg|_{s=\hat{s}} \\
&= \delta_{ab}\, \hat{s}^{(a)} \left(\frac{1}{\xi_d^{(a)}} + \frac{\delta_{\alpha\beta}}{\xi_\alpha^{(a)}}\right),
\end{aligned}$$

for $a, b \in \{1, \ldots, d+1\}$ and $\alpha, \beta \in \{1, \ldots, d-1\}$, where $\hat{s}^{(d+1)}$ and $\xi_d^{(a)}$ are formally defined as

$$\hat{s}^{(d+1)} := 1 - \sum_{a=1}^{d} \hat{s}^{(a)}, \qquad \xi_d^{(a)} := 1 - \sum_{\alpha=1}^{d-1} \xi_\alpha^{(a)}.$$

Secondly, the mixture connection $\tilde{\nabla}^{(m)}$ on Ξ is defined through the natural affine isomorphism between $M(\hat{s})$ and Ξ. Finally, the dual connection $\tilde{\nabla}^{(e)}$ is defined by the duality

$$\tilde{g}(\tilde{\nabla}_X^{(e)} Y, Z) := X\tilde{g}(Y, Z) - \tilde{g}(Y, \tilde{\nabla}_X^{(m)} Z).$$

Thus, the MLE ξ^* in the parameter space Ξ is interpreted as the $\tilde{\nabla}^{(m)}$-projection from $\hat{\xi}$ to the physical domain B with respect to the metric $\tilde{g}$.

4. Examples

In this section, we present some examples that demonstrate the implication of Proposition 2 as well as the general diagram given in Figure 2.

4.1. When $\dim \mathcal{H} = 2$

Let us first study the simplest case when $\mathcal{H} = \mathbb{C}^2$. A full set of MUBs is given by

$$\begin{aligned}
\left\{|1^{(1)}\rangle, |2^{(1)}\rangle\right\} &= \left\{\frac{1}{\sqrt{2}}\begin{pmatrix}1\\1\end{pmatrix}, \frac{1}{\sqrt{2}}\begin{pmatrix}1\\-1\end{pmatrix}\right\}, \\
\left\{|1^{(2)}\rangle, |2^{(2)}\rangle\right\} &= \left\{\frac{1}{\sqrt{2}}\begin{pmatrix}1\\-i\end{pmatrix}, \frac{1}{\sqrt{2}}\begin{pmatrix}1\\i\end{pmatrix}\right\}, \\
\left\{|1^{(3)}\rangle, |2^{(3)}\rangle\right\} &= \left\{\begin{pmatrix}1\\0\end{pmatrix}, \begin{pmatrix}0\\1\end{pmatrix}\right\}.
\end{aligned}$$

With these bases, the parameter representation (1) becomes

$$\rho = \frac{1}{2}\begin{pmatrix}1 + x_3 & x_1 - ix_2 \\ x_1 + ix_2 & 1 - x_3\end{pmatrix},$$

where $x = (x_1, x_2, x_3)$ is the standard Stokes parameter, which is related to $\xi = (\xi_1^{(1)}, \xi_1^{(2)}, \xi_1^{(3)})$ as $x_a = 2\xi_1^{(a)} - 1$ for $a = 1, 2, 3$.

Figure 3 demonstrates how the $\tilde{\nabla}^{(m)}$-projection is realized. Here, the trajectories of $\tilde{\nabla}^{(m)}$-projections that gives the MLE p^* are plotted only on the x_1x_2-plane. The left and right panels correspond to the cases when $N^{(1)} : N^{(2)} = 1 : 1$ and $N^{(1)} : N^{(2)} = 5 : 1$, respectively. The change of ξ_1-coordinate relative to the change of x_2-coordinate along each trajectory is less noticeable in the right panel than in the left panel. This is because a tomography with $N^{(1)}/N^{(2)} = 5$ provides us with more information about x_1-coordinate, relative to x_2-coordinate, as compared with the case when $N^{(1)}/N^{(2)} = 1$.

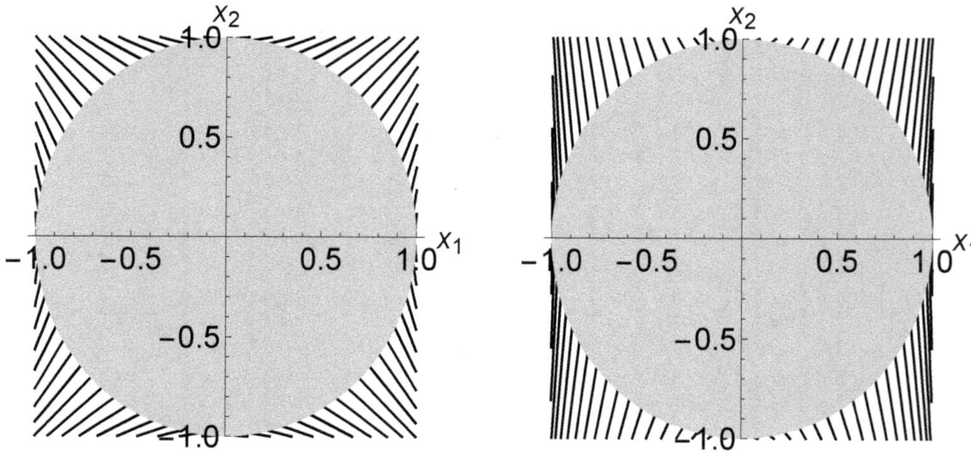

Figure 3. The trajectories of $\tilde{\nabla}^{(m)}$-projections on the Stokes parameter space when $N^{(1)} : N^{(2)} = 1 : 1$ (**left**) and $N^{(1)} : N^{(2)} = 5 : 1$ (**right**). The greyish disk represents the Bloch ball B.

4.2. *When* $\dim \mathcal{H} = 3$

The space $\mathcal{H} = \mathbb{C}^3$ admits a full set of MUBs; for example,

$$\left\{|1^{(1)}\rangle, |2^{(1)}\rangle, |3^{(1)}\rangle\right\} = \left\{\begin{pmatrix}1\\0\\0\end{pmatrix}, \begin{pmatrix}0\\1\\0\end{pmatrix}, \begin{pmatrix}0\\0\\1\end{pmatrix}\right\},$$

$$\left\{|1^{(2)}\rangle, |2^{(2)}\rangle, |3^{(2)}\rangle\right\} = \left\{\frac{1}{\sqrt{3}}\begin{pmatrix}1\\1\\1\end{pmatrix}, \frac{1}{\sqrt{3}}\begin{pmatrix}1\\\omega\\\omega^2\end{pmatrix}, \frac{1}{\sqrt{3}}\begin{pmatrix}1\\\omega^2\\\omega\end{pmatrix}\right\},$$

$$\left\{|1^{(3)}\rangle, |2^{(3)}\rangle, |3^{(3)}\rangle\right\} = \left\{\frac{1}{\sqrt{3}}\begin{pmatrix}\omega\\1\\1\end{pmatrix}, \frac{1}{\sqrt{3}}\begin{pmatrix}1\\\omega\\1\end{pmatrix}, \frac{1}{\sqrt{3}}\begin{pmatrix}1\\1\\\omega\end{pmatrix}\right\},$$

$$\left\{|1^{(4)}\rangle, |2^{(4)}\rangle, |3^{(4)}\rangle\right\} = \left\{\frac{1}{\sqrt{3}}\begin{pmatrix}\omega^2\\1\\1\end{pmatrix}, \frac{1}{\sqrt{3}}\begin{pmatrix}1\\\omega^2\\1\end{pmatrix}, \frac{1}{\sqrt{3}}\begin{pmatrix}1\\1\\\omega^2\end{pmatrix}\right\},$$

where $\omega = (-1 + i\sqrt{3})/2$ is a primitive third root of unity. With these bases, the parameter representation (1) becomes

$$\rho = \begin{pmatrix} \xi_1^{(1)} & a_{12} - ib_{12} & a_{13} - ib_{13} \\ a_{12} + ib_{12} & \xi_2^{(1)} & a_{23} - ib_{23} \\ a_{13} + ib_{13} & a_{23} + ib_{23} & 1 - \xi_1^{(1)} - \xi_2^{(1)} \end{pmatrix},$$

where

$$a_{12} = \frac{1}{2}\left(1 + \xi_1^{(2)} - \xi_1^{(3)} - \xi_2^{(3)} - \xi_1^{(4)} - \xi_2^{(4)}\right),$$
$$a_{13} = \frac{1}{2}\left(-1 + \xi_1^{(2)} + \xi_2^{(3)} + \xi_2^{(4)}\right),$$
$$a_{23} = \frac{1}{2}\left(-1 + \xi_1^{(2)} + \xi_1^{(3)} + \xi_1^{(4)}\right),$$
$$b_{12} = \frac{\sqrt{3}}{6}\left(1 - \xi_1^{(2)} - 2\xi_2^{(2)} + \xi_1^{(3)} - \xi_2^{(3)} - \xi_1^{(4)} + \xi_2^{(4)}\right),$$
$$b_{13} = \frac{\sqrt{3}}{6}\left(-1 + \xi_1^{(2)} + 2\xi_2^{(2)} + 2\xi_1^{(3)} + \xi_2^{(3)} - 2\xi_1^{(4)} - \xi_2^{(4)}\right),$$
$$b_{23} = \frac{\sqrt{3}}{6}\left(1 - \xi_1^{(2)} - 2\xi_2^{(2)} + \xi_1^{(3)} + 2\xi_2^{(3)} - \xi_1^{(4)} - 2\xi_2^{(4)}\right).$$

The physical domain B that corresponds to the state space $\mathcal{S}(\mathbb{C}^3)$ is a compact convex subset of the parameter space $\Xi\ (\subset \mathbb{R}^8)$, and the extreme points of B form an algebraic variety with respect to the parameters

$$\xi = (\xi_1^{(1)}, \xi_2^{(1)}, \xi_1^{(2)}, \xi_2^{(2)}, \xi_1^{(3)}, \xi_2^{(3)}, \xi_1^{(4)}, \xi_2^{(4)}).$$

A numerical example of a $\tilde{\nabla}^{(m)}$-projection that gives the MLE is illustrated in Figure 4, where no probe particle is lost, that is, when

$$\hat{s} = \left(\frac{1}{4}, \frac{1}{4}, \frac{1}{4}, \frac{1}{4}\right).$$

In Figure 4, the dot laid outside the greyish region indicates the empirical distribution, i.e., the temporary estimate

$$\hat{\xi} = (0.100, 0.100, 0.066, 0.333, 0.333, 0.333, 0.333, 0.333),$$

and the corresponding MLE is

$$\xi_* = (0.122, 0.122, 0.108, 0.329, 0.299, 0.327, 0.327, 0.299).$$

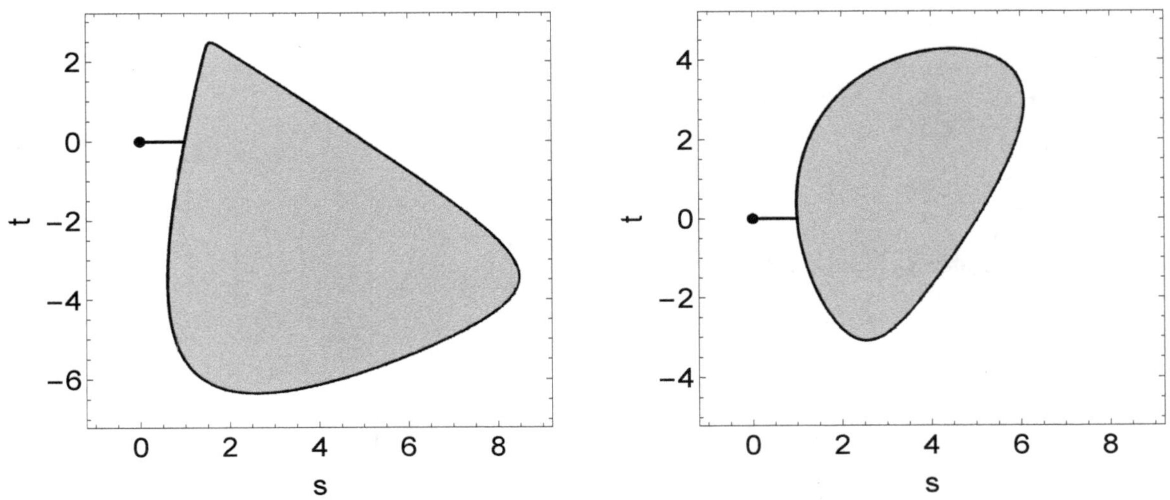

Figure 4. A trajectory of $\tilde{\nabla}^{(m)}$-projection displayed on randomly chosen two-dimensional affine subspaces of Ξ to which both the empirical distribution (marked as a dot) and the MLE belong. The greyish region represents the physical domain B.

Furthermore, the greyish region represents the physical domain B cut by a two-dimensional affine subspace of Ξ specified by the equation

$$\xi = (1-s)\,\hat{\xi} + s\,\xi_* + t\,v.$$

The vector v was chosen randomly under the condition that

$$v \perp \hat{\xi} - \xi_* \quad \text{and} \quad \|v\| = \|\hat{\xi} - \xi_*\|,$$

where the orthogonality $\perp$ and the norm $\|\cdot\|$ are understood relative to the standard Euclidean structure of $\mathbb{R}^8$. In Figure 4, the vector v was taken to be

$$v_1 = (-0.036, -0.038, 0.012, -0.026, -0.038, 0.011, 0.002, 0.005)$$

in the left panel, and

$$v_2 = (0.028, 0.000, -0.006, 0.034, -0.024, -0.022, 0.034, 0.030)$$

in the right panel.

Figure 4 also demonstrates that the sections of the physical domain B show a variety of shapes. Unfortunately, due to this asymmetry of B, we were unable to find a (nontrivial) two-dimensional affine subspace on which every $\tilde{\nabla}^{(m)}$-projection runs. Such a difficulty is in good contrast to the simplest case $\mathcal{H} \simeq \mathbb{C}^2$, where the set B is rotationally symmetric and the $\tilde{\nabla}^{(m)}$-projections can be displayed on any two-dimensional section of B that passes through the origin of B as Figure 3.

4.3. *When* $\dim \mathcal{H} \geq 4$

The space $\mathcal{H} = \mathbb{C}^4$ is also known to admit a full set of MUBs since $\dim \mathcal{H} = 4$ is the second power of the prime number 2; for example [22],

$$\left\{|1^{(1)}\rangle, |2^{(1)}\rangle, |3^{(1)}\rangle, |4^{(1)}\rangle\right\} = \left\{\begin{pmatrix}1\\0\\0\\0\end{pmatrix}, \begin{pmatrix}0\\1\\0\\0\end{pmatrix}, \begin{pmatrix}0\\0\\1\\0\end{pmatrix}, \begin{pmatrix}0\\0\\0\\1\end{pmatrix}\right\},$$

$$\left\{|1^{(2)}\rangle, |2^{(2)}\rangle, |3^{(2)}\rangle, |4^{(2)}\rangle\right\} = \left\{\frac{1}{2}\begin{pmatrix}1\\1\\1\\1\end{pmatrix}, \frac{1}{2}\begin{pmatrix}1\\1\\-1\\-1\end{pmatrix}, \frac{1}{2}\begin{pmatrix}1\\-1\\-1\\1\end{pmatrix}, \frac{1}{2}\begin{pmatrix}1\\-1\\1\\-1\end{pmatrix}\right\},$$

$$\left\{|1^{(3)}\rangle, |2^{(3)}\rangle, |3^{(3)}\rangle, |4^{(3)}\rangle\right\} = \left\{\frac{1}{2}\begin{pmatrix}1\\-1\\-i\\-i\end{pmatrix}, \frac{1}{2}\begin{pmatrix}1\\-1\\i\\i\end{pmatrix}, \frac{1}{2}\begin{pmatrix}1\\1\\i\\-i\end{pmatrix}, \frac{1}{2}\begin{pmatrix}1\\1\\-i\\i\end{pmatrix}\right\},$$

$$\left\{|1^{(4)}\rangle, |2^{(4)}\rangle, |3^{(4)}\rangle, |4^{(4)}\rangle\right\} = \left\{\frac{1}{2}\begin{pmatrix}1\\-i\\-i\\-1\end{pmatrix}, \frac{1}{2}\begin{pmatrix}1\\-i\\i\\1\end{pmatrix}, \frac{1}{2}\begin{pmatrix}1\\i\\i\\-1\end{pmatrix}, \frac{1}{2}\begin{pmatrix}1\\i\\-i\\1\end{pmatrix}\right\},$$

$$\left\{|1^{(5)}\rangle, |2^{(5)}\rangle, |3^{(5)}\rangle, |4^{(5)}\rangle\right\} = \left\{\frac{1}{2}\begin{pmatrix}1\\-i\\-1\\-i\end{pmatrix}, \frac{1}{2}\begin{pmatrix}1\\-i\\1\\i\end{pmatrix}, \frac{1}{2}\begin{pmatrix}1\\i\\-1\\i\end{pmatrix}, \frac{1}{2}\begin{pmatrix}1\\i\\1\\-i\end{pmatrix}\right\}.$$

It is straightforward to calculate the parameter representation (1) of a state $\rho \in \mathcal{S}(C^4)$; however, the corresponding density matrix is rather complicated, and we omit to display it here.

When $\mathcal{H} = \mathbb{C}^6$, or more generally, when $\dim \mathcal{H}$ is not a power of a prime, we do not know whether $\mathcal{H}$ admits a full set of MUBs. Let us touch upon a situation where a Hilbert space $\mathcal{H}$, if it exists, does not admit a full set of MUBs. In this case, there is no measurement basis $M^{(a)}$ that allows a parametrization ξ of the state space $\mathcal{S}(\mathcal{H})$ having a direct connection to the probability distribution of the outcomes as (2). Such a situation could be comparable to the case when the Gell-Mann matrices [23] are used as the measurement basis for estimating an unknown state on $\mathcal{H} = \mathbb{C}^3$. A state $\rho \in \mathcal{S}(\mathbb{C}^3)$ is represented as

$$\rho = \rho_x := \frac{1}{3}\left(I + \sqrt{3}\sum_{i=1}^{8} x_i\lambda_i\right),$$

where $\lambda_1, \ldots, \lambda_8$ are the Gell-Mann matrices, and $x = (x_1, \ldots, x_8)$ is a set of real parameters. The physical domain

$$B = \left\{x \in \mathbb{R}^8 \,\middle|\, \rho_x \geq 0\right\}$$

forms a compact convex subset of the unit ball in $\mathbb{R}^8$. With the state ρ_x, the probability distribution of obtaining the eigenvalues $(-1, 0, 1)$ of the observable λ_1 is

$$\left(\frac{1-\sqrt{3}x_1+x_8}{3}, \frac{1-2x_8}{3}, \frac{1+\sqrt{3}x_1+x_8}{3}\right),$$

while the probability distribution of obtaining the eigenvalues $(-1, 0, 1)$ of the observable λ_2 is

$$\left(\frac{1-\sqrt{3}x_2+x_8}{3}, \frac{1-2x_8}{3}, \frac{1+\sqrt{3}x_2+x_8}{3}\right).$$

Note that the probability of obtaining the eigenvalue 0 of λ_1 is identical to that of λ_2. However, in a randomized estimation scheme in which λ_i is measured N_i times, the frequency of obtaining the eigenvalue 0 of λ_1 would be different from that of λ_2. Consequently, one cannot assign a consistent temporary estimate $\hat{x}_8$ for the parameter x_8 in that case. Put differently, the empirical distribution $\hat{q}_N$ on the extended outcome space Ω does not in general have a coordinate representation (4). Thus, the existence of a full set of MUBs is crucial in our analysis.

5. Concluding Remarks

In the present paper, we explored an information geometrical structure of the randomized quantum state tomography, assuming that the Hilbert space under consideration admits a full set of MUBs. We first introduced a classical statistical model $\{p_{(s,\xi)}\}_{s,\xi}$ on an extended sample space Ω, and found that the probability simplex $\mathcal{P}(\Omega)$ was decomposed into mutually orthogonal dualistic foliation (Proposition 1). We then clarified that this geometrical structure had a statistical importance in estimating the coordinate ξ of an unknown quantum state $\rho(\xi)$ under the existence of the nuisance parameter s (Proposition 2). This result gave a generalized insight into the $\nabla^{(m)}$-projection interpretation for the MLE in that a similar interpretation was still valid for the randomized quantum state tomography by changing the standard Fisher metric into a deformed one. It also provided us with a new, convenient way of data processing in the actual quantum state tomography that may involve unexpected probe particle loss.

It should be noted that the existence of a full set of MUBs ensures the parametrization (1) of the quantum state space $\mathcal{S}(\mathcal{H})$. Such a parametrization is distinctive in that it enables a direct correspondence between the parameter space and the probability simplex, realizing the coordinate representation (4) of the empirical distribution $\hat{q}_N$. Thus, the use of a full set of MUBs is crucial in our analysis. Nevertheless, it is often the case that the Hilbert space under consideration takes the form $\mathcal{H} \simeq (\mathbb{C}^p)^{\otimes n}$ for $p = 2$ or 3 because qubits or qutrits are often regarded as building blocks of various quantum protocols. Therefore, the existence of a full set of MUBs would not be too strong a requirement in applications.

Author Contributions: The authors contributed equally to this work.

Acknowledgments: The authors are grateful to Ryo Okamoto and Shigeki Takeuchi for helpful discussions.

Abbreviations

The following abbreviations are used in this manuscript:

MUBs	mutually unbiased bases
PVM	projection-valued measure
MLE	maximum likelihood estimate

Appendix A. Information Geometry of MLE

Let $\mathcal{P}(\Omega)$ denote the set of probability distributions on a finite sample space Ω, i.e.,

$$\mathcal{P}(\Omega) := \left\{ p : \Omega \to \mathbb{R} \;\middle|\; p(\omega) > 0 \text{ for all } \omega \in \Omega, \text{ and } \sum_{\omega \in \Omega} p(\omega) = 1 \right\}.$$

This set may be identified with the $(|\Omega| - 1)$-dimensional (open) simplex, where $|\Omega|$ denotes the number of elements in Ω, and thus it is sometimes referred to as the *probability simplex* on Ω. The set $\mathcal{P}(\Omega)$ is also regarded as a *statistical manifold* endowed with the *dualistic structure* $(g, \nabla^{(e)}, \nabla^{(m)})$, where g is the *Fisher metric*, and $\nabla^{(e)}$ and $\nabla^{(m)}$ are the *exponential* and *mixture connections* [11–13].

Suppose that the state of the physical system at hand belongs to a (closed) subset $\mathcal{M}$ of $\mathcal{P}(\Omega)$, but we do not know which is the true state. We further assume that the probability distributions of $\mathcal{M}$ are faithfully parametrized by a finite dimensional parameter θ as

$$\mathcal{M} = \{ p_\theta(\omega) \mid \theta \in \Theta \}.$$

In this case, $\mathcal{M}$ is called a *parametric model*, and our task is to estimate the true value of the parameter θ that specifies the true state. Suppose that, by n independent experiments, we have obtained data $(x_1, x_2, \ldots, x_n) \in \Omega^n$. This information is compressed into the *empirical distribution*, an element of $\mathcal{P}(\Omega)$ defined by

$$\begin{aligned} \hat{q}_n(\omega) &:= \frac{\text{Number of occurrences of } \omega \text{ in data } (x_1, x_2, \ldots, x_n)}{n} \\ &= \frac{1}{n} \sum_{i=1}^{n} \delta_{x_i}(\omega) \end{aligned}$$

for each $\omega \in \Omega$, where $\delta_{x_i}(\omega)$ is the Kronecker delta. If $\hat{q}_n$ belongs to the model $\mathcal{M}$, then we have an estimate $\hat{\theta}_n$ that satisfies $p_{\hat{\theta}_n} = \hat{q}_n$. However, the empirical distribution $\hat{q}_n$ does not always belong to the model $\mathcal{M}$. When $\hat{q}_n \notin \mathcal{M}$, we need to find an alternative estimate from the data. A canonical method of finding an alternative estimate $p_{\hat{\theta}_n} \in \mathcal{M}$ is the *maximum likelihood method*, in which one seeks the maximizer of the likelihood function

$$\theta \longmapsto p_\theta(x_1) p_\theta(x_2) \ldots p_\theta(x_n),$$

within the domain Θ of the parameter θ, so that

$$\hat{\theta}_n := \underset{\theta\in\Theta}{\arg\max}\left\{p_\theta(x_1)p_\theta(x_2)\cdots p_\theta(x_n)\right\}.$$

We can rewrite this relation as follows.

$$\begin{aligned}\hat{\theta}_n &= \underset{\theta\in\Theta}{\arg\max}\ \frac{1}{n}\sum_{i=1}^{n}\log p_\theta(x_i)\\ &= \underset{\theta\in\Theta}{\arg\max}\ \sum_{\omega\in\Omega}\hat{q}_n(\omega)\log p_\theta(\omega)\\ &= \underset{\theta\in\Theta}{\arg\min}\ \sum_{\omega\in\Omega}\hat{q}_n(\omega)\left\{\log\hat{q}_n(\omega)-\log p_\theta(\omega)\right\}\\ &= \underset{\theta\in\Theta}{\arg\min}\ D(\hat{q}_n\|p_\theta),\end{aligned}$$

where

$$D(q\|p) := \sum_{\omega\in\Omega} q(\omega)\log\frac{q(\omega)}{p(\omega)}$$

is the *Kullback-Leibler divergence* from q to p. In other words, the maximum likelihood estimate (MLE) $p_{\hat{\theta}_n}$ is the point on $\mathcal{M}$ that is "closest" from the empirical distribution $\hat{q}_n$ as measured by the Kullback-Leibler divergence:

$$p_{\hat{\theta}_n} = \underset{p\in\mathcal{M}}{\arg\min}\ D(\hat{q}_n\|p).$$

Due to the generalized Pythagorean theorem, the MLE is geometrically understood as the $\nabla^{(m)}$-projection from $\hat{q}_n$ to $\mathcal{M}$ or its boundary, as illustrated in Figure A1.

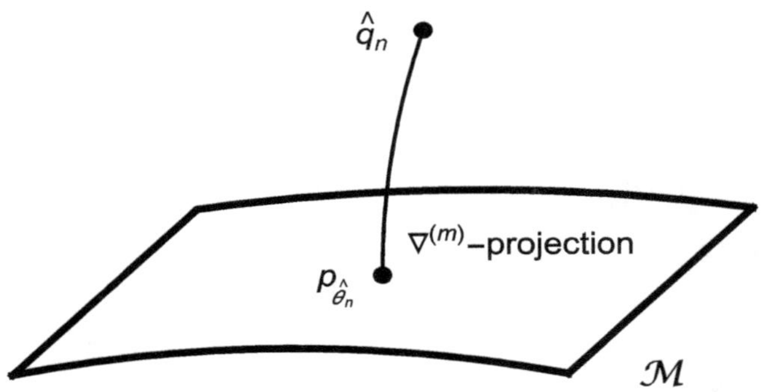

Figure A1. The maximum likelihood estimate $p_{\hat{\theta}_n}$ is the minimizer of the function $p \mapsto D(\hat{q}_n\|p)$ with respect to $p \in \mathcal{M}$, and is also understood as the $\nabla^{(m)}$-projection from the empirical distribution $\hat{q}_n$ to $\mathcal{M}$ or its boundary.

Appendix B. Parametrization of $\mathcal{S}(\mathcal{H})$

Suppose that the Hilbert space $\mathcal{H}\simeq\mathbb{C}^d$ under consideration admits a full set of MUBs

$$\left\{|\alpha^{(a)}\rangle\right\}_{\alpha\in\{1,\dots,d\}},\qquad (a=1,\dots,d+1).$$

For each $a\in\{1,\dots,d+1\}$, let

$$M^{(a)} := \left\{M_1^{(a)},\dots,M_d^{(a)}\right\} = \left\{|1^{(a)}\rangle\langle 1^{(a)}|,\dots,|d^{(a)}\rangle\langle d^{(a)}|\right\}.$$

Then, the operators

$$\left\{M_\alpha^{(a)}-\frac{I}{d}\right\}_{(a,\alpha)\in\{1,\dots,d+1\}\times\{1,\dots,d-1\}}$$

are linearly independent, spanning the space of selfadjoint operators with zero trace. This is easily seen from the orthogonality relation:

$$\mathrm{Tr}\left(M_\alpha^{(a)}-\frac{I}{d}\right)\left(M_\beta^{(b)}-\frac{I}{d}\right)=\delta_{ab}\left(\delta_{\alpha\beta}-\frac{1}{d}\right).$$

Thus, given $\rho\in\mathcal{S}(\mathcal{H})$, the operator $\rho-(I/d)$ is uniquely expanded as

$$\rho-\frac{I}{d}=\sum_{a=1}^{d+1}\sum_{\alpha=1}^{d-1}x_\alpha^{(a)}\left(M_\alpha^{(a)}-\frac{I}{d}\right),$$

where $x_\alpha^{(a)}$ are real numbers. We can regard $x_\alpha^{(a)}$ as a coordinate system of the state space $\mathcal{S}(\mathcal{H})$. When $d=2$, this is identical to the Stokes parametrization, up to a factor of 2.

Now, let us change the coordinate system $x_\alpha^{(a)}$ into $\xi_\alpha^{(a)}$ as

$$x_\alpha^{(a)}=\xi_\alpha^{(a)}+\left(\sum_{\beta=1}^{d-1}\xi_\beta^{(a)}\right)-1.$$

We then arrive at the parametrization (1), i.e.,

$$\rho=\sum_{a=1}^{d+1}\left\{\sum_{\alpha=1}^{d-1}\xi_\alpha^{(a)}M_\alpha^{(a)}+\left(1-\sum_{\beta=1}^{d-1}\xi_\beta^{(a)}\right)M_d^{(a)}\right\}-I.$$

This parametrization is useful in our analysis because it gives a direct connection to the probability distribution of outcomes of the measurement $M^{(a)}$ as

$$p_\alpha^{(a)}:=\mathrm{Tr}\,\rho M_\alpha^{(a)}=\begin{cases}\xi_\alpha^{(a)}, & \text{for } \alpha=1,\dots,d-1,\\ 1-\sum_{\beta=1}^{d-1}\xi_\beta^{(a)}, & \text{for } \alpha=d.\end{cases}$$

References

1. Nielsen, M.A.; Chuang, I.L. *Quantum Computation and Quantum Information*; Cambridge University Press: Cambridge, UK, 2000.
2. Lehmann, E.L.; Casella, G. *Theory of Point Estimation*, 2nd ed.; Springer: New York, NY, USA, 1998.
3. Helstrom, C.W. *Quantum Detection and Estimation Theory*; Academic Press: New York, NY, USA, 1976.
4. Holevo, A.S. *Probabilistic and Statistical Aspects of Quantum Theory*; North-Holland: Amsterdam, The Netherlands, 1982.
5. Hradil, Z. Quantum-State Estimation. *Phys. Rev. A* **1997**, *55*, R1561–R1564. [CrossRef]
6. Banaszek, K.; D'Ariano, G.M.; Paris, M.G.A.; Sacchi, M.F. Maximum-likelihood estimation of the density matrix. *Phys. Rev. A* **1999**, *61*, 010304. [CrossRef]
7. Hradil, Z.; Summhammer, J.; Badurek, G.; Rauch, H. Reconstruction of the spin state. *Phys. Rev. A* **2000**, *62*, 014101. [CrossRef]
8. James, D.F.V.; Kwiat, P.G.; Munro, W.J.; White, A.G. Measurement of qubits. *Phys. Rev. A* **2001**, *64*, 052312. [CrossRef]
9. De Burgh, M.D.; Langford, N.K.; Doherty, A.C.; Gilchrist, A. Choice of measurement sets in qubit tomography. *Phys. Rev. A* **2008**, *78*, 052122. [CrossRef]
10. Blune-Kohout, R. Optimal, reliable estimation of quantum states. *New J. Phys.* **2010**, *12*, 043034. [CrossRef]
11. Amari, S.-I.; Nagaoka, H. *Methods of Information Geometry*; Translations of Mathematical Monographs 191; AMS and Oxford: Charles Street, RI, USA, 2000.

12. Amari, S.-I. *Differential-Geometrical Methods in Statistics*; Lecture Notes in Statistics 28; Springer: Berlin, Germany, 1985.
13. Murray, M.K.; Rice, J.W. *Differential Geometry and Statistics*; Chapman & Hall: London, UK, 1993.
14. Fujiwara, A.; Yamagata, K. Data processing for qubit state tomography: An information geometric approach. *arXiv* **2016**, arXiv:1608.07983.
15. Fraïsse, J.M.E.; Braun, D. Quantum channel-estimation with particle loss: GHZ versus W states. *Quantum Meas. Quantum Metrol.* **2016**, *3*, 53. [CrossRef]
16. Durt, T.; Englert, B.-G.; Bengtsson, I.; Życzkowski, K. On mutually unbiased bases. *Int. J. Quantum Inf.* **2010**, *8*, 535–640. [CrossRef]
17. Yuan, H.; Zhou, Z.; Guo, G. Quantum state tomography via mutually unbiased measurements in driven cavity QED systems. *New J. Phys.* **2016**, *18*, 043013. [CrossRef]
18. Wootters, W.K.; Fields, B.D. Optimal state-determination by mutually unbiased measurements. *Ann. Phys.* **1989**, *191*, 363–381. [CrossRef]
19. Bengtsson, I. Three ways to look at mutually unbiased bases. *arXiv* **2006**, arXiv:quant-ph/0610216.
20. Ivonovic, I.D. Geometrical description of quantal state determination. *J. Phys. A* **1981**, *14*, 3241–3245. [CrossRef]
21. Yamagata, K. Efficiency of quantum state tomography for qubits. *Int. J. Quantum Inform.* **2011**, *9*, 1167–1183. [CrossRef]
22. Klappenecker, A.; Rötteler, M. Constructions of mutually unbiased bases. *arXiv* **2003**, arXiv:quant-ph/0309120.
23. Gell-Mann, M. Symmetries of baryons and mesons. *Phys. Rev.* **1962**, *125*, 1067. [CrossRef]

3

Geometry Induced by a Generalization of Rényi Divergence

David C. de Souza [1,†], Rui F. Vigelis [2,*,†] and Charles C. Cavalcante [3,†]

1 Instituto Federal do Ceará, Campus Maracanaú, Fortaleza 61939-140, Brazil; davidcs@ifce.edu.br
2 Computer Engineering School, Campus Sobral, Federal University of Ceará, Sobral 62010-560, Brazil
3 Department of Teleinformatics Engineering, Federal University of Ceará, Fortaleza 60455-900, Brazil; charles@ufc.br
* Correspondence: rfvigelis@ufc.br
† These authors contributed equally to this work.

Academic Editors: Frédéric Barbaresco and Frank Nielsen

Abstract: In this paper, we propose a generalization of Rényi divergence, and then we investigate its induced geometry. This generalization is given in terms of a φ-function, the same function that is used in the definition of non-parametric φ-families. The properties of φ-functions proved to be crucial in the generalization of Rényi divergence. Assuming appropriate conditions, we verify that the generalized Rényi divergence reduces, in a limiting case, to the φ-divergence. In generalized statistical manifold, the φ-divergence induces a pair of dual connections $D^{(-1)}$ and $D^{(1)}$. We show that the family of connections $D^{(\alpha)}$ induced by the generalization of Rényi divergence satisfies the relation $D^{(\alpha)} = \frac{1-\alpha}{2}D^{(-1)} + \frac{1+\alpha}{2}D^{(1)}$, with $\alpha \in [-1,1]$.

Keywords: Rényi divergence; φ-function; φ-divergence; φ-family; statistical manifold; information geometry

1. Introduction

Information geometry, the study of statistical models equipped with a differentiable structure, was pioneered by the work of Rao [1], and gained maturity with the work of Amari and many others [2–4]. It has been successfully applied in many different areas, such as statistical inference, machine learning, signal processing or optimization [4,5]. In appropriate statistical models, the differentiable structure is induced by a (statistical) divergence. The Kullback–Leibler divergence induces a Riemannian metric, called the Fisher–Rao metric, and a pair of dual connections, the exponential and mixture connections. A statistical model endowed with the Fisher–Rao metric is called a (classical) statistical manifold. Amari also considered a family of α-divergences that induce a family of α-connections.

Much research in recent years has focused on the geometry of non-standard statistical models [6–8]. These models are defined in terms of a deformed exponential (also called ϕ-exponential). In particular, κ-exponential models and q-exponential families are investigated in [9,10]. Non-parametric (or infinite-dimensional) φ-families were introduced by the authors in [11,12], which generalize exponential families in the non-parametric setting [13–16]. Based on the similarity between exponential and φ-families, we defined the so-called φ-divergence, with respect to which the Kullback–Leibler divergence is a particular case. Statistical models equipped with a geometric structure induced by φ-divergences, which are called generalized statistical manifolds, are investigated in [17,18]. With respect to these connections, parametric φ-families are dually flat.

The φ-divergence is intrinsically related to the (ρ,τ)-model of Zhang, which was proposed in [19,20], extended to the infinite-dimension setting in [21], and explained in more details in [22,23].

For instance, the metric induced by φ-divergence and the (ρ, τ)-generalization of the Fisher–Rao metric, for the choices $\rho = \varphi^{-1}$ and $f = \rho^{-1}$, differ by a conformal factor.

Among many attempts to generalize Kullback–Leibler divergence, Rényi divergence [24] is one of the most successful, having found many applications [25]. In the present paper, we propose a generalization of Rényi divergence, which we use to define a family of α-connections. This generalization is based on an interpretation of Rényi divergence as a kind of normalizing function. To generalize Rényi divergence, we considered functions satisfying some suitable conditions. To a function for which these conditions hold, we give the name of φ-function. In a limiting case, the generalized Rényi divergence reduces to the φ-divergence. In [17,18], the φ-divergence gives rise to a pair of dual connections $D^{(-1)}$ and $D^{(1)}$. We show that the connection $D^{(\alpha)}$ induced by the generalization of Rényi divergence satisfies the convex combination $D^{(\alpha)} = \frac{1-\alpha}{2} D^{(-1)} + \frac{1+\alpha}{2} D^{(1)}$.

Eguchi in [26] investigated a geometry based on a normalizing function similar to the one used in the generalization of Rényi divergence. In [26], results were derived supposing that this normalizing function exists; conditions for its existence were not given. In the present paper, the existence of the normalizing function is ensured by conditions involved in the definition of φ-functions.

The rest of the paper is organized as follows. In Section 2, φ-functions are introduced and some properties are discussed. The Rényi divergence is generalized in Section 3. We investigate in Section 4 the geometry induced by the generalization of Rényi divergence. Section 4.2 provides evidence of the role of the generalized Rényi divergence in φ-families.

2. φ-Functions

Rényi divergence is defined in terms of the exponential function (to be more precise, the logarithm). A way of generalizing Rényi divergence is to replace the exponential function by another function, which satisfies some suitable conditions. To a function for which these conditions hold, we give the name φ-function. In this section, we define and investigate some properties of φ-functions.

Let (T, Σ, μ) be a measure space. Although we do not restrict our analysis to a particular measure space, the reader can think of T as the set of real numbers $\mathbb{R}$, Σ as the Borel σ-algebra on $\mathbb{R}$, and μ as the Lebesgue measure. We can also consider T to be a discrete set, a case in which μ is the counting measure.

We say that $\varphi\colon \mathbb{R} \to (0, \infty)$ is a *φ-function* if the following conditions are satisfied:

(a1) $\varphi(\cdot)$ is convex;
(a2) $\lim_{u\to-\infty} \varphi(u) = 0$ and $\lim_{u\to\infty} \varphi(u) = \infty$;
(a3) there exists a measurable function $u_0\colon T \to (0, \infty)$ such that

$$\int_T \varphi(c + \lambda u_0) d\mu < \infty, \qquad \text{for all } \lambda > 0, \tag{1}$$

for each measurable function $c\colon T \to \mathbb{R}$ satisfying $\int_T \varphi(c) d\mu = 1$.

Thanks to condition (a3), we can generalize Rényi divergence using φ-functions. These conditions appeared first at [12] where the authors constructed non-parametric φ-families of probability distributions. We remark that if T is finite, condition (a3) is always satisfied.

Examples of functions $\varphi\colon \mathbb{R} \to (0, \infty)$ satisfying (a1)–(a3) abound. An example of great relevance is the exponential function $\varphi(u) = \exp(u)$, which satisfies conditions (a1)–(a3) with $u_0 = \mathbf{1}_T$. Another example of φ-function is the Kaniadakis' κ-exponential [12,27,28].

Example 1. *The Kaniadakis' κ-exponential* $\exp_\kappa\colon \mathbb{R} \to (0, \infty)$ *for* $\kappa \in [-1, 1]$ *is defined as*

$$\exp_\kappa(u) = \begin{cases} (\kappa u + \sqrt{1 + \kappa^2 u^2})^{1/\kappa}, & \text{if } \kappa \neq 0, \\ \exp(u), & \text{if } \kappa = 0, \end{cases}$$

whose inverse is the so called the Kaniadakis' κ-logarithm $\log_k\colon (0,\infty) \to \mathbb{R}$, *which is given by*

$$\log_\kappa(u) = \begin{cases} \frac{u^\kappa - u^{-\kappa}}{2\kappa}, & \text{if } \kappa \neq 0, \\ \ln(u), & \text{if } \kappa = 0. \end{cases}$$

It is clear that $\exp_\kappa(\cdot)$ *satisfies (a1) and (a2). Let* $u_0\colon T \to (0,\infty)$ *be any measurable function for which* $\int_T \exp_\kappa(u_0) d\mu < \infty$. *We will show that* u_0 *satisfies expression (1). For any* $u \in \mathbb{R}$ *and* $\alpha \geq 1$, *we can write*

$$\begin{aligned} \exp_\kappa(\alpha u) &= \alpha^{1/|\kappa|}(|\kappa|u + \sqrt{1/\alpha^2 + |\kappa|^2 u^2})^{1/|\kappa|} \\ &\leq \alpha^{1/|\kappa|}(|\kappa|u + \sqrt{1 + |\kappa|^2 u^2})^{1/|\kappa|} \\ &= \alpha^{1/|\kappa|} \exp_\kappa(u), \end{aligned}$$

where we used that $\exp_\kappa(\cdot) = \exp_{-\kappa}(\cdot)$. *Then, we conclude that* $\int_T \exp_\kappa(\alpha u_0) d\mu < \infty$ *for all* $\alpha \geq 0$. *Fix any measurable function* $c\colon T \to \mathbb{R}$ *such that* $\int_T \varphi(c) d\mu = 1$. *For each* $\lambda > 0$, *we have*

$$\begin{aligned} \int_T \exp_\kappa(c + \lambda u_0) d\mu &\leq \frac{1}{2} \int_T \exp_\kappa(2c) d\mu + \frac{1}{2} \int_T \exp_\kappa(2\lambda u_0) d\mu \\ &\leq 2^{1/|\kappa|-1} \int_T \exp_\kappa(c) d\mu + 2^{1/|\kappa|-1} \int_T \exp_\kappa(\lambda u_0) d\mu \\ &< \infty, \end{aligned}$$

which shows that $\exp_\kappa(\cdot)$ *satisfies (a3). Therefore, the Kaniadakis' κ-exponential* $\exp_\kappa(\cdot)$ *is an example of φ-function.*

The restriction that $\int_T \varphi(c) d\mu = 1$ can be weakened, as asserted in the next result.

Lemma 1. *Let* $\widetilde{c}\colon T \to \mathbb{R}$ *be any measurable function such that* $\int_T \varphi(\widetilde{c}) d\mu < \infty$. *Then,* $\int_T \varphi(\widetilde{c} + \lambda u_0) d\mu < \infty$ *for all* $\lambda > 0$.

Proof. Notice that if $\int_T \varphi(\widetilde{c}) d\mu \geq 1$, then $\int_T \varphi(\widetilde{c} - \alpha u_0) d\mu = 1$ for some $\alpha > 1$. From the definition of u_0, it follows that $\int_T \varphi(\widetilde{c} + \lambda u_0) d\mu = \int_T \varphi(c + (\alpha + \lambda) u_0) d\mu < \infty$, where $c = \widetilde{c} - \alpha u_0$. Now assume that $\int_T \varphi(\widetilde{c}) d\mu < 1$. Consider any measurable set $A \subseteq T$ with measure $0 < \mu(A) < \mu(T)$. Let $u\colon T \to [0,\infty)$ be a measurable function supported on A satisfying $\varphi(\widetilde{c} + u)\mathbf{1}_A = [\varphi(\widetilde{c}) + \alpha]\mathbf{1}_A$, where $\alpha = (1 - \int_T \varphi(\widetilde{c}) d\mu)/\mu(A)$. Defining $c = (\widetilde{c} + u)\mathbf{1}_A + \widetilde{c}\mathbf{1}_{T \setminus A}$, we see that $\int_T \varphi(c) d\mu = 1$. By the definition of u_0, we can write

$$\int_T \varphi(\widetilde{c} + \lambda u_0) d\mu \leq \int_T \varphi(c + \lambda u_0) d\mu < \infty, \qquad \text{for any } \lambda > 0,$$

which is the desired result. □

As a consequence of Lemma 1, condition (a3) can be replaced by the following one:

(a3′) There exists a measurable function $u_0\colon T \to (0,\infty)$ such that

$$\int_T \varphi(c + \lambda u_0) d\mu < \infty, \qquad \text{for all } \lambda > 0, \tag{2}$$

for each measurable function $c\colon T \to \mathbb{R}$ for which $\int_T \varphi(c) d\mu < \infty$.

Without the equivalence between conditions (a3) and (a3′), we could not generalize Rényi divergence in the manner we propose. In fact, φ-functions could be defined directly in terms of (a3′), without mentioning (a3). We chose to begin with (a3) because this condition appeared initially in [12].

Not all functions $\varphi\colon \mathbb{R} \to (0,\infty)$, for which conditions (a1) and (a2) hold, satisfy condition (a3). Such a function is given below.

Example 2. *Assume that the underlying measure μ is σ-finite and non-atomic. This is the case of the Lebesgue measure. Let us consider the function*

$$\varphi(u) = \begin{cases} e^{(u+1)^2/2}, & u \geq 0, \\ e^{(u+1/2)}, & u \leq 0, \end{cases} \tag{3}$$

which clearly is convex, and satisfies the limits $\lim_{u\to-\infty}\varphi(u) = 0$ and $\lim_{u\to\infty}\varphi(u) = \infty$. Given any measurable function $u_0\colon T \to (0,\infty)$, we will find a measurable function $c\colon T \to \mathbb{R}$ with $\int_T \varphi(c)d\mu < \infty$, for which expression (2) is not satisfied.

For each $m \geq 1$, we define

$$v_m(t) := \Big(m\frac{\log(2)}{u_0(t)} - \frac{u_0(t)}{2} - 1\Big)\mathbf{1}_{E_m}(t),$$

where $E_m = \{t \in T : m\frac{\log(2)}{u_0(t)} - \frac{u_0(t)}{2} - 1 > 0\}$. Because $v_m \uparrow \infty$, we can find a sub-sequence $\{v_{m_n}\}$ such that

$$\int_{E_{mn}} e^{(v_{m_n}+u_0+1)^2/2} d\mu \geq 2^n.$$

According to (Lemma 8.3 in [29]) , there exists a sub-sequence $w_k = v_{m_{n_k}}$ and pairwise disjoint sets $A_k \subseteq E_{m_{n_k}}$ for which

$$\int_{A_k} e^{(w_k+u_0+1)^2/2} d\mu = 1.$$

Let us define $c = \overline{c}\mathbf{1}_{T\setminus A} + \sum_{k=1}^{\infty} w_k \mathbf{1}_{A_k}$, where $A = \bigcup_{k=1}^{\infty} A_k$ and $\overline{c}$ is any measurable function such that $\varphi(\overline{c}(t)) > 0$ for $t \in T \setminus A$ and $\int_{T\setminus A}\varphi(\overline{c})d\mu < \infty$. Observing that

$$e^{(w_k(t)+u_0(t)+1)^2/2} = 2^{m_{n_k}} e^{(w_k(t)+1)^2/2}, \qquad \textit{for } t \in A_k,$$

we get

$$\int_{A_k} e^{(w_k+1)^2/2} d\mu = \frac{1}{2^{m_{n_k}}}, \qquad \textit{for every } m \geq 1.$$

Then, we can write

$$\begin{aligned} \int_T \varphi(c)d\mu &= \int_{T\setminus A} \varphi(\overline{c})d\mu + \sum_{k=1}^{\infty}\int_{A_k} e^{(w_k+1)^2/2} d\mu \\ &= \int_{T\setminus A} \varphi(\overline{c})d\mu + \sum_{k=1}^{\infty} \frac{1}{2^{m_{n_k}}} < \infty. \end{aligned}$$

On the other hand,

$$\begin{aligned} \int_T \varphi(c+u_0)d\mu &= \int_{T\setminus A} \varphi(\overline{c})d\mu + \sum_{k=1}^{\infty}\int_{A_k} e^{(u_0+w_k+1)^2/2} d\mu \\ &= \int_{T\setminus A} \varphi(\overline{c})d\mu + \sum_{k=1}^{\infty} 1 = \infty, \end{aligned}$$

which shows that (2) is not satisfied.

3. Generalization of Rényi Divergence

In this section, we provide a generalization of Rényi divergence, which is given in terms of a φ-function. This generalization also depends on a parameter $\alpha \in [-1,1]$; for $\alpha = \pm 1$, it is defined as a limit. Supposing that the underlying φ-function is continuously differentiable, we show that this limit exists and results in the φ-divergence [12]. In what follows, all probability distributions are assumed to have positive density. In other words, they belong to the collection

$$\mathcal{P}_\mu = \left\{ p \in L^0 : \int_T p\, d\mu = 1 \text{ and } p > 0 \right\},$$

where L^0 is the space of all real-valued, measurable functions on T, with equality μ-a.e. (μ-almost everywhere).

The Rényi divergence of order $\alpha \in (-1,1)$ between two probability distributions p and q in $\mathcal{P}_\mu$ is defined as

$$\mathcal{D}_{\mathrm{R}}^{(\alpha)}(p \parallel q) = \frac{4}{\alpha^2 - 1} \log\left(\int_T p^{\frac{1-\alpha}{2}} q^{\frac{1+\alpha}{2}} d\mu \right). \tag{4}$$

For $\alpha = \pm 1$, the Rényi divergence is defined by taking a limit:

$$\mathcal{D}_{\mathrm{R}}^{(-1)}(p \parallel q) = \lim_{\alpha \downarrow -1} \mathcal{D}_{\mathrm{R}}^{(\alpha)}(p \parallel q), \tag{5}$$

$$\mathcal{D}_{\mathrm{R}}^{(1)}(p \parallel q) = \lim_{\alpha \uparrow 1} \mathcal{D}_{\mathrm{R}}^{(\alpha)}(p \parallel q). \tag{6}$$

Under some conditions, the limits in (5) and (6) are finite-valued, and converge to the Kullback–Leibler divergence. In other words,

$$\mathcal{D}_{\mathrm{R}}^{(-1)}(p \parallel q) = \mathcal{D}_{\mathrm{R}}^{(1)}(q \parallel p) = \mathcal{D}_{\mathrm{KL}}(p \parallel q) < \infty,$$

where $\mathcal{D}_{\mathrm{KL}}(p \parallel q)$ denotes the Kullback–Leibler divergence between p and q, which is given by

$$\mathcal{D}_{\mathrm{KL}}(p \parallel q) = \int_T p \log\left(\frac{p}{q}\right) d\mu.$$

These conditions are stated in Proposition 1, given in the end of this section, for the case involving the generalized Rényi divergence.

The Rényi divergence in its standard form is given by

$$\mathcal{D}^{(\alpha)}(p \parallel q) = \frac{1}{1-\alpha} \log\left(\int_T p^\alpha q^{1-\alpha} d\mu \right), \qquad \text{for } \alpha \in (0,1). \tag{7}$$

Expression (4) is related to this form by

$$\mathcal{D}_{\mathrm{R}}^{(\alpha)}(p \parallel q) = \frac{2}{1-\alpha} \mathcal{D}^{((1-\alpha)/2)}(p \parallel q).$$

Beyond the change of variables, which results in α ranging in $[-1,1]$, expressions (4) and (7) differ by the factor $2/(1-\alpha)$. We opted to insert the term $2/(1-\alpha)$ so that some kind of symmetry could be maintained when the limits $\alpha \downarrow -1$ and $\alpha \uparrow 1$ are considered. In addition, the geometry induced by the version (4) conforms with Amari's notation [5].

The Rényi divergence $\mathcal{D}_{\mathrm{R}}^{(\alpha)}(\cdot \parallel \cdot)$ can be defined for every $\alpha \in \mathbb{R}$. However, for $\alpha \notin (-1,1)$, the expression (4) may not be finite-valued for every p and q in $\mathcal{P}_\mu$. To avoid some technicalities, we just consider $\in [\ 1,1]$

Given p and q in $\mathcal{P}_\mu$, let us define

$$\kappa(\alpha) = -\log\Big(\int_T p^{\frac{1-\alpha}{2}} q^{\frac{1+\alpha}{2}} d\mu\Big), \qquad \text{for } \alpha \in [-1,1],$$

which can be used to express the Rényi divergence as

$$\mathcal{D}_{\mathrm{R}}^{(\alpha)}(p \parallel q) = \frac{4}{1-\alpha^2}\kappa(\alpha), \qquad \text{for } \alpha \in (-1,1).$$

The function $\kappa(\alpha)$, which depends on p and q, can be defined as the unique non-negative real number for which

$$\int_T \exp\Big(\frac{1-\alpha}{2}\ln(p) + \frac{1+\alpha}{2}\ln(q) + \kappa(\alpha)\Big) d\mu = 1. \tag{8}$$

The function $\kappa(\alpha)$ makes the role of a normalizing term. The generalization of Rényi divergence, which we propose, is based on the interpretation of $\kappa(\alpha)$ given in (8). Instead of the exponential function, we consider a φ-function in (8).

Fix any φ-function $\varphi\colon \mathbb{R} \to (0,\infty)$. Given any p and q in $\mathcal{P}_\mu$, we take $\kappa(\alpha) = \kappa(\alpha; p, q) \geq 0$ so that

$$\int_T \varphi\Big(\frac{1-\alpha}{2}\varphi^{-1}(p) + \frac{1+\alpha}{2}\varphi^{-1}(q) + \kappa(\alpha)u_0\Big) d\mu = 1, \tag{9}$$

or, in other words, the term inside the integral is a probability distribution in $\mathcal{P}_\mu$. The existence and uniqueness of $\kappa(\alpha)$ as defined in (9) is guaranteed by condition (a3').

We define a generalization of the Rényi divergence of order $\alpha \in (-1,1)$ as

$$\mathcal{D}_\varphi^{(\alpha)}(p \parallel q) = \frac{4}{1-\alpha^2}\kappa(\alpha). \tag{10}$$

For $\alpha = \pm 1$, this generalization is defined as a limit:

$$\mathcal{D}_\varphi^{(-1)}(p \parallel q) = \lim_{\alpha \downarrow -1} \mathcal{D}_\varphi^{(\alpha)}(p \parallel q), \tag{11}$$

$$\mathcal{D}_\varphi^{(1)}(p \parallel q) = \lim_{\alpha \uparrow 1} \mathcal{D}_\varphi^{(\alpha)}(p \parallel q). \tag{12}$$

The cases $\alpha = \pm 1$ are related to a generalization of the Kullback–Leibler divergence, the so-called φ-divergence, which was introduced by the authors in [12]. The φ-divergence is given by (It was pointed out to us by an anonymous referee that this form of divergence is a special case of the (ρ,τ)-divergence for $\rho = \varphi^{-1}$ and $f = \rho^{-1}$ (see Section 3.5 in [19]) apart from a conformal factor, which is the denominator of (13)):

$$\mathcal{D}_\varphi(p \parallel q) = \frac{\int_T \frac{\varphi^{-1}(p) - \varphi^{-1}(q)}{(\varphi^{-1})'(p)} d\mu}{\int_T \frac{u_0}{(\varphi^{-1})'(p)} d\mu}. \tag{13}$$

Under some conditions, the limit in (11) or (12) is finite-valued and converges to the φ-divergence:

$$\mathcal{D}_\varphi^{(-1)}(p \parallel q) = \mathcal{D}_\varphi^{(1)}(q \parallel p) = \mathcal{D}_\varphi(p \parallel q) < \infty. \tag{14}$$

To show (14), we make use of the following result.

Lemma 2. *Assume that $\varphi(\cdot)$ is continuously differentiable. If for $\alpha_0, \alpha_1 \in \mathbb{R}$, the expression*

$$\int_T \varphi\Big(\frac{1-\alpha}{2}\varphi^{-1}(p) + \frac{1+\alpha}{2}\varphi^{-1}(q)\Big) d\mu < \infty \tag{15}$$

is satisfied for all $\alpha \in [\alpha_0, \alpha_1]$, *then the derivative of* $\kappa(\alpha)$ *exists at any* $\alpha \in (\alpha_0, \alpha_1)$, *and is given by*

$$\frac{\partial \kappa}{\partial \alpha}(\alpha) = -\frac{1}{2} \frac{\int_T [\varphi^{-1}(q) - \varphi^{-1}(p)]\varphi'(c_\alpha) d\mu}{\int_T \varphi'(c_\alpha) u_0 d\mu}, \tag{16}$$

where $c_\alpha = \frac{1-\alpha}{2}\varphi^{-1}(p) + \frac{1+\alpha}{2}\varphi^{-1}(q) + \kappa(\alpha)u_0$.

Proof. For $\alpha \in (\alpha_0, \alpha_1)$ and $\kappa > 0$, define

$$g(\alpha, \kappa) = \int_T \varphi\Big(\frac{1-\alpha}{2}\varphi^{-1}(p) + \frac{1+\alpha}{2}\varphi^{-1}(q) + \kappa u_0\Big) d\mu.$$

The function $\kappa(\alpha)$ is defined implicitly by $g(\alpha, \kappa(\alpha)) = 1$. If we show that

(i) the function $g(\alpha, \kappa)$ is continuous in a neighborhood of $(\alpha, \kappa(\alpha))$,
(ii) the partial derivatives $\frac{\partial g}{\partial \alpha}$ and $\frac{\partial g}{\partial \kappa}$ exist and are continuous at $(\alpha, \kappa(\alpha))$,
(iii) and $\frac{\partial g}{\partial \kappa}(\alpha, \kappa(\alpha)) > 0$,

then by the Implicit Function Theorem $\kappa(\alpha)$ is differentiable at $\alpha \in (\alpha_0, \alpha_1)$, and

$$\frac{\partial \kappa}{\partial \alpha}(\alpha) = -\frac{(\partial g/\partial \alpha)(\alpha, \kappa(\alpha))}{(\partial g/\partial \kappa)(\alpha, \kappa(\alpha))}. \tag{17}$$

We begin by verifying that $g(\alpha, \kappa)$ is continuous. For fixed $\alpha \in (\alpha_0, \alpha_1)$ and $\kappa > 0$, set $\kappa_0 = 2\kappa$. Denoting $A = \{t \in T : \varphi^{-1}(q(t)) > \varphi^{-1}(p(t))\}$, we can write

$$\begin{aligned}\varphi\Big(\frac{1-\beta}{2}\varphi^{-1}(p) + \frac{1+\beta}{2}\varphi^{-1}(q) + \lambda u_0\Big) &\le \varphi\Big(\varphi^{-1}(p) + \frac{1+\beta}{2}[\varphi^{-1}(q) - \varphi^{-1}(p)] + \kappa_0 u_0\Big) \\ &\le \varphi\Big(\varphi^{-1}(p) + \frac{1+\alpha_1}{2}[\varphi^{-1}(q) - \varphi^{-1}(p)] + \kappa_0 u_0\Big)\mathbf{1}_A \\ &\quad + \varphi\Big(\varphi^{-1}(p) + \frac{1+\alpha_0}{2}[\varphi^{-1}(q) - \varphi^{-1}(p)] + \kappa_0 u_0\Big)\mathbf{1}_{T\setminus A},\end{aligned} \tag{18}$$

for every $\beta \in (\alpha_0, \alpha_1)$ and $\lambda \in (0, \kappa_0)$. Because the function on the right-hand side of (18) is integrable, we can apply the Dominated Convergence Theorem to conclude that

$$\lim_{(\beta,\lambda)\to(\alpha,\kappa)} g(\beta, \lambda) = g(\alpha, \kappa).$$

Now, we will show that the derivative of $g(\alpha, \kappa)$ with respect to α exists and is continuous. Consider the difference

$$\frac{g(\gamma, \lambda) - g(\beta, \lambda)}{\gamma - \beta} = \int_T \frac{1}{\gamma - \beta}\Big[\varphi\Big(c_\beta + \frac{\gamma-\beta}{2}[\varphi^{-1}(q) - \varphi^{-1}(p)] + \lambda u_0\Big) - \varphi(c_\beta + \lambda u_0)\Big] d\mu, \tag{19}$$

where $c_\beta = \frac{1-\beta}{2}\varphi^{-1}(p) + \frac{1+\beta}{2}\varphi^{-1}(q)$. Represent by $f_{\beta,\gamma,\lambda}$ the function inside the integral sign in (19). For fixed $\alpha \in (\alpha_0, \alpha_1)$ and $\kappa > 0$, denote $\overline{\alpha}_0 = (\alpha_0 + \alpha)/2$, $\overline{\alpha}_1 = (\alpha + \alpha_1)/2$, and $\kappa_0 = 2\kappa$. Because $\varphi(\cdot)$ is convex and increasing, it follows that

$$|f_{\beta,\gamma,\lambda}| \le f_{\overline{\alpha}_1,\alpha_1,\kappa_0}\mathbf{1}_A - f_{\overline{\alpha}_0,\alpha_0,\kappa_0}\mathbf{1}_{T\setminus A} =: f, \qquad \text{for all } \beta, \gamma \in (\overline{\alpha}_0, \overline{\alpha}_1) \text{ and } \lambda \in (0, \kappa_0),$$

where $A = \{t \in T : \varphi^{-1}(q(t)) > \varphi^{-1}(p(t))\}$. Observing that f is integrable, we can use the Dominated Convergence Theorem to get

$$\lim_{\gamma\to\beta} \int_T f_{\beta,\gamma,\lambda} d\mu = \int_T (\lim_{\gamma\to\beta} f_{\beta,\gamma,\lambda}) d\mu,$$

and then

$$\frac{\partial g}{\partial \alpha}(\beta,\lambda) = \frac{1}{2}\int_T [\varphi^{-1}(q) - \varphi^{-1}(p)]\varphi'(c_\beta + \lambda u_0)d\mu. \tag{20}$$

For $\beta \in (\overline{\alpha}_0, \overline{\alpha}_1)$ and $\lambda \in (0, \kappa_0)$, the function inside the integral sign in (20) is dominated by f. As a result, a second use of the Dominated Convergence Theorem shows that $\frac{\partial g}{\partial \alpha}$ is continuous at (α, κ):

$$\lim_{(\beta,\lambda)\to(\alpha,\kappa)} \frac{\partial g}{\partial \alpha}(\beta,\lambda) = \frac{\partial g}{\partial \alpha}(\alpha,\kappa).$$

Using similar arguments, one can show that $\frac{\partial g}{\partial \kappa}(\alpha,\kappa)$ exists and is continuous at any $\alpha \in (\alpha_0, \alpha_1)$ and $\kappa > 0$, and is given by

$$\frac{\partial g}{\partial \kappa}(\alpha,\kappa) = \int_T u_0 \varphi'(c_\alpha + \kappa u_0)d\mu. \tag{21}$$

Clearly, expression (21) implies that $\frac{\partial g}{\partial \kappa}(\alpha,\kappa) > 0$ for all $\alpha \in (0, \alpha_0)$ and $\kappa > 0$.

We proved that items (i)–(iii) are satisfied. As consequence, the derivative of $\kappa(\alpha)$ exists at any $\alpha \in (\alpha_0, \alpha_1)$. Expression (16) for the derivative of $\kappa(\alpha)$ follows from (17), (20) and (21). □

As an immediate consequence of Lemma 2, we get the proposition below.

Proposition 1. *Assume that $\varphi(\cdot)$ is continuously differentiable.*

(a) *If, for some $\alpha_0 < -1$, expression (15) is satisfied for all $\alpha \in [\alpha_0, -1)$, then*

$$\mathcal{D}_\varphi^{(-1)}(p \parallel q) = \lim_{\alpha \downarrow -1} \mathcal{D}_\varphi^{(\alpha)}(p \parallel q) = 2\frac{\partial \kappa}{\partial \alpha}(-1) = \mathcal{D}_\varphi(p \parallel q) < \infty.$$

(b) *If, for some $\alpha_1 > 1$, expression (15) is satisfied for all $\alpha \in (1, \alpha_1]$, then*

$$\mathcal{D}_\varphi^{(1)}(p \parallel q) = \lim_{\alpha \uparrow 1} \mathcal{D}_\varphi^{(\alpha)}(p \parallel q) = -2\frac{\partial \kappa}{\partial \alpha}(1) = \mathcal{D}_\varphi(q \parallel p) < \infty.$$

4. Generalized Statistical Manifolds

Statistical manifolds consist of a collection of probability distributions endowed with a metric and α-connections, which are defined in terms of the derivative of $l(t;\theta) = \log p(t;\theta)$. In a generalized statistical manifold, the metric and connection are defined in terms of $f(t;\theta) = \varphi^{-1}(p(t;\theta))$. Instead of the logarithm, we consider the inverse $\varphi^{-1}(\cdot)$ of a φ-function. Generalized statistical manifolds were introduced by the authors in [17,18]. Among examples of the generalized statistical manifold, (parametric) φ-families of probability distributions are of greatest importance. The non-parametric counterpart was investigated in [11,12]. The metric in φ-families can be defined as the Hessian of a function; i.e., φ-families are Hessian manifolds [30]. In [17,18], the φ-divergence gives rise to a pair of dual connections $D^{(-1)}$ and $D^{(1)}$; and then for $\alpha \in (-1, 1)$ the α-connection $D^{(\alpha)}$ is defined as the convex combination $D^{(\alpha)} = \frac{1-\alpha}{2}D^{(-1)} + \frac{1+\alpha}{2}D^{(1)}$. In the present paper, we show that the connection induced by $\mathcal{D}_\varphi^{(\alpha)}(\cdot \parallel \cdot)$, the generalization of Rényi divergence, corresponds to $D^{(\alpha)}$.

4.1. Definitions

Let $\varphi\colon \mathbb{R} \to (0,\infty)$ be a φ-function. A *generalized statistical manifold* $\mathcal{P} = \{p(t;\theta) : \theta \in \Theta\}$ is a collection of probability distributions $p_\theta(t) := p(t;\theta)$, indexed by parameters $\theta = (\theta^1, \dots, \theta^n) \in \Theta$ in a one-to-one relation, such that

(m1) Θ is a domain (open and connected set) in $\mathbb{R}^n$;
(m2) $p(t;\theta)$ is differentiable with respect to θ;

(m3) the matrix $g = (g_{ij})$ defined by

$$g_{ij} = -E'_\theta\Big[\frac{\partial^2 \varphi^{-1}(p_\theta)}{\partial\theta^i \partial\theta^j}\Big], \tag{22}$$

is positive definite at each $\theta \in \Theta$, where

$$E'_\theta[\cdot] = \frac{\int_T (\cdot)\varphi'(\varphi^{-1}(p_\theta))d\mu}{\int_T u_0 \varphi'(\varphi^{-1}(p_\theta))d\mu}; \tag{23}$$

(m4) the operations of integration with respect to μ and differentiation with respect to θ^i commute in all calculations found below, which are related to the metric and connections.

The matrix $g = (g_{ij})$ equips $\mathcal{P}$ with a metric. By the chain rule, the tensor related to $g = (g_{ij})$ is invariant under change of coordinates. The (classical) statistical manifold is a particular case in which $\varphi(u) = \exp(u)$ and $u_0 = \mathbf{1}_T$.

We introduce a notation similar to Equation (23) that involves higher order derivatives of $\varphi(\cdot)$. For each $n \geq 1$, we define

$$E_\theta^{(n)}[\cdot] = \frac{\int_T (\cdot)\varphi^{(n)}(\varphi^{-1}(p_\theta))d\mu}{\int_T u_0 \varphi'(\varphi^{-1}(p_\theta))d\mu}. \tag{24}$$

We also use $E'_\theta[\cdot]$, $E''_\theta[\cdot]$ and $E'''_\theta[\cdot]$ to denote $E_\theta^{(n)}[\cdot]$ for $n = 1, 2, 3$, respectively. The notation (24) appears in expressions related to the metric and connections.

Using property (m4), we can find an alternate expression for g_{ij} as well as an identification involving tangent spaces. The matrix $g = (g_{ij})$ can be equivalently defined by

$$g_{ij} = E''_\theta\Big[\frac{\partial\varphi^{-1}(p_\theta)}{\partial\theta^i}\frac{\partial\varphi^{-1}(p_\theta)}{\partial\theta^j}\Big]. \tag{25}$$

As a consequence of this equivalence, the tangent space $T_{p_\theta}\mathcal{P}$ can be identified with $\widetilde{T}_{p_\theta}\mathcal{P}$, the vector space spanned by $\frac{\partial\varphi^{-1}(p_\theta)}{\partial\theta^i}$, and endowed with the inner product $\langle\widetilde{X},\widetilde{Y}\rangle_\theta := E''_\theta[\widetilde{X}\widetilde{Y}]$. The mapping

$$\sum_i a_i \frac{\partial}{\partial\theta^i} \mapsto \sum_i a_i \frac{\partial\varphi^{-1}(p_\theta)}{\partial\theta^i}$$

defines an isometry between $T_{p_\theta}\mathcal{P}$ and $\widetilde{T}_{p_\theta}\mathcal{P}$.

To verify (25), we differentiate $\int_T p_\theta d\mu = 1$, with respect to θ^i, to get

$$0 = \frac{\partial}{\partial\theta^i}\int_T p_\theta d\mu = \int_T \frac{\partial}{\partial\theta^i}\varphi(\varphi^{-1}(p_\theta))d\mu = \int_T \frac{\partial\varphi^{-1}(p_\theta)}{\partial\theta^i}\varphi'(\varphi^{-1}(p_\theta))d\mu. \tag{26}$$

Now, differentiating with respect to θ^j, we obtain

$$0 = \int_T \frac{\partial^2\varphi^{-1}(p_\theta)}{\partial\theta^i\partial\theta^j}\varphi'(\varphi^{-1}(p_\theta))d\mu + \int_T \frac{\partial\varphi^{-1}(p_\theta)}{\partial\theta^i}\frac{\partial\varphi^{-1}(p_\theta)}{\partial\theta^j}\varphi''(\varphi^{-1}(p_\theta))d\mu,$$

and then (25) follows. In view of (26), we notice that every vector $\widetilde{X}$ belonging to $\widetilde{T}_{p_\theta}\mathcal{P}$ satisfies $E'_\theta[\widetilde{X}] = 0$.

The metric $g = (g_{ij})$ gives rise to a Levi–Civita connection ∇ (i.e., a torsion-free, metric connection), whose corresponding Christoffel symbols Γ_{ijk} are given by

$$\Gamma_{ijk} := \frac{1}{2}\Big(\frac{\partial g_{ki}}{\partial\theta^j} + \frac{\partial g_{kj}}{\partial\theta^i} - \frac{\partial g_{ij}}{\partial\theta^k}\Big). \tag{27}$$

Using expression (25) to calculate the derivatives in (27), we can express

$$\Gamma_{ijk} = E''_\theta\Big[\frac{\partial^2\varphi^{-1}(p_\theta)}{\partial\theta^i\partial\theta^j}\frac{\partial\varphi^{-1}(p_\theta)}{\partial\theta^k}\Big] + \frac{1}{2}E'''_\theta\Big[\frac{\partial\varphi^{-1}(p_\theta)}{\partial\theta^i}\frac{\partial\varphi^{-1}(p_\theta)}{\partial\theta^j}\frac{\partial\varphi^{-1}(p_\theta)}{\partial\theta^k}\Big]$$
$$-\frac{1}{2}E''_\theta\Big[\frac{\partial\varphi^{-1}(p_\theta)}{\partial\theta^i}\frac{\partial\varphi^{-1}(p_\theta)}{\partial\theta^k}\Big]E''_\theta\Big[u_0\frac{\partial\varphi^{-1}(p_\theta)}{\partial\theta^j}\Big] - \frac{1}{2}E''_\theta\Big[\frac{\partial\varphi^{-1}(p_\theta)}{\partial\theta^j}\frac{\partial\varphi^{-1}(p_\theta)}{\partial\theta^k}\Big]E''_\theta\Big[u_0\frac{\partial\varphi^{-1}(p_\theta)}{\partial\theta^i}\Big]$$
$$+\frac{1}{2}E''_\theta\Big[\frac{\partial\varphi^{-1}(p_\theta)}{\partial\theta^i}\frac{\partial\varphi^{-1}(p_\theta)}{\partial\theta^j}\Big]E''_\theta\Big[u_0\frac{\partial\varphi^{-1}(p_\theta)}{\partial\theta^k}\Big].$$

As we will show later, the Levi–Civita connection ∇ corresponds to the connection derived from the divergence $\mathcal{D}_\varphi^{(\alpha)}(\cdot \,\|\, \cdot)$ with $\alpha = 0$.

4.2. φ-Families

Let $c\colon T \to \mathbb{R}$ be a measurable function for which $p = \varphi(c)$ is a probability density in $\mathcal{P}_\mu$. Fix measurable functions $u_1, \dots, u_n\colon T \to \mathbb{R}$. A *(parametric) φ-family* $\mathcal{F}_p = \{p_\theta : \theta \in \Theta\}$, centered at $p = \varphi(c)$, is a set of probability distributions in $\mathcal{P}_\mu$, whose members can be written in the form

$$p_\theta := \varphi\Big(c + \sum_{i=1}^n \theta^i u_i - \psi(\theta)u_0\Big), \qquad \text{for each } \theta = (\theta^i) \in \Theta, \tag{28}$$

where $\psi\colon \Theta \to [0,\infty)$ is a *normalizing function*, which is introduced so that expression (28) defines a probability distribution belonging to $\mathcal{P}_\mu$.

The functions $u_1, \dots, u_n$ are not arbitrary. They are chosen to satisfy the following assumptions:

(i) $u_0, u_1, \dots, u_n$ are linearly independent,
(ii) $\int_T u_i\varphi'(c)d\mu = 0$, and
(iii) there exists $\varepsilon > 0$ such that $\int_T \varphi(c + \lambda u_i)d\mu < \infty$, for all $\lambda \in (-\varepsilon, \varepsilon)$.

Moreover, the domain $\Theta \subseteq \mathbb{R}^n$ is defined as the set of all vectors $\theta = (\theta^i)$ for which

$$\int_T \varphi\Big(c + \lambda\sum_{i=1}^n \theta^i u_i\Big)d\mu < \infty, \qquad \text{for some } \lambda > 1.$$

Condition (i) implies that the mapping defined by (28) is one-to-one. Assumption (ii) makes of ψ a non-negative function. Indeed, by the convexity of $\varphi(\cdot)$, along with (ii), we can write

$$\int_T \varphi(c)d\mu = \int_T \Big[\varphi(c) + \Big(\sum_{i=1}^n \theta^i u_i\Big)\varphi'(c)\Big]d\mu \le \int_T \varphi\Big(c + \sum_{i=1}^n \theta^i u_i\Big)d\mu,$$

which implies $\psi(\theta) \ge 0$. By condition (iii), the domain Θ is an open neighborhood of the origin. If the set T is finite, condition (iii) is always satisfied. One can show that the domain Θ is open and convex. Moreover, the normalizing function ψ is also convex (or strictly convex if $\varphi(\cdot)$ is strictly convex). Conditions (ii) and (iii) also appears in the definition of non-parametric φ-families. For further details, we refer to [11,12].

In a φ-family $\mathcal{F}_p$, the matrix (g_{ij}) given by (22) or (25) can be expressed as the Hessian of ψ. If $\varphi(\cdot)$ is strictly convex, then(g_{ij}) is positive definite. From

$$\frac{\partial\varphi^{-1}(p_\theta)}{\partial\theta^i} = u_i - \frac{\partial\psi}{\partial\theta^i}, \qquad -\frac{\partial^2\varphi^{-1}(p_\theta)}{\partial\theta^i\partial\theta^j} = -\frac{\partial^2\psi}{\partial\theta^i\partial\theta^j},$$

it follows that $g_{ij} = \frac{\partial^2\psi}{\partial\theta^i\partial\theta^j}$.

The next two results show how the generalization of Rényi divergence and the φ-divergence are related to the normalizing function in φ-families.

Proposition 2. *In a φ-family $\mathcal{F}_p$, the generalization of Rényi divergence for $\alpha \in (-1,1)$ can be expressed in terms of the normalizing function ψ as follows:*

$$\mathcal{D}_{\varphi}^{(\alpha)}(p_\theta \parallel p_\vartheta) = \frac{2}{1+\alpha}\psi(\theta) + \frac{2}{1-\alpha}\psi(\vartheta) - \frac{4}{1-\alpha^2}\psi\Big(\frac{1-\alpha}{2}\theta + \frac{1+\alpha}{2}\vartheta\Big), \tag{29}$$

for all $\theta, \vartheta \in \Theta$.

Proof. Recall the definition of $\kappa(\alpha)$ as the real number for which

$$\int_T \varphi\Big(\frac{1-\alpha}{2}\varphi^{-1}(p_\theta) + \frac{1+\alpha}{2}\varphi^{-1}(p_\vartheta) + \kappa(\alpha)u_0\Big)d\mu = 1.$$

Using expression (28) for probability distributions in $\mathcal{F}_p$, we can write

$$\begin{aligned}\frac{1-\alpha}{2}\varphi^{-1}(p_\theta) &+ \frac{1+\alpha}{2}\varphi^{-1}(p_\vartheta) + \kappa(\alpha)u_0 \\ &= c + \sum_{i=1}^{n}\Big(\frac{1-\alpha}{2}\theta^i + \frac{1+\alpha}{2}\vartheta^i\Big)u_i - \Big(\frac{1-\alpha}{2}\psi(\theta) + \frac{1+\alpha}{2}\psi(\vartheta) - \kappa(\alpha)\Big)u_0 \\ &= c + \sum_{i=1}^{n}\Big(\frac{1-\alpha}{2}\theta^i + \frac{1+\alpha}{2}\vartheta^i\Big)u_i - \psi\Big(\frac{1-\alpha}{2}\theta + \frac{1+\alpha}{2}\vartheta\Big)u_0.\end{aligned}$$

The last equality is a consequence of the domain Θ being convex. Thus, it follows that

$$\kappa(\alpha) = \frac{1-\alpha}{2}\psi(\theta) + \frac{1+\alpha}{2}\psi(\vartheta) - \psi\Big(\frac{1-\alpha}{2}\theta + \frac{1+\alpha}{2}\vartheta\Big).$$

By the definition of $\mathcal{D}_{\varphi}^{(\alpha)}(\cdot \parallel \cdot)$, we get (29). □

Proposition 3. *In a φ-family $\mathcal{F}_p$, the φ-divergence is related to the normalizing function ψ by the equality*

$$\mathcal{D}_\varphi(p_\theta \parallel p_\vartheta) = \psi(\vartheta) - \psi(\theta) - \nabla\psi(\theta)\cdot(\vartheta - \theta), \tag{30}$$

for all $\theta, \vartheta \in \Theta$.

Proof. To show (30), we use

$$\frac{\partial\psi}{\partial\theta^i}(\theta) = \frac{\int_T u_i\varphi'(\varphi^{-1}(p_\theta))d\mu}{\int_T u_0\varphi'(\varphi^{-1}(p_\theta))d\mu},$$

which is a consequence of (Lemma 10 in [12]). In view of $(\varphi^{-1})'(u) = 1/\varphi'(\varphi^{-1}(u))$, expression (13) with $p = p_\theta$ and $q = p_\vartheta$ results in

$$\mathcal{D}_\varphi(p_\theta \parallel p_\vartheta) = \frac{\int_T[\varphi^{-1}(p_\theta) - \varphi^{-1}(p_\vartheta)]\varphi'(\varphi^{-1}(p_\theta))d\mu}{\int_T u_0\varphi'(\varphi^{-1}(p_\theta))d\mu}. \tag{31}$$

Inserting into (31) the difference

$$\begin{aligned}\varphi^{-1}(p_\theta) - \varphi^{-1}(p_\vartheta) &= \Big(c + \sum_{i=1}^{n}\theta^i u_i - \psi(\theta)u_0\Big) - \Big(c + \sum_{i=1}^{n}\vartheta^i u_i - \psi(\vartheta)u_0\Big) \\ &= \psi(\vartheta)u_0 - \psi(\theta)u_0 - \sum_{i=1}^{n}(\vartheta^i - \theta^i)u_i,\end{aligned}$$

we get expression (30). □

In Proposition 2, the expression on the right-hand side of Equation (29) defines a divergence on its own, which was investigated by Jun Zhang in [19]. Proposition 3 asserts that the φ-divergence $\mathcal{D}_\varphi(p_\theta \parallel p_\vartheta)$ coincides with the Bregman divergence [31,32] associated with the normalizing function ψ for points ϑ and θ in Θ. Because ψ is convex and attains a minimum at $\theta = 0$, it follows that $\frac{\partial \psi}{\partial \theta^i}(\theta) = 0$ at $\theta = 0$. As a result, equality (30) reduces to $\mathcal{D}_\varphi(p \parallel p_\theta) = \psi(\theta)$.

4.3. Geometry Induced by $\mathcal{D}_\varphi^{(\alpha)}(\cdot \parallel \cdot)$

In this section, we assume that $\varphi(\cdot)$ is continuously differentiable and strictly convex. The latter assumption guarantees that

$$\mathcal{D}_\varphi^{(\alpha)}(p \parallel q) = 0 \quad \text{if and only if} \quad p = q. \tag{32}$$

The generalized Rényi divergence induces a metric $g = (g_{ij})$ in generalized statistical manifolds $\mathcal{P}$. This metric is given by

$$g_{ij} = -\Big[\Big(\frac{\partial}{\partial \theta^i}\Big)_p \Big(\frac{\partial}{\partial \theta^j}\Big)_q \mathcal{D}_\varphi^\alpha(p \parallel q)\Big]_{q=p}. \tag{33}$$

To show that this expression defines a metric, we have to verify that g_{ij} is invariant under change of coordinates, and (g_{ij}) is positive definite. The first claim follows from the chain rule. The positive definiteness of (g_{ij}) is a consequence of Proposition 4, which is given below.

Proposition 4. *The metric induced by $\mathcal{D}_\varphi^{(\alpha)}(\cdot \parallel \cdot)$ coincides with the metric given by (22) or (25).*

Proof. Fix any $\alpha \in (-1, 1)$. Applying the operator $(\frac{\partial}{\partial \theta^j})_{p_\vartheta}$ to

$$\int_T \varphi(c_\alpha) d\mu = 1,$$

where $c_\alpha = \frac{1-\alpha}{2}\varphi^{-1}(p_\theta) + \frac{1+\alpha}{2}\varphi^{-1}(p_\vartheta) + \kappa(\alpha) u_0$, we obtain

$$\int_T \Big(\frac{1+\alpha}{2}\frac{\partial \varphi^{-1}(p_\vartheta)}{\partial \theta^j} + \Big(\frac{\partial}{\partial \theta^j}\Big)_{p_\vartheta} \kappa(\alpha) u_0\Big) \varphi'(c_\alpha) d\mu = 0,$$

which results in

$$\Big(\frac{\partial}{\partial \theta^j}\Big)_{p_\vartheta} \kappa(\alpha) = -\frac{1+\alpha}{2}\frac{\int_T \frac{\partial \varphi^{-1}(p_\vartheta)}{\partial \theta^j}\varphi'(c_\alpha) d\mu}{\int_T u_0 \varphi'(c_\alpha) d\mu}.$$

By the standard differentiation rules, we can write

$$\begin{aligned}\Big(\frac{\partial}{\partial \theta^i}\Big)_{p_\theta}\Big(\frac{\partial}{\partial \theta^j}\Big)_{p_\vartheta} \kappa(\alpha) = &-\frac{1+\alpha}{2}\frac{\int_T [\frac{1-\alpha}{2}\frac{\partial \varphi^{-1}(p_\theta)}{\partial \theta^i} + (\frac{\partial}{\partial \theta^i})_{p_\theta}\kappa(\alpha) u_0]\frac{\partial \varphi^{-1}(p_\vartheta)}{\partial \theta^j}\varphi''(c_\alpha) d\mu}{\int_T u_0 \varphi'(c_\alpha) d\mu} \\ &+ \frac{1+\alpha}{2}\frac{\int_T \frac{\partial \varphi^{-1}(p_\vartheta)}{\partial \theta^j}\varphi'(c_\alpha) d\mu}{\int_T u_0 \varphi'(c_\alpha) d\mu}\frac{\int_T u_0 [\frac{1-\alpha}{2}\frac{\partial \varphi^{-1}(p_\theta)}{\partial \theta^i} + (\frac{\partial}{\partial \theta^i})_{p_\theta}\kappa(\alpha) u_0]\varphi''(c_\alpha) d\mu}{\int_T u_0 \varphi'(c_\alpha) d\mu}.\end{aligned} \tag{34}$$

Noticing that $\int_T \frac{\partial \varphi^{-1}(p_\vartheta)}{\partial \theta^j}\varphi'(c_\alpha) d\mu = 0$ for $p_\vartheta = p_\theta$, the second term on the right-hand side of Equation (34) vanishes, and then

$$\Big[\Big(\frac{\partial}{\partial \theta^i}\Big)_{p_\theta}\Big(\frac{\partial}{\partial \theta^j}\Big)_{p_\vartheta} \kappa(\alpha)\Big]_{p_\vartheta = p_\theta} = -\frac{1-\alpha^2}{4}\frac{\int_T \frac{\partial \varphi^{-1}(p_\theta)}{\partial \theta^i}\frac{\partial \varphi^{-1}(p_\theta)}{\partial \theta^j}\varphi''(\varphi^{-1}(p_\theta)) d\mu}{\int_T u_0 \varphi'(\varphi^{-1}(p_\theta)) d\mu}.$$

If we use the notation introduced in (24), we can write

$$g_{ij} = -\Big[\Big(\frac{\partial}{\partial\theta^i}\Big)_{p_\theta}\Big(\frac{\partial}{\partial\theta^j}\Big)_{p_\vartheta}\mathcal{D}_\varphi^{(\alpha)}(p_\theta \parallel p_\vartheta)\Big]_{p_\vartheta=p_\theta} = E''_\theta\Big[\frac{\partial\varphi^{-1}(p_\theta)}{\partial\theta^i}\frac{\partial\varphi^{-1}(p_\theta)}{\partial\theta^j}\Big].$$

It remains to show the case $\alpha = \pm 1$. Comparing (13) and (23), we can write

$$\mathcal{D}_\varphi(p_\theta \parallel p_\vartheta) = E'_\theta[\varphi^{-1}(p_\theta) - \varphi^{-1}(p_\vartheta)]. \tag{35}$$

We use the equivalent expressions

$$g_{ij} = \Big[\Big(\frac{\partial^2}{\partial\theta^i\partial\theta^j}\Big)_p\mathcal{D}_\varphi^\alpha(p \parallel q)\Big]_{q=p} = \Big[\Big(\frac{\partial^2}{\partial\theta^i\partial\theta^j}\Big)_q\mathcal{D}_\varphi^\alpha(p \parallel q)\Big]_{q=p},$$

which follows from condition (32), to infer that

$$g_{ij} = \Big[\Big(\frac{\partial^2}{\partial\theta^i\partial\theta^j}\Big)_{p_\vartheta}\mathcal{D}_\varphi(p_\theta \parallel p_\vartheta)\Big]_{p_\theta=p_\vartheta} = -E'_\theta\Big[\frac{\partial^2\varphi^{-1}(p_\theta)}{\partial\theta^i\partial\theta^j}\Big]. \tag{36}$$

Because $\mathcal{D}_\varphi^{(-1)}(p \parallel q) = \mathcal{D}_\varphi^{(1)}(q \parallel p) = \mathcal{D}_\varphi(p \parallel q)$, we conclude that the metric defined by (22) coincides with the metric induced by $\mathcal{D}_\varphi^{(-1)}(\cdot \parallel \cdot)$ and $\mathcal{D}_\varphi^{(1)}(\cdot \parallel \cdot)$. □

In generalized statistical manifolds, the generalized Rényi divergence $\mathcal{D}_\varphi^{(\alpha)}(\cdot \parallel \cdot)$ induces a connection $D^{(\alpha)}$, whose Christoffel symbols $\Gamma_{ijk}^{(\alpha)}$ are given by

$$\Gamma_{ijk}^{(\alpha)} = -\Big[\Big(\frac{\partial^2}{\partial\theta^i\partial\theta^j}\Big)_p\Big(\frac{\partial}{\partial\theta^k}\Big)_q\mathcal{D}_\varphi^{(\alpha)}(p \parallel q)\Big]_{q=p}.$$

Because $\mathcal{D}_\varphi^{(\alpha)}(p \parallel q) = \mathcal{D}_\varphi^{(-\alpha)}(q \parallel p)$, it follows that $D^{(\alpha)}$ and $D^{(-\alpha)}$ are mutually dual for any $\alpha \in [-1,1]$. In other words, $\Gamma_{ijk}^{(\alpha)}$ and $\Gamma_{ijk}^{(-\alpha)}$ satisfy the relation $\frac{\partial g_{jk}}{\partial\theta^i} = \Gamma_{ijk}^{(\alpha)} + \Gamma_{ikj}^{(-\alpha)}$. A development involving expression (35) results in

$$\Gamma_{ijk}^{(1)} = E''_\theta\Big[\frac{\partial^2\varphi^{-1}(p_\theta)}{\partial\theta^i\partial\theta^j}\frac{\partial\varphi^{-1}(p_\theta)}{\partial\theta^k}\Big] - E'_\theta\Big[\frac{\partial^2\varphi^{-1}(p_\theta)}{\partial\theta^i\partial\theta^j}\Big]E''_\theta\Big[u_0\frac{\partial\varphi^{-1}(p_\theta)}{\partial\theta^k}\Big], \tag{37}$$

and

$$\begin{aligned}\Gamma_{ijk}^{(-1)} = {} & E''_\theta\Big[\frac{\partial^2\varphi^{-1}(p_\theta)}{\partial\theta^i\partial\theta^j}\frac{\partial\varphi^{-1}(p_\theta)}{\partial\theta^k}\Big] + E'''_\theta\Big[\frac{\partial\varphi^{-1}(p_\theta)}{\partial\theta^i}\frac{\partial\varphi^{-1}(p_\theta)}{\partial\theta^j}\frac{\partial\varphi^{-1}(p_\theta)}{\partial\theta^k}\Big] \\ & - E''_\theta\Big[\frac{\partial\varphi^{-1}(p_\theta)}{\partial\theta^j}\frac{\partial\varphi^{-1}(p_\theta)}{\partial\theta^k}\Big]E''_\theta\Big[u_0\frac{\partial\varphi^{-1}(p_\theta)}{\partial\theta^i}\Big] \\ & - E''_\theta\Big[\frac{\partial\varphi^{-1}(p_\theta)}{\partial\theta^i}\frac{\partial\varphi^{-1}(p_\theta)}{\partial\theta^k}\Big]E''_\theta\Big[u_0\frac{\partial\varphi^{-1}(p_\theta)}{\partial\theta^j}\Big].\end{aligned} \tag{38}$$

For $\alpha \in (-1,1)$, the Christoffel symbols $\Gamma_{ijk}^{(\alpha)}$ can be written as a convex combination of $\Gamma_{ijk}^{(-1)}$ and $\Gamma_{ijk}^{(-1)}$, as asserted in the next result.

Proposition 5. *The Christoffel symbols $\Gamma_{ijk}^{(\alpha)}$ induced by the divergence $\mathcal{D}_\varphi^{(\alpha)}(\cdot \parallel \cdot)$ satisfy the relation*

$$\Gamma_{ijk}^{(\alpha)} = \frac{1-\alpha}{2}\Gamma_{ijk}^{(-1)} + \frac{1+\alpha}{2}\Gamma_{ijk}^{(1)}, \qquad \text{for } \alpha \in [-1,1]. \tag{39}$$

Proof. For $\alpha = \pm 1$, equality (39) follows trivially. Thus, we assume $\alpha \in (-1,1)$. By (34), we can write

$$\begin{aligned}\Big(\frac{\partial}{\partial\theta^i}\Big)_{p_\theta}\Big(\frac{\partial}{\partial\theta^k}\Big)_{p_\vartheta}\kappa(\alpha) &= -\frac{1+\alpha}{2}\frac{\int_T [\frac{1-\alpha}{2}\frac{\partial\varphi^{-1}(p_\theta)}{\partial\theta^i} + (\frac{\partial}{\partial\theta^i})_{p_\theta}\kappa(\alpha)u_0]\frac{\partial\varphi^{-1}(p_\vartheta)}{\partial\theta^k}\varphi''(c_\alpha)d\mu}{\int_T u_0\varphi'(c_\alpha)d\mu} \\ &\quad + \frac{1+\alpha}{2}\frac{\int_T \frac{\partial\varphi^{-1}(p_\vartheta)}{\partial\theta^k}\varphi'(c_\alpha)d\mu}{\int_T u_0\varphi'(c_\alpha)d\mu}\frac{\int_T u_0[\frac{1-\alpha}{2}\frac{\partial\varphi^{-1}(p_\theta)}{\partial\theta^i} + (\frac{\partial}{\partial\theta^i})_{p_\theta}\kappa(\alpha)u_0]\varphi''(c_\alpha)d\mu}{\int_T u_0\varphi'(c_\alpha)d\mu}.\end{aligned} \tag{40}$$

Applying $(\frac{\partial}{\partial\theta^j})_{p_\theta}$ to the first term on the right-hand side of (40), and then equating $p_\vartheta = p_\theta$, we obtain

$$\begin{aligned}&-\frac{1-\alpha^2}{4}E''_\theta\Big[\frac{\partial^2\varphi^{-1}(p_\theta)}{\partial\theta^i\partial\theta^j}\frac{\partial\varphi^{-1}(p_\theta)}{\partial\theta^k}\Big] - \frac{1+\alpha}{2}\Big(\frac{\partial^2}{\partial\theta^i\partial\theta^j}\Big)_{p_\theta}\kappa(\alpha)E''_\theta\Big[u_0\frac{\partial\varphi^{-1}(p_\theta)}{\partial\theta^k}\Big] \\ &\quad - \frac{1-\alpha^2}{4}\frac{1-\alpha}{2}E'''_\theta\Big[\frac{\partial\varphi^{-1}(p_\theta)}{\partial\theta^i}\frac{\partial\varphi^{-1}(p_\theta)}{\partial\theta^j}\frac{\partial\varphi^{-1}(p_\theta)}{\partial\theta^k}\Big] \\ &\quad + \frac{1-\alpha^2}{4}\frac{1-\alpha}{2}E''_\theta\Big[\frac{\partial\varphi^{-1}(p_\theta)}{\partial\theta^i}\frac{\partial\varphi^{-1}(p_\vartheta)}{\partial\theta^k}\Big]E''_\theta\Big[u_0\frac{\partial\varphi^{-1}(p_\vartheta)}{\partial\theta^j}\Big].\end{aligned} \tag{41}$$

Similarly, if we apply $(\frac{\partial}{\partial\theta^j})_{p_\theta}$ to the second term on the right-hand side of (40), and make $p_\vartheta = p_\theta$, we get

$$\frac{1-\alpha^2}{4}\frac{1-\alpha}{2}E''_\theta\Big[\frac{\partial\varphi^{-1}(p_\theta)}{\partial\theta^j}\frac{\partial\varphi^{-1}(p_\theta)}{\partial\theta^k}\Big]E''_\theta\Big[u_0\frac{\partial\varphi^{-1}(p_\theta)}{\partial\theta^i}\Big]. \tag{42}$$

Collecting (41) and (42), we can write

$$\begin{aligned}\Gamma^{(\alpha)}_{ijk} &= -\frac{4}{1-\alpha^2}\Big[\Big(\frac{\partial^2}{\partial\theta^i\partial\theta^j}\Big)_{p_\theta}\Big(\frac{\partial}{\partial\theta^k}\Big)_{p_\vartheta}\kappa(\alpha)\Big]_{p_\theta=p_\vartheta} \\ &= E''_\theta\Big[\frac{\partial^2\varphi^{-1}(p_\theta)}{\partial\theta^i\partial\theta^j}\frac{\partial\varphi^{-1}(p_\theta)}{\partial\theta^k}\Big] + \frac{1-\alpha}{2}E'''_\theta\Big[\frac{\partial\varphi^{-1}(p_\theta)}{\partial\theta^i}\frac{\partial\varphi^{-1}(p_\theta)}{\partial\theta^j}\frac{\partial\varphi^{-1}(p_\theta)}{\partial\theta^k}\Big] \\ &\quad - \frac{1-\alpha}{2}E''_\theta\Big[\frac{\partial\varphi^{-1}(p_\theta)}{\partial\theta^j}\frac{\partial\varphi^{-1}(p_\theta)}{\partial\theta^k}\Big]E''_\theta\Big[u_0\frac{\partial\varphi^{-1}(p_\theta)}{\partial\theta^i}\Big] \\ &\quad - \frac{1-\alpha}{2}E''_\theta\Big[\frac{\partial\varphi^{-1}(p_\theta)}{\partial\theta^i}\frac{\partial\varphi^{-1}(p_\vartheta)}{\partial\theta^k}\Big]E''_\theta\Big[u_0\frac{\partial\varphi^{-1}(p_\vartheta)}{\partial\theta^j}\Big] \\ &\quad - \frac{1+\alpha}{2}E'_\theta\Big[\frac{\partial^2\varphi^{-1}(p_\theta)}{\partial\theta^i\partial\theta^j}\Big]E''_\theta\Big[u_0\frac{\partial\varphi^{-1}(p_\theta)}{\partial\theta^k}\Big],\end{aligned} \tag{43}$$

where we used

$$\begin{aligned}\Big(\frac{\partial^2}{\partial\theta^i\partial\theta^j}\Big)_{p_\theta}\kappa(\alpha) &= \frac{1-\alpha^2}{4}\Big[\Big(\frac{\partial^2}{\partial\theta^i\partial\theta^j}\Big)_{p_\theta}\mathcal{D}^{(\alpha)}_\varphi(p_\theta \parallel p_\vartheta)\Big]_{p_\vartheta=p_\theta} \\ &= \frac{1-\alpha^2}{4}g_{ij} = -\frac{1-\alpha^2}{4}E'_\theta\Big[\frac{\partial^2\varphi^{-1}(p_\theta)}{\partial\theta^i\partial\theta^i}\Big].\end{aligned}$$

Expression (39) follows from (37), (38) and (43). □

5. Conclusions

In [17,18], the authors introduced a pair of dual connections $D^{(-1)}$ and $D^{(1)}$ induced by φ-divergence. The main motivation of the present work was to find a (non-trivial) family of α-divergences, whose induced α-connections are convex combinations of $D^{(-1)}$ and $D^{(1)}$. As a result of our efforts, we proposed a generalization of Rényi divergence. The connection $D^{(\alpha)}$ induced by the generalization of Rényi divergence satisfies the relation $D^{(\alpha)} = \frac{1-\alpha}{2}D^{(-1)} + \frac{1+\alpha}{2}D^{(1)}$. To generalize Rényi divergence, we made use of properties of φ-functions. This makes evident the importance of φ-functions in the geometry of non-standard models. In standard statistical manifolds, even though Amari's α-divergence and Rényi divergence (with $\alpha \in [-1,1]$) do not coincide, they induce the same family of α-connections. This striking result requires further investigation. Future work should focus on how the generalization of Rényi divergence is related to Zhang's (ρ, τ)-divergence, and also how the present proposal is related to the model presented in [33].

Acknowledgments: The authors are indebted to the anonymous reviewers for their valuable comments and corrections, which led to a great improvement of this paper. Charles C. Cavalcante also thanks the CNPq (Proc. 309055/2014-8) for partial funding.

Author Contributions: All authors contributed equally to the design of the research. The research was carried out by all authors. Rui F. Vigelis and Charles C. Cavalcante gave the central idea of the paper and managed the organization of it. Rui F. Vigelis wrote the paper. All the authors read and approved the final manuscript.

References

1. Rao, C.R. Information and the accuracy attainable in the estimation of statistical parameters. *Bull. Calcutta Math. Soc.* **1945**, *37*, 81–91.
2. Amari, S.-I. Differential geometry of curved exponential families—Curvatures and information loss. *Ann. Stat.* **1982**, *10*, 357–385.
3. Amari, S.-I. *Differential-Geometrical Methods in Statistics*; Springer: Berlin/Heidelberg, Germany, 1985; Volume 28.
4. Amari, S.-I.; Nagaoka, H. *Methods of Information Geometry (Translations of Mathematical Monographs)*; American Mathematical Society: Providence, RI, USA, 2000; Volume 191.
5. Amari, S.-I. *Information Geometry and Its Applications*; Applied Mathematical Sciences Series; Springer: Berlin/Heidelberg, Germany, 2016; Volume 194.
6. Amari, S.-I.; Ohara, A.; Matsuzoe, H. Geometry of deformed exponential families: Invariant, dually-flat and conformal geometries. *Physica A* **2012**, *391*, 4308–4319.
7. Matsuzoe, H. Hessian structures on deformed exponential families and their conformal structures. *Differ. Geom. Appl.* **2014**, *35* (Suppl.), 323–333.
8. Naudts, J. Estimators, escort probabilities, and ϕ-exponential families in statistical physics. *J. Inequal. Pure Appl. Math.* **2004**, *5*, 102.
9. Pistone, G. κ-exponential models from the geometrical viewpoint. *Eur. Phys. J. B* **2009**, *70*, 29–37.
10. Amari, S.-I.; Ohara, A. Geometry of q-exponential family of probability distributions. *Entropy* **2011**, *13*, 1170–1185.
11. Vigelis, R.F.; Cavalcante, C.C. The Δ_2-Condition and φ-Families of Probability Distributions. In *Geometric Science of Information*; Springer: Berlin/Heidelberg, Germany, 2013; Volume 8085, pp. 729–736.
12. Vigelis, R.F.; Cavalcante, C.C. On φ-families of probability distributions. *J. Theor. Probab.* **2013**, *26*, 870–884.
13. Cena, A.; Pistone, G. Exponential statistical manifold. *Ann. Inst. Stat. Math.* **2007**, *59*, 27–56.
14. Grasselli, M.R. Dual connections in nonparametric classical information geometry. *Ann. Inst. Stat. Math.* **2010**, *62*, 873–896.
15. Pistone, G.; Sempi, C. An infinite-dimensional geometric structure on the space of all the probability measures equivalent to a given one. *Ann. Stat.* **1995**, *23*, 1543–1561.
16. Santacroce, M.; Siri, P.; Trivellato, B. New results on mixture and exponential models by Orlicz spaces. *Bernoulli* **2016**, *22*, 1431–1447.
17. Vigelis, R.F.; Cavalcante, C.C. Information Geometry: An Introduction to New Models for Signal Processing. In *Signals and Images*; CRC Press: Boca Raton, FL, USA, 2015; pp. 455–491.
18. Vigelis, R.F.; de Souza, D.C.; Cavalcante, C.C. New Metric and Connections in Statistical Manifolds. In *Geometric Science of Information*; Springer: Berlin/Heidelberg, Germany, 2015; Volume 9389, pp. 222–229.
19. Zhang, J. Divergence function, duality, and convex analysis. *Neural Comput.* **2004**, *16*, 159–195.
20. Zhang, J. Referential Duality and Representational Duality on Statistical Manifolds. In Proceedings of the 2nd International Symposium on Information Geometry and Its Applications, Pescara, Italy, 12–16 December 2005; pp. 58–67.
21. Zhang, J. Nonparametric information geometry: From divergence function to referential-representational biduality on statistical manifolds. *Entropy* **2013**, *15*, 5384–5418.
22. Zhang, J. Divergence Functions and Geometric Structures They Induce on a Manifold. In *Geometric Theory of Information*; Springer: Berlin/Heidelberg, Germany, 2014; pp. 1–30.

23. Zhang, J. On monotone embedding in information geometry. *Entropy* **2015**, *17*, 4485–4489.
24. Rényi, A. On measures of entropy and information. In *Proceedings of 4th Berkeley Symposium on Mathematical Statistics and Probability*; University California Press: Berkeley, CA, USA, 1961; Volume I, pp. 547–561.
25. Van Erven, T.; Harremoës, P. Rényi divergence and Kullback-Leibler divergence. *IEEE Trans. Inform. Theory* **2014**, *60*, 3797–3820.
26. Eguchi, S.; Komori, O. Path Connectedness on a Space of Probability Density Functions. In *Geometric Science of Information*; Springer: Berlin/Heidelberg, Germany, 2015; Volume 9389, pp. 615–624.
27. Kaniadakis, G.; Lissia, M.; Scarfone, A.M. Deformed logarithms and entropies. *Physica A* **2004**, *340*, 41–49.
28. Kaniadakis, G. Theoretical foundations and mathematical formalism of the power-law tailed statistical distributions. *Entropy* **2013**, *15*, 3983–4010.
29. Musielak, J. *Orlicz Spaces and Modular Spaces*; Springer: Berlin/Heidelberg, Germany, 1983; Volume 1034.
30. Shima, H. *The Geometry of Hessian Structures*; World Scientific: Singapore, 2007.
31. Banerjee, A.; Merugu, S.; Dhillon, I.S.; Ghosh, J. Clustering with Bregman Divergences. *J. Mach. Learn. Res.* **2005**, *6*, 1705–1749.
32. Bregman, L.M. The relaxation method of finding the common point of convex sets and its application to the solution of problems in convex programming. *USSR Comput. Math. Math. Phys.* **1967**, *7*, 200–217.
33. Zhanga, J.; Hästö, P. Statistical manifold as an affine space: A functional equation approach. *J. Math. Psychol.* **2006**, *50*, 60–65.

4

Kernel Density Estimation on the Siegel Space with an Application to Radar Processing †

Emmanuel Chevallier [1,*], Thibault Forget [2,3], Frédéric Barbaresco [2] and Jesus Angulo [3]

1 Department of Computer Science and Applied Mathematics, Weizmann Institute of Science, Rehovot 7610001, Israel

2 Thales Air Systems, Surface Radar Business Line, Advanced Radar Concepts Business Unit, Voie Pierre-Gilles de Gennes, Limours 91470, France; thibault.forget@mines-paristech.fr (T.F.); frederic.barbaresco@thalesgroup.com (F.B.)

3 CMM-Centre de Morphologie Mathématique, MINES ParisTech, PSL-Research University, Paris 75006, France; jesus.angulo@mines-paristech.fr

* Correspondence: emmanuelchevallier1@gmail.com.

† This paper is an extended version of our paper published in the 2nd conference on Geometric Science of Information, Paris, France, 28–30 October 2015.

Academic Editors: Arye Nehorai, Satyabrata Sen and Murat Akcakaya

Abstract: This paper studies probability density estimation on the Siegel space. The Siegel space is a generalization of the hyperbolic space. Its Riemannian metric provides an interesting structure to the Toeplitz block Toeplitz matrices that appear in the covariance estimation of radar signals. The main techniques of probability density estimation on Riemannian manifolds are reviewed. For computational reasons, we chose to focus on the kernel density estimation. The main result of the paper is the expression of Pelletier's kernel density estimator. The computation of the kernels is made possible by the symmetric structure of the Siegel space. The method is applied to density estimation of reflection coefficients from radar observations.

Keywords: kernel density estimation; Siegel space; symmetric spaces; radar signals

1. Introduction

Various techniques can be used to estimate the density of probability measure in the Euclidean spaces, such as histograms, kernel methods, or orthogonal series. These methods can sometimes be adapted to densities in Riemannian manifolds. The computational cost of the density estimation depends on the isometry group of the manifold. In this paper, we study the special case of the Siegel space. The Siegel space is a generalization of the hyperbolic space. It has a structure of symmetric Riemannian manifold, which enables the adaptation of different density estimation methods at a reasonable cost. Convergence rates of the density estimation using kernels and orthogonal series were gradually generalized to Riemannian manifolds (see [1–3]).

The Siegel space appears in radar processing in the study of Toeplitz block Toeplitz matrices, whose blocks represent covariance matrices of a radar signal (see [4–6]). The Siegel also appears in statistical mechanics, see [7] and was recently used in image processing (see [8]). Information geometry is now a standard framework in radar processing (see [4–6,9–13]). The information geometry on positive definite Teoplitz block Teoplitz matrices is directly related to the metric on the Siegel space (see [14]). Indeed, Toeplitz block Toeplitz matrices can be represented by a symmetric positive definite matrix and a point laying in a product of Siegel disks. The metric considered on Toeplitz block Toeplitz matrices is induced by the product metric between a metric on the symmetric positive definite matrices and the Siegel disks metrics (see [4–6,9,14]).

One already encounters the problem of density estimation in the hyperbolic space for electrical impedance [15], networks [16] and radar signals [17]. In [18], a generalization of the Gaussian law on the hyperbolic space was proposed. Apart from [19], where authors propose a generalization of the Gaussian law, probability density estimation on the Siegel space has not yet been addressed.

The contributions of the paper are the following. We review the main non parametric density estimation techniques on the Siegel disk. We provide some rather simple explicit expressions of the kernels defined by Pelletier in [1]. These expressions make the kernel density estimation the most adapted method. We present visual results of estimated densities in the simple case where the Siegel disk reduces to the Poincaré disk.

The paper begins with an introduction to the Siegel space in Section 2. Section 3 reviews the main non-parametric density estimation techniques on the Siegel space. Section 3.3 contains the original results of the paper. Section 4 presents an application to radar data estimation.

2. The Siegel Space

This section presents facts about the Siegel space. The interested reader can find more details in [20,21]. The necessary background on Lie groups and symmetric space can be found in [22].

2.1. The Siegel Upper Half Space

The Siegel upper half space is a generalization of the Poincaré upper half space (see [23]) for a description of the hyperbolic space. Let $Sym(n)$ be the space of real symmetric matrices of size $n \times n$ and $Sym_+(n)$ the set of real symmetric positive definite matrices of size $n \times n$. The Siegel upper half space is defined by

$$\mathcal{H}_n = \{Z = X + iY | X \in Sym(n), Y \in Sym_+(n)\}.$$

$\mathcal{H}_n$ is equipped with the following metric:

$$ds = 2tr(Y^{-1}dZY^{-1}d\overline{Z}).$$

The set of real symplectic matrices $Sp(n,\mathbb{R})$ is defined by

$$g \in Sp(n,\mathbb{R}) \Leftrightarrow g^t J g = J,$$

where

$$J = \begin{pmatrix} 0 & I_n \\ -I_n & 0 \end{pmatrix},$$

and I_n is the $n \times n$ identity matrix. $Sp(n,\mathbb{R})$ is a subgroup of $SL_{2n}(\mathbb{R})$, the set of $2n \times 2n$ invertible matrices of determinant 1. Let $g = \begin{pmatrix} A & B \\ C & D \end{pmatrix} \in Sp(n,\mathbb{R})$. The metric ds is invariant under the following action of $Sp(n,\mathbb{R})$,

$$g.Z = (AZ+B)(CZ+D)^{-1}.$$

This action is transitive, i.e.,

$$\forall Z \in \mathcal{H}_n, \exists g \in Sp(n,\mathbb{R}), g.iI = Z.$$

The stabilizer K of iI is the set of elements g of $Sp(n,\mathbb{R})$ whose action leaves iI fixed. K is a subgroup of $Sp(n,\mathbb{R})$ called the isotropy group. We can verify that

$$K = \left\{ \begin{pmatrix} A & B \\ -B & A \end{pmatrix}, A + iB \in SU(n) \right\}.$$

A symmetric space is a Riemannian manifold, where the reversal of the geodesics is well defined and is an isometry. Formally, $exp_p(u) \mapsto exp_p(-u)$ is an isometry for each p on the manifold, where u is a vector in the tangent space at p, and exp_p the Riemannian exponential application at p. In other words, the symmetry around each point is an isometry. $\mathcal{H}_n$ is a symmetric space (see [20]). The structure of a symmetric space can be studied through its isometry group and the Lie algebra of its isometry group. The present work will make use of the Cartan and Iwasawa decompositions of the Lie algebra of $Sp(n,\mathbb{R})$ (see [22]). Let $\mathfrak{sp}(n,\mathbb{R})$ be the Lie algebra of $Sp(n,\mathbb{R})$. Given A, B and C three real $n \times n$ matrices, let denote $\begin{pmatrix} A & B \\ C & -A^t \end{pmatrix} = (A,B,C)$. We have

$$\mathfrak{sp}(n,\mathbb{R}) = \{(A,B,C) | B \text{ and } C \text{ symmetric}\}.$$

The Cartan decomposition of $\mathfrak{sp}(n,\mathbb{R})$ is given by

$$\mathfrak{sp}(n,\mathbb{R}) = \mathfrak{t} \oplus \mathfrak{p},$$

where

$$\mathfrak{t} = \{(A,B,-B) | B \text{ symmetric and } A \text{ skew-symmetric}\},$$

$$\mathfrak{p} = \{(A,B,B) | A, B, \text{ symmetric}\}. \tag{1}$$

The Iwasawa decomposition is given by

$$\mathfrak{sp}(n,\mathbb{R}) = \mathfrak{t} \oplus \mathfrak{a} \oplus \mathfrak{n},$$

where

$$\mathfrak{a} = \{(H,0,0) | H \text{ diagonal}\},$$

$$\mathfrak{n} = \{(A,B,0) | A \text{ upper triangular with 0 on the diagonal }, B \text{ symmetric}\}.$$

It can be shown that

$$\mathfrak{p} = \cup_{k \in K} Ad_k(\mathfrak{a}), \tag{2}$$

where Ad is the adjoint representation of $Sp(n,\mathbb{R})$.

2.2. The Siegel Disk

The Siegel disk $\mathcal{D}_n$ is the set of complex matrices $\{Z | I - Z^*Z \geq 0\}$, where $\geq$ stands for the Loewner order (see [24] for details on the Loewner order). Recall that for A and B two Hermitian matrices, $A \geq B$ with respect to the Loewner order means that $A - B$ is positive definite. The transformation

$$Z \in \mathcal{H}_n \mapsto (Z - iI)(Z + iI)^{-1} \in \mathcal{D}_n$$

is an isometry between the Siegel upper half space and the Siegel disk. Let $C = \begin{pmatrix} I & -iI \\ I & iI \end{pmatrix}$. The application $g \in Sp(n,\mathbb{R}) \mapsto CgC^{-1}$ identifies the set of isometries of $\mathcal{H}_n$ and of $\mathcal{D}_n$. Thus, it can be shown that a matrix $g = \begin{pmatrix} A & B \\ \overline{A} & \overline{B} \end{pmatrix} \in Sp(n,\mathbb{C})$ acts isometrically on $\mathcal{D}_n$ by

$$g.Z = (AZ + B)(\overline{A}Z + \overline{B})^{-1},$$

where $\overline{A}$ stands for the conjugate of A. The point iI in $\mathcal{H}_n$ is identified with the null matrix noted 0 in $\mathcal{D}_n$. Let $Z \in \mathcal{D}_n$. There exists P a diagonal matrix with decreasing positive real entries and U a unitary matrix such that $Z = UPU^t$. Let $\tau_1 \geq ... \geq \tau_n$ be such that

$$P = \begin{pmatrix} th(\tau_1) & & \\ & \ddots & \\ & & th(\tau_n) \end{pmatrix}.$$

Let

$$A_0 = \begin{pmatrix} ch(\tau_1) & & \\ & \ddots & \\ & & ch(\tau_n) \end{pmatrix}, B_0 = \begin{pmatrix} sh(\tau_1) & & \\ & \ddots & \\ & & sh(\tau_n) \end{pmatrix}$$

and

$$g_Z = \begin{pmatrix} U & 0 \\ 0 & \overline{U} \end{pmatrix} . \begin{pmatrix} A_0 & B_0 \\ A_0 & B_0 \end{pmatrix}.$$

It can be shown that

$$g_Z \in Sp(n, \mathbb{C}) \text{ and } g_Z.0 = Z. \tag{3}$$

We provide now a correspondence between the elements of $\mathcal{D}_n$ and the elements of $\mathfrak{p}$ defined in Equation (1). Let

$$H_Z = \begin{pmatrix} \tau_1 & & & & & \\ & \ddots & & & & \\ & & \tau_n & & & \\ & & & -\tau_1 & & \\ & & & & \ddots & \\ & & & & & -\tau_n \end{pmatrix} \in \mathfrak{a}, \tag{4}$$

and

$$a_Z = \begin{pmatrix} e^{\tau_1} & & & & & \\ & \ddots & & & & \\ & & e^{\tau_n} & & & \\ & & & e^{-\tau_1} & & \\ & & & & \ddots & \\ & & & & & e^{-\tau_n} \end{pmatrix} \in A = exp(\mathfrak{a}).$$

It can be shown that there exists $k \in K$ such that

$$Cexp(Ad_k(H_Z))C^{-1}.0 = Z,$$

or equivalently

$$Cka_ZkC^{-1}.0 = Z.$$

Recall that Equation (2) gives $Ad_k(H) \in \mathfrak{p}$ and $kak \in exp(\mathfrak{p})$. The distance between Z and 0 in $\mathcal{D}_n$ is given by

$$d(0, Z) = \left(2\sum \tau_i^2\right)^{1/2} \tag{5}$$

(see p. 292 in [20]).

3. Non Parametric Density Estimation on the Siegel Space

Let Ω be a space, endowed with a σ-algebra and a probability measure p. Let X be a random variable $\Omega \to \mathcal{D}_n$. The Riemannian measure of $\mathcal{D}_n$ is called vol and the measure on $\mathcal{D}_n$ induced by X is noted μ_X. We assume that μ_X has a density, noted f, with respect to vol, and that the support of X is a compact set noted $Supp$. Let $(x_1, ..., x_k) \in \mathcal{D}_n^k$ be a set of draws of X.

The Dirac measure at a point $a \in \mathcal{D}_n$ is denoted δ_a. Let $\mu_k = \frac{1}{k}\sum_{i=1}^{k} \delta_{x_i}$ denotes the empirical measure of the set of draws. This section presents four non-parametric techniques of estimation of the density f from the set of draws $(x_1, ..., x_k)$. The estimated density at x in $\mathcal{D}_n$ is noted $\hat{f}_k(x) = \hat{f}(x, x_1, ..., x_k)$. The relevance of a density estimation technique depends on several aspects. When the space allows it, the estimation technique should equally consider each direction and location. This leads to an isotropy and a homogeneity condition. In the kernel method, for instance, a kernel density function K_{x_i} is placed at each observation x_i. Firstly, in order to treat directions equally, the function K_{x_i} should be invariant under the isotropy group of x_i; Secondly, for another observation x_j, functions K_{x_i} and K_{x_j} should be similar up to the isometries that send x_i on x_j. These considerations strongly depend on the geometry of the space: if the space is not homogeneous and the isotropy group is empty, these indifference principles have no meaning. Since the Siegel space is symmetric, it is homogeneous and has a non empty isotropy group. Thus, the density estimation technique should be chosen accordingly.

The convergence of the different estimation techniques is widely studied. Results were first obtained in the Euclidean case, and are gradually extended to the probability densities on manifold (see [1,2,15,25]).

The last relevant aspect is computational. Each estimation technique has its own computational framework that presents pros and cons given the different applications. For instance, the estimation by orthogonal series needs an initial pre-processing, but provides a fast evaluation of the estimated density in compact manifolds.

3.1. Histograms

The histogram is the simplest density estimation method. Given a partition of the space $\mathcal{D}_n = \cup_i A_i$, the estimated density is given by

$$\hat{f}(x \in A_i) = \frac{1}{vol(A_i)} \sum_{j=1}^{k} 1_{A_i}(x_j),$$

where 1_{A_i} stands for the indicator function of A_i. Following the considerations of the previous sections, the elements of the partition should firstly be as isotropic as possible, and secondly as similar as possible to each other. Regarding the problem of histograms, the case of the Siegel space is similar to the case of the hyperbolic space. There exist various uniform polygonal tilings on the Siegel space that could be used to compute histograms. However, there are ratio $\lambda \in \mathbb{R}$ for which there is no homothety. Thus, it is not always possible to adapt the size of the bins to a given set of draws of the random variable. Modifying the size of the bins can require a change of the structure of the tiling. This is why the study of histograms has not been deepened.

3.2. Orthogonal Series

The estimation of the density f can be made out of the estimation of the scalar product between f and a set of "orthonormal" functions $\{e_j\}$. The most standard choice for $\{e_j\}$ is the eigenfunctions of the Laplacian. When the variable X takes its values in $\mathbb{R}^n$, this estimation technique becomes the characteristic function method. When the underlying space is compact, the spectrum of the Laplacian operator is countable, while when the space is non-compact, the spectrum is uncountable. In the first case, the estimation of the density f is made through the estimation of a sum, while in the second case is made through the estimation of an integral. In practice, the second situation presents a larger computational complexity. Unfortunately, eigenfunctions of the Laplacian operator are known on $\mathcal{D}_n$ but not on compact sub-domains. This is why the study of this method has not been deepened.

3.3. Kernels

Let $\mathcal{K} : \mathbb{R}_+ \to \mathbb{R}_+$ be a map which verifies the following properties:

(i) $\int_{\mathbb{R}^d} \mathcal{K}(||x||)dx = 1$;
(ii) $\int_{\mathbb{R}^d} x\mathcal{K}(||x||)dx = 0$;
(iii) $\mathcal{K}(x > 1) = 0$;
(iv) $sup(\mathcal{K}(x)) = \mathcal{K}(0)$.

Let $p \in \mathcal{D}_n$. Generally, given a point p on a Riemannian manifold, exp_p defines an injective application only on a neighborhood of 0. On the Siegel space, exp_p is injective on the whole space. When the tangent space $T_p\mathcal{D}_n$ is endowed with the local scalar product,

$$||u|| = d(p, exp_p(u)),$$

where $||.||$ is the Euclidean distance associated with the local scalar product and $d(.,.)$ is the Riemannian distance. The corresponding Lebesgue measure on $T_p\mathcal{D}_n$ is noted Leb_p. Let $exp_p^*(Leb_p)$ denote the push-forward measure of Lep_p by exp_p. The function θ_p defined by:

$$\theta_p : q \mapsto \theta_p(q) = \frac{dvol}{dexp_p^*(Leb_p)}(q) \tag{6}$$

is the density of the Riemannian measure on $\mathcal{D}_n$ with respect to the Lebesgue measure Leb_p after the identification of $\mathcal{D}_n$ and $T_p\mathcal{D}_n$ induced by exp_p (see Figure 1).

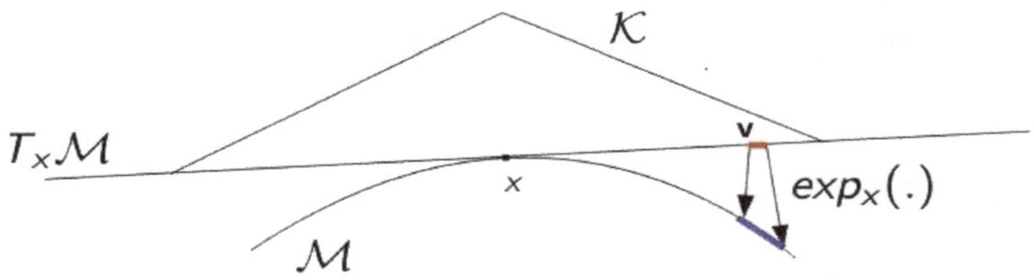

Figure 1. $\mathcal{M}$ is a Riemannian manifold, and $T_x\mathcal{M}$ is its tangent space at x. The exponential application induces a volume change θ_x between $T_x\mathcal{M}$ and $\mathcal{M}$.

Given $\mathcal{K}$ and a positive radius r, the estimator of f proposed by [1] is defined by:

$$\hat{f}_k = \frac{1}{k}\sum_i \frac{1}{r^n}\frac{1}{\theta_{x_i}(x)}\mathcal{K}\left(\frac{d(x,x_i)}{r}\right). \tag{7}$$

The corrective factor $\theta_{x_i}(x)^{-1}$ is necessary since the kernel $\mathcal{K}$ originally integrates to one with respect to the Lebesgue measure and not with respect to the Riemannian measure. It can be noticed that this estimator is the usual kernel estimator in the case of Euclidean space. When the curvature of the space is negative, which is the case of the Siegel space, the distribution placed over each sample x_i has x_i as intrinsic mean. The following theorem provides convergence rate of the estimator. It is a direct adaptation of Theorem 3.1 of [1].

Theorem 1. *Let $(\mathcal{M}, g)$ be a Riemannian manifold of dimension n and μ its Riemannian volume measure. Let X be a random variable taking its values in a compact subset C of $(\mathcal{M}, g)$. Let $0 < r \leq r_{inj}$, where r_{inj} is the infimum of the injectivity radius on C. Assume the law of X has a twice differentiable density f with respect to* *the Riemannian volume measure. Let $\hat{f}_k$ be the estimator defined in Equation (7). There exists a constant C_f such that*

$$\int_{x\in\mathcal{M}} E_{x_1,\dots,x_k}[(f(x)-\hat{f}_k(x))^2]d\mu \leq C_f(\frac{1}{kr^n}+r^4). \tag{8}$$

If $r \sim k^{\frac{-1}{n+4}}$,

$$\int_{x\in\mathcal{M}} E_{x_1,\dots,x_k}[(f(x)-\hat{f}_k(x))^2]d\mu = O(k^{-\frac{4}{n+4}}). \tag{9}$$

Proof. See Appendix A. □

It can be checked that on the Siegel space $r_{inj} = +\infty$ and that, for an isometry α, we have:

$$\hat{f}_k(x, x_1, \dots, x_k) = \hat{f}_k(\alpha(x), \alpha(x_1), \dots, \alpha(x_k)).$$

Each location and direction are processed as similarly as possible. This density estimator can be used for data classification on Riemannian manifolds, see [26].

In order to obtain the explicit expression of the estimator, one must have the explicit expression of the Riemannian exponential, of its inverse, and of the function θ_p (see Equations (6) and (7)). These expressions are difficult and sometimes impossible to obtain for general Riemannian manifolds. In the case of the Siegel space, the symmetric structure makes the computation possible. Since the space is homogeneous, the computation can be made at the origin $iI \in \mathcal{H}_n$ or $0 \in \mathcal{D}_n$ and transported to the whole space. In the present work, the random variable lays in $\mathcal{D}_n$. However, in the literature, the Cartan and Iwasawa decompositions are usually given for the isometry group of $\mathcal{H}_n$. Thus, our computation starts in $\mathcal{H}_n$ before moving to $\mathcal{D}_n$.

The Killing form on the Lie algebra $\mathfrak{sp}(n,\mathbb{R})$ of the isometry group of $\mathcal{H}_n$ induces a scalar product on $\mathfrak{p}$. This scalar product can be transported on $exp(\mathfrak{p})$ by left multiplication. This operation gives $exp(\mathfrak{p})$ a Riemannian structure. It can be shown that on this Riemannian manifold, the Riemannian exponential at the identity coincides with the group exponential. Furthermore,

$$\begin{array}{cccc} \phi & : & exp(\mathfrak{p}) & \rightarrow & \mathcal{H}_n \\ & & g & \mapsto & g.iI \end{array} \tag{10}$$

is a bijective isometry, up to a scaling factor. Since the volume change θ_p is invariant under rescaling of the metric, this scaling factor has no impact. Thus, $\mathcal{H}_n$ can be identified with $exp(\mathfrak{p})$ and $exp_{iI\in\mathcal{H}_n}$ with $exp_{|\mathfrak{p}}$. The expression of the Riemannian exponential is difficult to obtain in general; however, it boils down to the group exponential in the case of symmetric spaces. This is the main element of the computation of θ_p. The Riemannian volume measure on $exp(\mathfrak{p})$ is noted vol'. Let

$$\begin{array}{cccc} \psi & : & K\times\mathfrak{a} & \rightarrow & \mathfrak{p} \\ & & (k,H) & \mapsto & Ad_k(H). \end{array}$$

Let $\mathfrak{a}^+$ be the diagonal matrices with strictly decreasing positive eigenvalues. Let Λ^+ be the set of positive roots of $\mathfrak{sp}(n,\mathbb{R})$ as described in p. 282 in [20] ,

$$\Lambda^+ = \{e_i + e_j, i \leq j\} \cup \{e_i - e_j, i < j\},$$

where $e_i(H)$ is the i-th diagonal term of the diagonal matrix H. Let $C_c(E)$ be the set of continuous compactly supported functions on the space E. In [27], at page 73, it is given that for all $t \in C_c(\mathfrak{p})$, there exists $c_1 > 0$ such that

$$\int_{\mathfrak{p}} t(Y)dY = c_1 \int_K \int_{\mathfrak{a}^+} t(\psi(k,H)) \prod_{\lambda\in\Lambda^+} \lambda(H)dkdH, \tag{11}$$

where dY is a Lebesgue measure on the coefficients of Y. Let $\tilde{\mathfrak{p}} = \psi(K \times \mathfrak{a}^+)$. $\lambda \in \Lambda^+$ never vanishes on $\mathfrak{a}^+$ and $\mathfrak{p} \setminus \tilde{\mathfrak{p}}$ has a null measure. Thus,

$$\int_{\tilde{\mathfrak{p}}} t(Y) \prod_{\lambda \in \Lambda^+} \frac{1}{\lambda(H_Y)} dY = c_1 \int_{K'} \int_{\mathfrak{a}^+} t(Ad_k(H)) dk dH, \tag{12}$$

where H_Y is the point in $\mathfrak{a}^+$ such that there exists k in K such that $\psi(k, H_Y) = Y$. Calculation in p. 73 in [27] also gives that for all $t \in C_c(\mathfrak{p})$, there exists $c_2 > 0$, such that

$$\int_{Sp(n,\mathbb{R})} t(g) dg = c_2 \int_K \int_{\mathfrak{a}^+} \int_K t(k_2.exp(Ad_{k_1}(H))) J(H) dk_1 dH dk_2, \tag{13}$$

where dg is the Haar measure on $Sp(n, \mathbb{R})$ and

$$\begin{aligned} J(H) &= \prod_{\lambda \in \Lambda^+} e^{\lambda(H)} - e^{-\lambda(H)} \\ &= 2^{|\Lambda^+|} \prod_{\lambda \in \Lambda^+} \sinh(\lambda(H)). \end{aligned}$$

Thus, for all $t \in C_c(Sp(n, \mathbb{R})/K)$,

$$\int_{Sp(n,\mathbb{R})/K} t(x) dx = c_2 \int_K \int_{\mathfrak{a}^+} t(exp(Ad_k(H))) J(H) dk dH, \tag{14}$$

where dx is the invariant measure on $Sp(n, \mathbb{R})/K$. After identifying $Sp(n, \mathbb{R})/K$ and $exp(\mathfrak{p})$, the Riemannian measure on $exp(\mathfrak{p})$ coincides with the invariant measure on $Sp(n, \mathbb{R})/K$. Thus, for all $t \in C_c(exp(\mathfrak{p}))$,

$$\int_{exp(\mathfrak{p})} t(x) dvol' = c_2 \int_K \int_{\mathfrak{a}^+} t(exp(Ad_k(H))) J(H) dk dH. \tag{15}$$

Using the notation H_Y of Equation (12),

$$\int_{\tilde{\mathfrak{p}}} t(exp(Y)) J(H_Y) \prod_{\lambda \in \Lambda^+} \frac{1}{\lambda(H_Y)} dY = c_1 \int_K \int_{\mathfrak{a}^+} t(exp(Ad_k(H))) J(H) dk dH. \tag{16}$$

Combining Equations (15) and (16), we obtain that there exists c_3 such that

$$\int_{\tilde{\mathfrak{p}}} t(exp(Y)) \prod_{\lambda \in \Lambda^+} \frac{\sinh(\lambda(H_Y))}{\lambda(H_Y)} dY = c_3 \int_{exp(\mathfrak{p})} t(x) dvol'. \tag{17}$$

The term $\frac{\sinh(\lambda(H))}{\lambda(H)}$ can be extended by continuity on $\mathfrak{a}$; thus,

$$\int_{\mathfrak{p}} t(exp(Y)) \prod_{\lambda \in \Lambda^+} \frac{\sinh(\lambda(H_Y))}{\lambda(H_Y)} dY = c_3 \int_{exp(\mathfrak{p})} t(x) dvol'. \tag{18}$$

Let dY be the Lebesgue measure corresponding to the metric. Then, the exponential application does not introduce a volume change at $0 \in \mathfrak{p}$. Since $H_0 = 0$ and $\frac{\sinh(\lambda(H))}{\lambda(H)} \underset{H \to 0}{\longrightarrow} 1$, we have $c_3 = 1$. Let log denote the inverse of the exponential application. We have

$$\frac{dlog^*(vol')}{dY} = \prod_{\lambda \in \Lambda^+} \frac{\sinh(\lambda(H_Y))}{\lambda(H_Y)}.$$

Since ϕ from Equation (10) is an isometry up to a scaling factor, if $Y \in \mathfrak{p}$ and $C\phi(exp(Y))C^{-1} = exp_0(u \in T_0\mathcal{D}_n)$, then

$$\frac{dlog^*(vol)}{dLeb_0}(u) = \frac{dlog^*(vol')}{dY}(Y),$$

where Leb_0 refers to the Lebesgue measure on the tangent space $T_0\mathcal{D}_n$ as in Equation (6). Given $Z \in \mathcal{D}_n$, H_Z from Equation (4) verifies $C\phi(exp(Ad_k(H_Z)))C^{-1} = Z$ for some k in K. Thus,

$$\theta_0(Z) = \frac{dlog^*(vol')}{dY}(Ad_k(H_Z)) = \prod_{\lambda\in\Lambda^+} \frac{\sinh(\lambda(H_Z))}{\lambda(H_Z)}.$$

We have then

$$\theta_0(Z) = \prod_{i<j} \frac{\sinh(\tau_i - \tau_j)}{\tau_i - \tau_j} \prod_{i\leq j} \frac{\sinh(\tau_i + \tau_j)}{\tau_i + \tau_j},$$

where the (τ_i) are described in Section 2.2. Given $Z_1, Z_2 \in \mathcal{D}_n$,

$$\theta_{Z_1}(Z_2) = \theta_0(g_{Z_1}^{-1}.Z_2),$$

where $g_{Z_1}^{-1}$ is defined in Equation (3). It is thus possible to use the density estimator defined in Equation (7). Indeed,

$$\frac{1}{\theta_{Z_1}(Z_2)}\mathcal{K}\left(\frac{d(Z_1,Z_2)}{r}\right) = \prod_{i<j} \frac{\tau_i - \tau_j}{\sinh(\tau_i - \tau_j)} \prod_{i\leq j} \frac{\tau_i + \tau_j}{\sinh(\tau_i + \tau_j)} \mathcal{K}\left(\frac{(2\sum \tau_i^2)^{1/2}}{r}\right), \tag{19}$$

where the (τ_i) are the diagonal elements of $H_{g_{Z_1}^{-1}.Z_2}$. Recall that when $n = 1$, the Siegel disk corresponds to the Poincaré disk. Thus, we retrieve the expression of the kernel for the hyperbolic space,

$$\frac{1}{\theta_{Z_1}(Z_2)}\mathcal{K}\left(\frac{d(Z_1,Z_2)}{r}\right) = \frac{2\tau}{\sinh(2\tau)}\mathcal{K}\left(\frac{(2\tau^2)^{1/2}}{r}\right). \tag{20}$$

4. Application to Radar Processing

4.1. Radar Data

In space time adaptative radar processing (STAP), the signal is formed by a succession of matrices X representing the realization of a temporal and spatial process. Let $\mathcal{B}_{n,m}^+$ be the set of positive definite block Teoplitz matrices composed of $n \times n$ blocks of $m \times m$ matrices (PD BT). For a stationary signal, the autocorrelation matrix R is PD BT (see [5,6,14]). Authors of [5,6,14] proposed a generalization of Verblunsky coefficients and defined a parametrization of PD BT matrices,

$$\begin{array}{ccc} \mathcal{B}_{n,m}^+ & \rightarrow & Sym^+ \times \mathbb{D}_n^{m-1} \\ \mathcal{R} & \mapsto & (P_0, \Omega_1, ..., \Omega_{m-1}), \end{array} \tag{21}$$

in which the metric induced by the Kähler potential is the product metric of an affine invariant metric on Sym^+ and $m-1$ times the metric of the Siegel disk, up to a scaling factor. When the signal is not Gaussian, reflection/Verblunsky coefficients in Poincaré or Siegel Disks should be normalized as described in [28] by a normalized Burg algorithm. Among other references, positive definite block Teoplitz matrices have been studied in the context of STAP-radar processing in [4–6].

4.2. Marginal Densities of Reflection Coefficients

In this section, we show density estimation results of the marginal parameters Ω_k. For the sake of visualization, only the Siegel disk $\mathbb{D}_1$ is considered. Recall that $\mathbb{D}_1$ coincides with the Poincaré disk.

The results are partly extracted from the conference paper [17]. Data used in the experimental tests are radar observations from THALES X-band Radar, recorded during 2014 field trials campaign at Toulouse Blagnac Airport for European FP7 UFO study (Ultra-Fast wind sensOrs for wake-vortex hazards mitigation) (see [29,30]). Data are representative of Turbulent atmosphere monitored by radar. Figure 2 illustrates the density estimation of six coefficients on the Poincaré unit disk under a rainy environment. The densities are individually re-scaled for visualization purposes. For each environment, the dataset is composed of 120 draws. The densities of the coefficients Ω_k are representative of the background. This information on the background is expected to ease the detection of interesting targets.

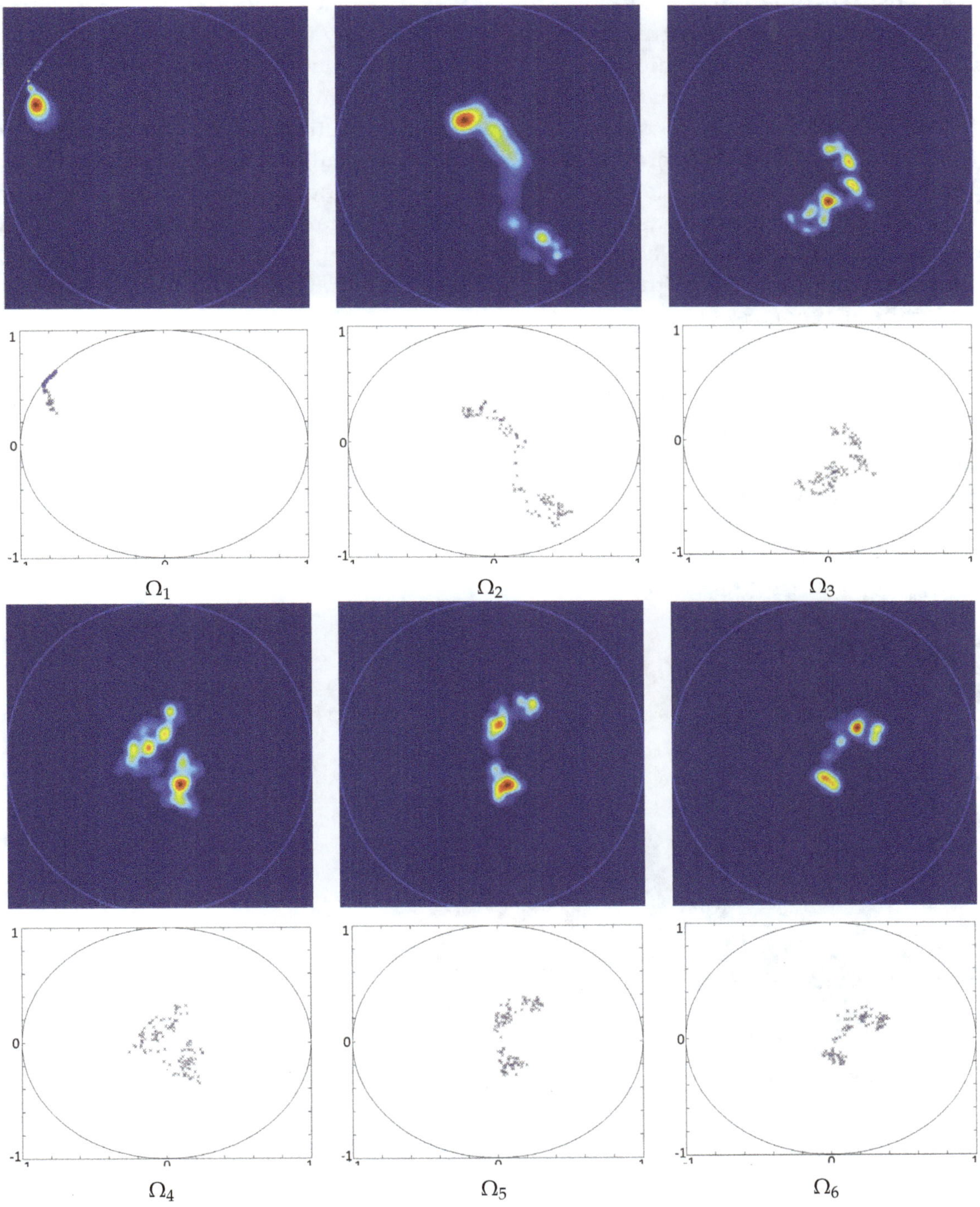

Figure 2. Estimation of the density of six coefficients Ω_k under rainy conditions. The expression of the used kernel is $K(x) = \frac{3}{\pi}(1 - x^2)^2 \mathbf{1}_{x<1}$. Densities are rescaled for visual purposes.

4.3. Radar Clutter Segmentation

Clutter refers to background Doppler signal related to meteorological conditions (e.g., wind in wooded areas, currents and breaking waves on water), which hinders detection of small and slow targets. At each range, a set of reflection coefficients are computed from the Doppler spectrum (see [31]). This set of coefficients is a point in the Poincaré poly-disk. From this set of points in the poly-disk, it is possible to estimate the underlying density. Segmenting clutter, i.e., determining zones of homogeneous Doppler characteristics (see Figure 3), enables the improvement of detection algorithms on each zone. The mean-shift algorithm enables segmentation of the space according to the kernel density estimation of a set of points. It was introduced by Fukunaga and Hostetler in 1975 (see [32]). It corresponds to a gradient ascent of the density estimator (see [33]) for a study of the statistical consistency of the gradient lines estimation. Each data point moves to a local mode of the density estimator, which yields as many clusters as modes. This algorithm has been generalized on manifolds in [34], and applied to radar images in [35]. It can thus be used to segment the set of points in the Poincaré poly-disk. Unfortunately, the mean-shift algorithm requires working with a kernel depending only on the distance to its barycenter, which is not the case of the kernel defined in Equation (19). Thus, the computations are performed without the use of the corrective term θ_p. It is possible to solve this problem by replacing the corrective term by its average at a given radius, which leads to a kernel depending only on the distance to its barycenter. Our future work will focus on the computation of these averages. Let

$$\hat{f}_r^{\mathcal{K}}(x) = \frac{c_d}{k} \sum_{i=1}^{k} \frac{1}{r^n} \mathcal{K}\left(\frac{d(x_i, x)^2}{r^2} \right),$$

where c_n is a normalization constant. Let $g = -k'$.

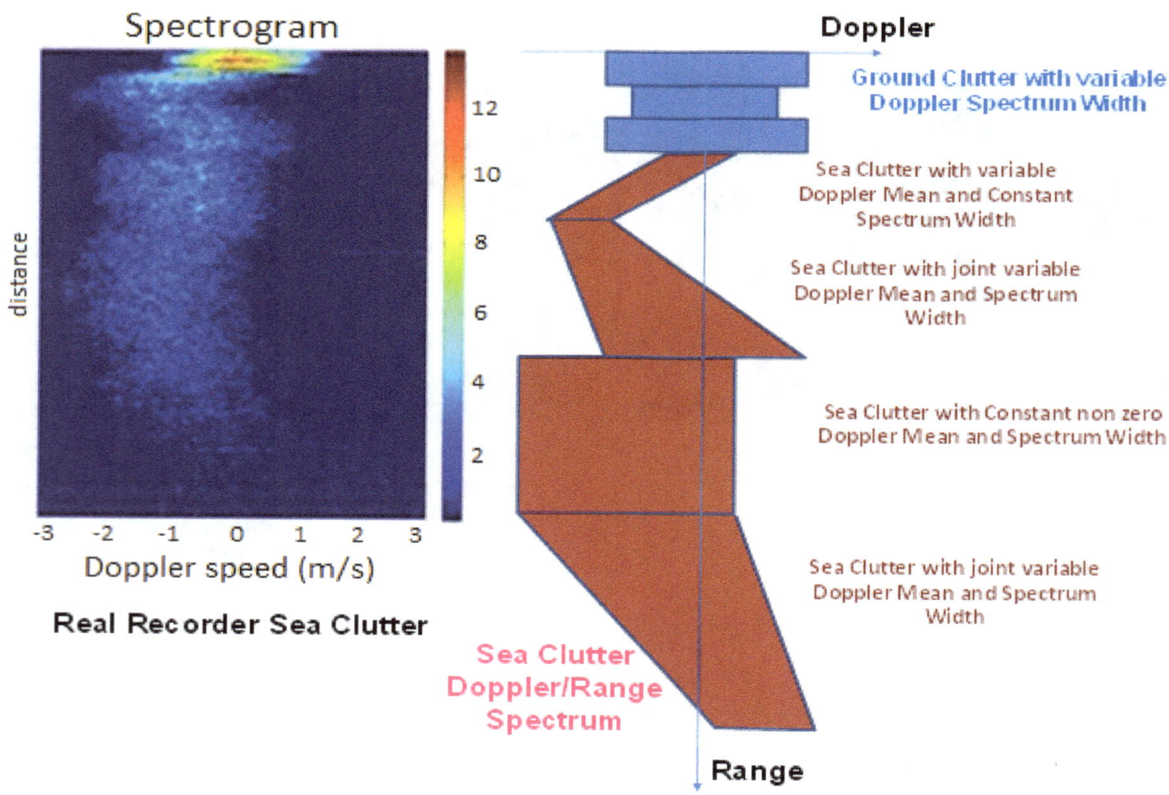

Figure 3. Mean and width variability of sea clutter Doppler spectrum.

The mean-shift is defined by

$$m(x) = \sum_{i=1}^{k} \frac{\frac{1}{r^{n+2}} g\left(\frac{d(x,x_i)^2}{r^2}\right)}{\sum_{i=1}^{k} \frac{1}{r^{n+2}} g\left(\frac{d(x,x_i)^2}{r^2}\right)} log_x(x_i) \propto \frac{\nabla f_r^{\mathcal{K}}}{f_r^g},$$

where $m(x)$ is in the tangent space at x. The algorithm moves from x to $exp_x(m(x))$ until convergence to a local maximum. The points of the space are segmented according to the local maxima to which they converge.

In order to assess the quality of unsupervised classification, we use the notion of Silhouette, see [36], which computes for each point a proximity criterion with respect to other points of the same cluster and other points of different clusters (see Figure 4). Let x be in the cluster A. We respectively define $a(x) = min_{y \in A} d(x, y)$ and $b(x) = min_{y \neq A} d(x, y)$, the minimum distance to points of the same (resp. other) class(es). The Silhouette of x is

$$\frac{a(x) - b(x)}{max\{a(x), b(x)\}},$$

which takes values between -1 and 1, respectively, when the data point is considered "badly" and "well" clustered. The average of all the silhouettes provides an indication of the relevance of the classification. One can represent graphically the silhouette profile by plotting for each class horizontal segments of the length of the silhouette value (see Figure 5).

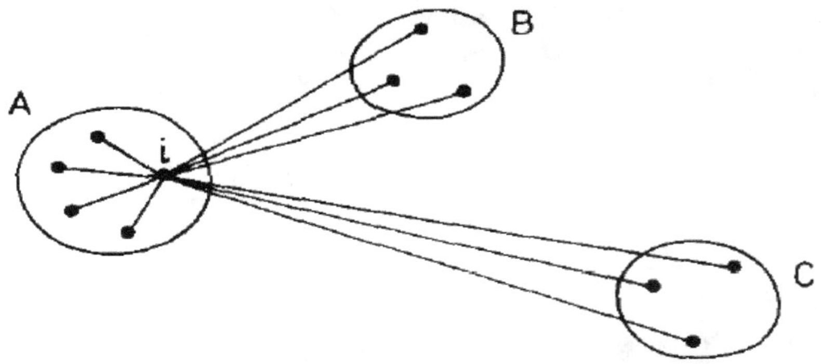

Figure 4. Intra and inter cluster distances.

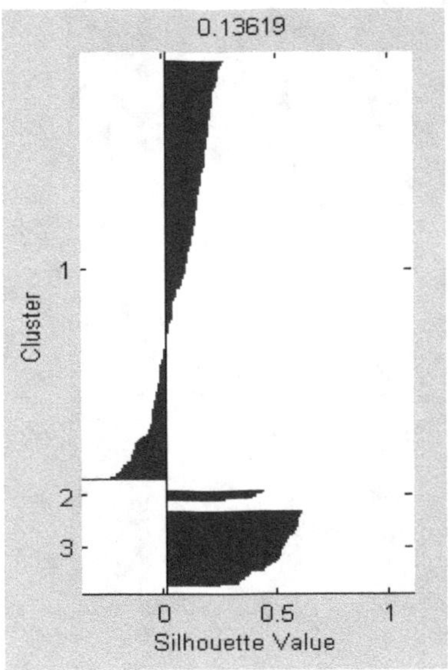

Figure 5. Example of silhouette.

In order to test the Riemannian Mean Shift performance, we generate simple synthetic radar clutter data. Given 250 range cells, we generate 125 cells of ground clutter (wind) centered at 0 m·s^{-1}, of spectral width 5 m·s^{-1}, to which we add 125 cells of rain clutter, centered at 5 m·s^{-1}, of spectral width 10 m·s^{-1}. This clutter is sampled 10 times and the segmentation is performed on each simulation (see Figures 6–8).

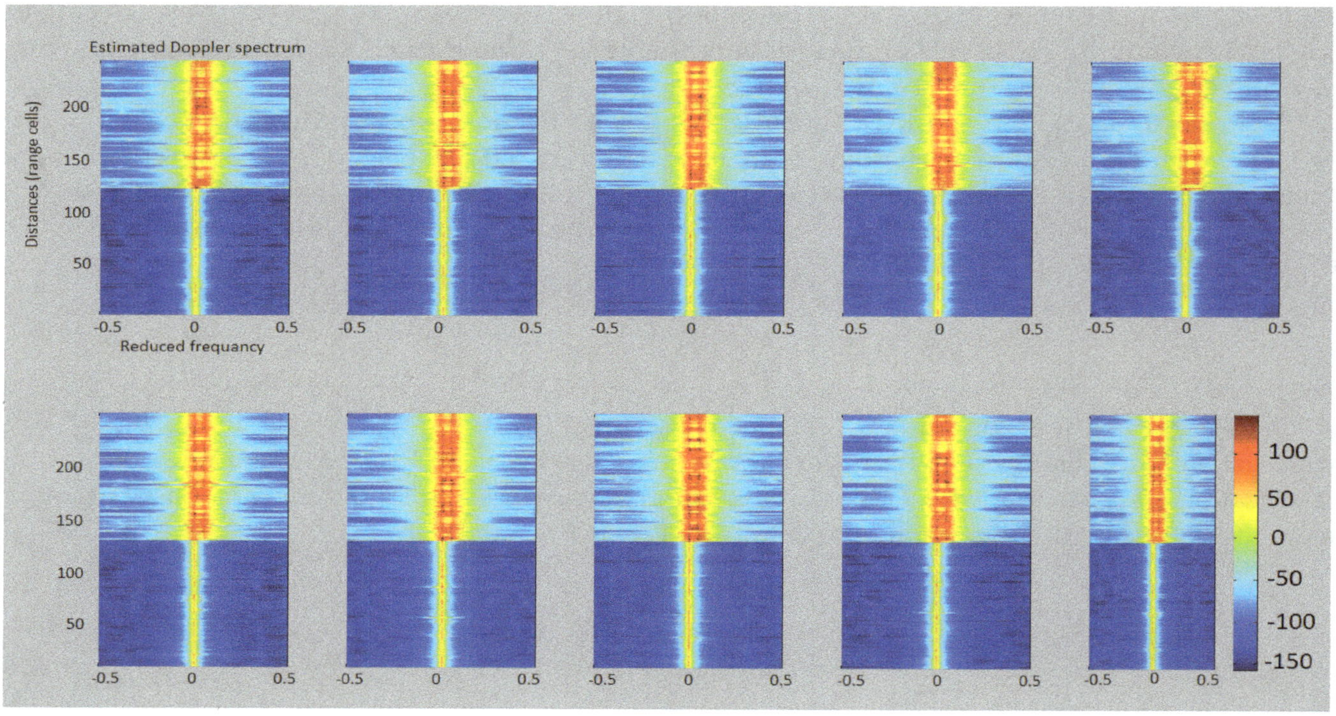

Figure 6. Autoregressive spectra.

Figure 7. Classification results (one color per cluster).

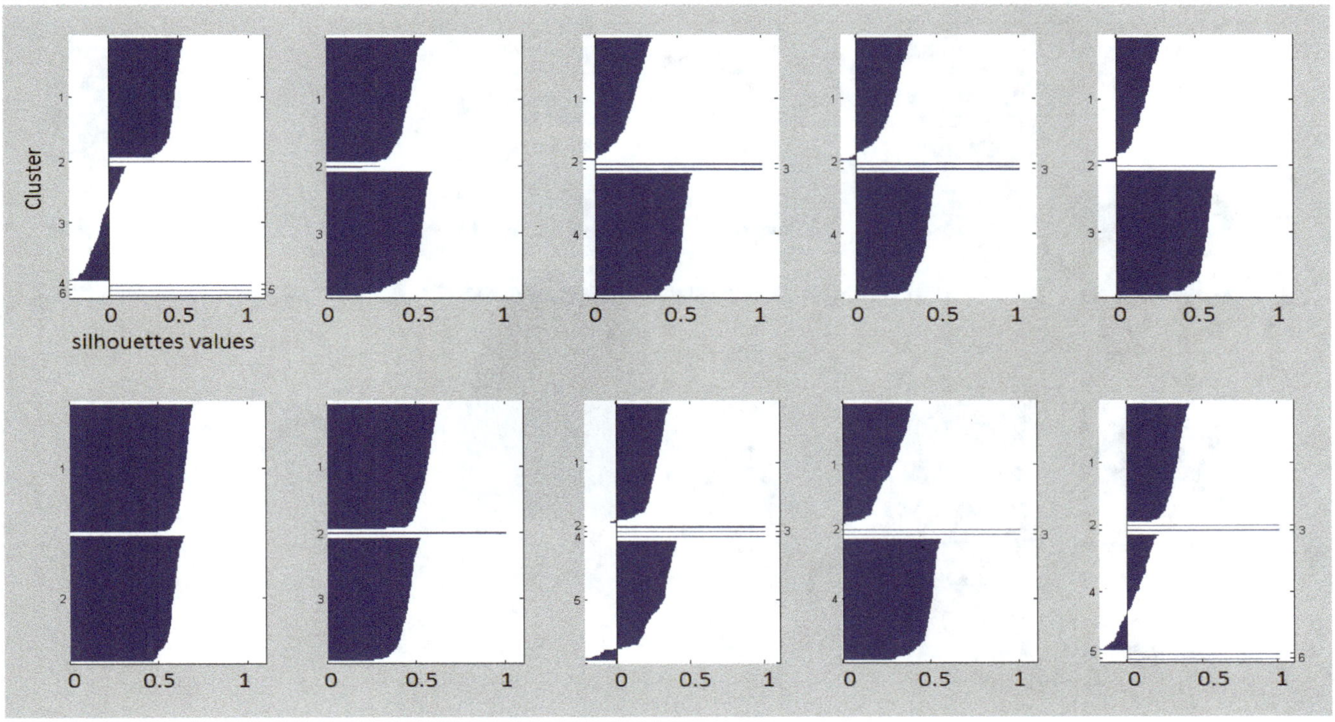

Figure 8. Silhouettes.

It can be seen that, apart from a few outliers, the two clutters are well classified and that the algorithm was able to distinguish between two zones of different Doppler characteristics.

We then test our algorithm on real sea clutter data (see Figures 9–11).

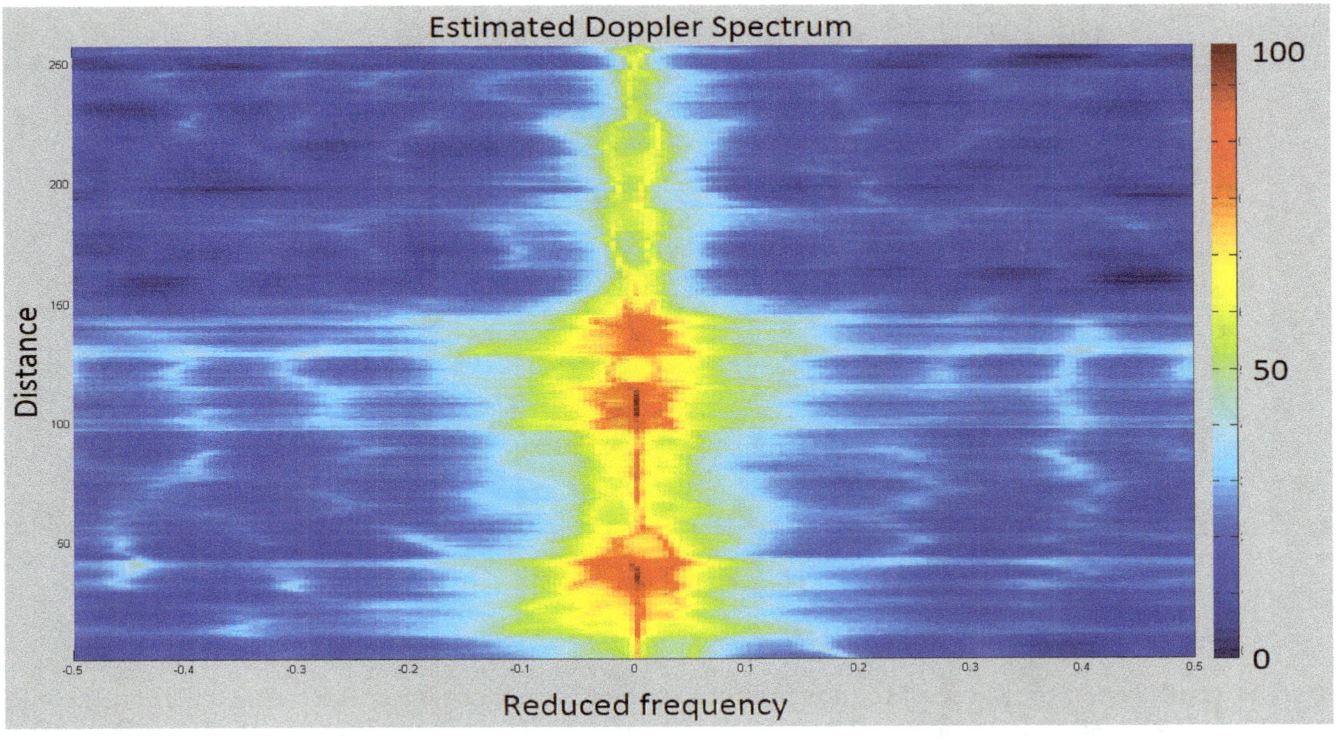

Figure 9. Autoregressive spectrum.

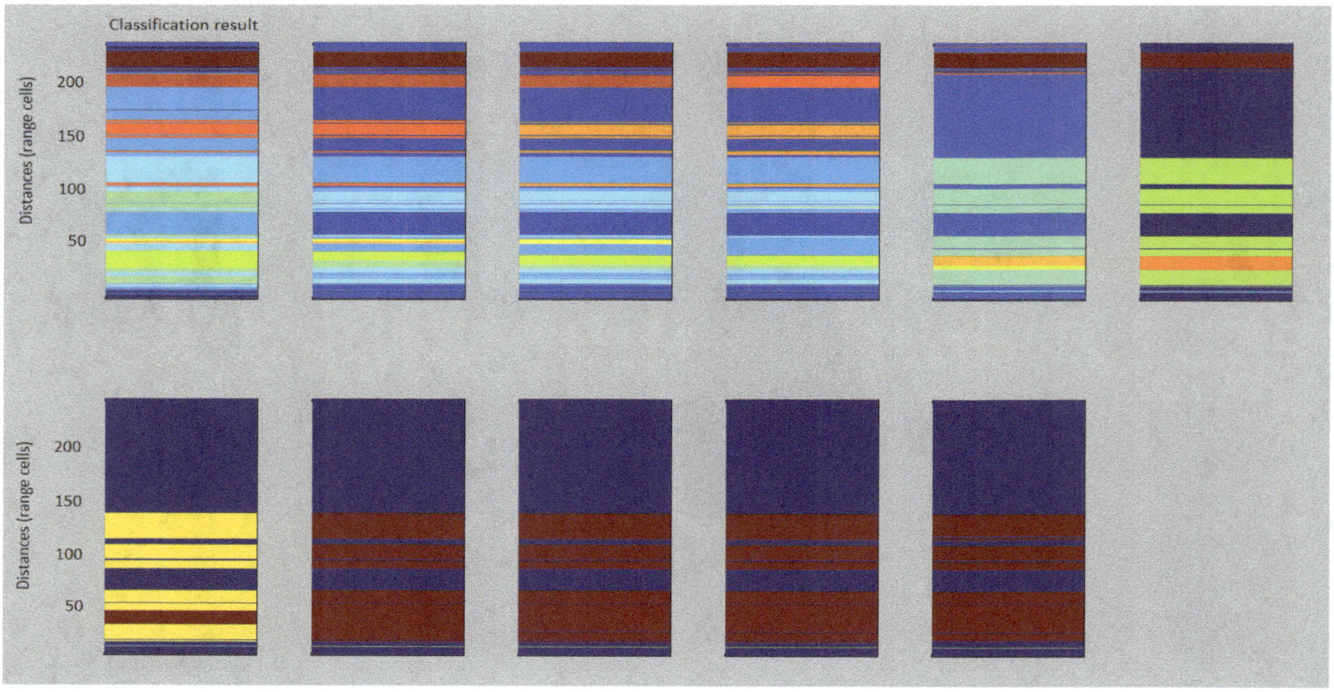

Figure 10. Classification results for varying radii size in the density estimator (10 to 20 closest neighbours).

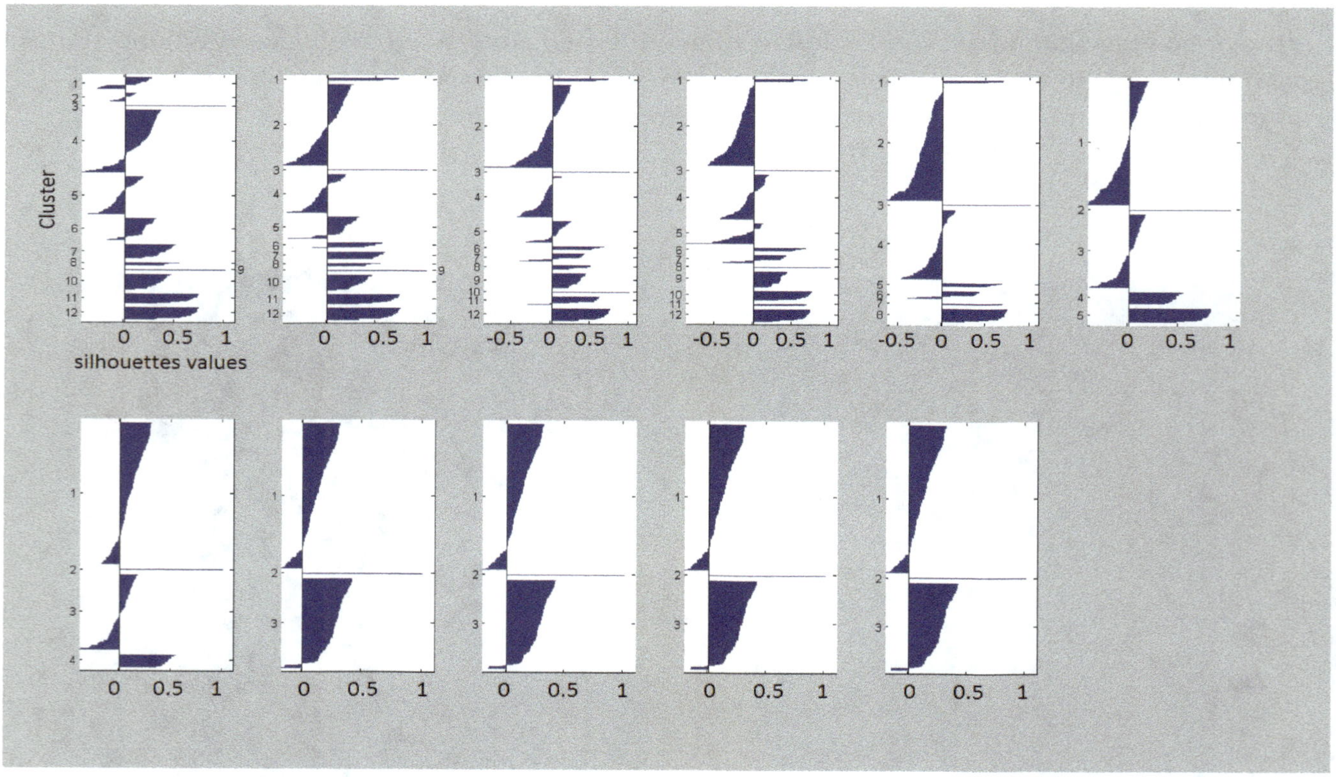

Figure 11. Silhouettes.

The results are more difficult to interpret in that case. The Doppler spectra are varying quite a lot along the range axis. Even though it looks over-segmented, the first classification (kernel size defined by the distance to the 10th closest neighbor point) displays the highest average silhouette value.

5. Conclusions

Three non parametric density estimation techniques have been considered. The main advantage of histograms in the Euclidean context is their simplicity of use. This makes histograms an interesting tool despite the fact that they do not present optimal convergence rates. On the Siegel space, histograms lose their simplicity advantage. They were thus not deeply studied. The orthogonal series density estimation also presents technical disadvantages on the Siegel space. Indeed, the series become integrals, which make the numerical computation of the estimator more difficult than in the Euclidean case. On the other hand, the use of the kernel density estimator does not present major differences with the Euclidean case. The convergence rate obtained in [1] can be extended to compactly supported random variables on non compact Riemannian manifolds. Furthermore, the corrective term whose computation is required to use Euclidean kernels on Riemannian manifolds turns out to have a reasonably simple expression. Our future efforts will concentrate on the use of kernel density estimation on the Siegel space in radar signal processing. As the experimental section suggests, we strongly believe that the estimation of the densities of the Ω_k will provide an interesting description of the different backgrounds. This non-parametric method of density estimation should be compared with parametric ones, as "Maximum Entropy Density" (Gibbs density) on homogenesous manifold as proposed in [37] based on the works of Jean-Marie Souriau. As proposed in [38], a median-shift approach might also be investigated.

Acknowledgments: The authors would like to thank Salem Said, Michal Zidor and Dmitry Gourevitch for the help they provided in the understanding of symmetric spaces and the Siegel space.

Author Contributions: Emmanuel Chevallier carried out the mathematical development. Thibault Forget has set up the Radar clutter segmentation. Frédéric Barbaresco has introduced Poincaré/Siegel Half space and Poincaré/Siegel Disk parameterization for Radar Doppler and Space-Time Adaptive Processing based on Metric spaces deduced from Information Geometry. This parameterization has been re-used in this paper. Jesus Angulo was the Ph.D. supervisor of Emmanuel Chevallier and participates in the supervision of master thesis of Thibault Forget, both thesis are at the origin of this study. All authors have read and approved the final manuscript.

Appendix Demonstration of Theorem 1

Lemma A1. *Let (M, g) be a Riemannian manifold, let C be a compact subset of M and let U be a relatively compact open subset of M containing C. Then, there is a compact Riemannian manifold (M', g') such that U is an open subset of M', the inclusion $i : U \hookrightarrow M'$ is a diffeomorphism onto its image and $g' = g$ on $\overline{U}$.*

Proof. We can assume that M is not compact. Let $f : M \to \mathbb{R}$ be a smooth function on M which tends to $+\infty$ at infinity. Since $\overline{U}$ is compact, $f^{-1}(]-\infty, a[)$ contains $\overline{U}$ for a large enough. By Sard Theorem, there exists a value $a \in \mathbb{R}$ such that $f^{-1}(a)$ contains no critical point of f and such that $f^{-1}(]-\infty, a[)$ contains $\overline{U}$. It follows that $N = f^{-1}(]-\infty, a])$ is a submanifold with boundary of M. Since f tends to $+\infty$ at infinity, N is compact as well as its boundary $\partial N = f^{-1}(\{a\})$.

Call M' the double of N. It is a compact manifold which contains N such that the inclusion $i : N \hookrightarrow M'$ is a diffeomorphism onto its image (see [39], Theorem 5.9 and Definition 5.10). Choose any metric g_0 on M'. Consider two open subsets W_1 and W_2 in M' and two smooth functions $f_1, f_2 : M' \to [0, 1]$ such that

$$\overline{U} \subset W_1 \subset \overline{W}_1 \subset W_2 \subset \overline{W}_2 \subset \operatorname{int} N,$$

the interior of N,

$$f_1(x) = 1$$

on $\overline{W}_1$, vanishes outside of W_2, and

$$f_2(x) = 1$$

outside W_1, and vanishes in $\overline{U}$. Define g' on M' by

$$g' = f_1 g + f_2 g_0$$

on N and

$$g' = f_2 g_0$$

outside of N. Since $f_1 + f_2 > 0$, g' is positive definite everywhere on M'. Since f_1 vanishes outside of W_2, g' is smooth on M'. Finally, since $f_1 = 1$ and $f_2 = 0$ on $\overline{U}$, $g' = g$ on $\overline{U}$. □

We can now prove Theorem 1. Let X be a random variable as in Theorem 1. Following the notations of the theorem and the lemma, let $U = \{x \in \mathcal{M}, d(x, C) < r_{inj}\}$. U is open, relatively compact and contains C. Let $(\mathcal{M}', g')$ be as in the lemma. Let $\hat{f}$ and $\hat{f}'$ be the kernel density estimators defined on M and M', respectively. Theorem 3.1 of [1] provides the desired results for $\hat{f}'$. For $r \leq r_{inj}$, the support and the values on the support of $\hat{f}'$ and $\hat{f}$ coincide. Thus, the desired result also holds for $\hat{f}$.

References

1. Pelletier, B. Kernel density estimation on Riemannian manifolds. *Stat. Probab. Lett.* **2005**, *73*, 297–304.
2. Hendriks, H. Nonparametric estimation of a probability density on a Riemannian manifold using Fourier expansions. *Ann. Stat.* **1990**, *18*, 832–849.
3. Asta, D.M. Kernel Density Estimation on Symmetric Spaces. In *Geometric Science of Information*; Springer: Berlin/Heidelberg, Germany, 2015; Volume 9389, pp. 779–787.
4. Barbaresco, F. Robust statistical radar processing in Fréchet metric space: OS-HDR-CFAR and OS-STAP processing in siegel homogeneous bounded domains. In Proceedings of the 2011 12th International Radar Symposium (IRS), Leipzig, Germany, 7–9 Septerber 2011.
5. Barbaresco, F. Information Geometry of Covariance Matrix: Cartan-Siegel Homogeneous Bounded Domains, Mostow/Berger Fibration and Fréchet Median. In *Matrix Information Geometry*; Bhatia, R., Nielsen, F., Eds.; Springer: Berlin/Heidelberg, Germany, 2012; pp. 199–256.
6. Barbaresco, F. Information geometry manifold of Toeplitz Hermitian positive definite covariance matrices: Mostow/Berger fibration and Berezin quantization of Cartan-Siegel domains. *Int. J. Emerg. Trends Signal Process.* **2013**, *1*, 1–87.
7. Berezin, F.A. Quantization in complex symmetric spaces. *Izv. Math.* **1975**, *9*, 341–379.
8. Lenz, R. Siegel Descriptors for Image Processing. *IEEE Signal Process. Lett.* **2016**, *25*, 625–628.
9. Barbaresco, F. Robust Median-Based STAP in Inhomogeneous Secondary Data: Frechet Information Geometry of Covariance Matrices. In Proceedings of the 2nd French-Singaporian SONDRA Workshop on EM Modeling, New Concepts and Signal Processing For Radar Detection and Remote Sensing, Cargese, France, 25–28 May 2010.
10. Degurse, J.F.; Savy, L.; Molinie, J.P.; Marcos, S. A Riemannian Approach for Training Data Selection in Space-Time Adaptive Processing Applications. In Proceedings of the 2013 14th International Radar Symposium (IRS), Dresden, Germany, 19–21 June 2013; Volume 1, pp. 319–324.
11. Degurse, J.F.; Savy, L.; Marcos, S. Information Geometry for radar detection in heterogeneous environments. In Proceedings of the 33rd International Workshop on Bayesian Inference and Maximum Entropy Methods in Science and Engineering, Amboise, France, 21–26 September 2014.
12. Barbaresco, F. Koszul Information Geometry and Souriau Geometric Temperature/Capacity of Lie Group Thermodynamics. *Entropy* **2014**, *16*, 4521–4565.
13. Barbaresco, F. New Generation of Statistical Radar Processing based on Geometric Science of Information: Information Geometry, Metric Spaces and Lie Groups Models of Radar Signal Manifolds. In Proceedings of the 4th French-Singaporian Radar Workshop SONDRA, Lacanau, France, 23 May 2016.
14. Jeuris, B.; Vandebril, R. The Kahler mean of Block-Toeplitz matrices with Toeplitz structured block. *SIAM J. Matrix Anal. Appl.* **2015**, *37*, 1151–1175.
15. Huckemann, S.; Kim, P.; Koo, J.; Munk, A. Mobius deconvolution on the hyperbolic plan with application to impedance density estimation. *Ann. Stat.* **2010**, *38*, 2465–2498.
16. Asta, D.; Shalizi, C. Geometric network comparison. 2014, arXiv:1411.1350.

17. Chevallier, E.; Barbaresco, F.; Angulo, J. Probability density estimation on the hyperbolic space applied to radar processing. In *Geometric Science of Information*; Springer: Berlin/Heidelberg, Germany, 2015; pp. 753–761.
18. Said, S.; Bombrun, L.; Berthoumieu, Y. New Riemannian Priors on the Univariate Normal Model. *Entropy* **2014**, *16*, 4015–4031.
19. Said, S.; Hatem, H.; Bombrun, L.; Baba, C.;Vemuri, B.C. Gaussian distributions on Riemannian symmetric spaces: Statistical learning with structured covariance matrices. 2016, arXiv:1607.06929.
20. Terras, A. *Harmonic Analysis on Symmetric Spaces and Applications II*; Springer: Berlin/Heidelberg, Germany, 2012.
21. Siegel, C.L. Symplectic geometry. *Am. J. Math.* **1943**, *65*, doi:10.2307/2371774.
22. Helgason, S. *Differential Geometry, Lie Groups, and Symmetric Spaces*; Academic Press: Cambridge, MA, USA, 1979.
23. Cannon, J.W.; Floyd, W.J.; Kenyon, R.; Parry, W.R. Hyperbolic geometry. In *Flavors of Geometry*; Cambridge University Press: Cambridge, UK, 1997; Volume 31, pp. 59–115.
24. Bhatia, R. Matrix Analysis. In *Graduate Texts in Mathematics-169*; Springer: Berlin/Heidelberg, Germany, 1997.
25. Kim, P.; Richards, D. Deconvolution density estimation on the space of positive definite symmetric matrices. In *Nonparametric Statistics and Mixture Models: A Festschrift in Honor of Thomas P. Hettmansperger*; World Scientific Publishing: Singapore, 2008; pp. 147–168.
26. Loubes, J.-M.; Pelletier, B. A kernel-based classifier on a Riemannian manifold. *Stat. Decis.* **2016**, *26*, 35–51.
27. Gangolli, R.; Varadarajan, V.S. *Harmonic Analysis of Spherical Functions on Real Reductive Groups*; Springer: Berlin/Heidelberg, Germany, 1988.
28. Decurninge, A.; Barbaresco, F. Robust Burg Estimation of Radar Scatter Matrix for Mixtures of Gaussian Stationary Autoregressive Vectors. 2016, arxiv:1601.02804.
29. Barbaresco, F. Eddy Dissipation Rate (EDR) retrieval with ultra-fast high range resolution electronic-scanning X-band airport radar: Results of European FP7 UFO Toulouse Airport trials. In Proceedings of the 2015 16th International Radar Symposium, Dresden, Germany, 24–26 June 2015.
30. Oude Nijhuis, A.C.P.; Thobois, L.P.; Barbaresco, F. *Monitoring of Wind Hazards and Turbulence at Airports with Lidar and Radar Sensors and Mode-S Downlinks: The UFO Project*; Bulletin of the American Meteorological Society, 2016, submitted for publication.
31. Barbaresco, F.; Forget, T.; Chevallier, E.; Angulo, J. Doppler spectrum segmentation of radar sea clutter by mean-shift and information geometry metric. In Proceedings of the 17th International Radar Symposium (IRS), Krakow, Poland, 10–12 May 2016; pp. 1–6.
32. Fukunaga, K.; Hostetler, L.D. The Estimation of the Gradient of a Density Function, with Applications in Pattern Recognition. *Proc. IEEE Trans. Inf. Theory* **1975**, *21*, 32–40.
33. Arias-Castro, E.; Mason, D.; Pelletier, B. On the estimation of the gradient lines of a density and the consistency of the mean-shift algorithm. *J. Mach. Learn. Res.* **2000**, *17*, 1–28.
34. Subbarao, R.; Meer, P. Nonlinear Mean Shift over Riemannian Manifolds. *Int. J. Comput. Vis.* **2009**, *84*, doi:10.1007/s11263-008-0195-8.
35. Wang, Y.H.; Han, C.Z. PolSAR Image Segmentation by Mean Shift Clustering in the Tensor Space. *Acta Autom. Sin.* **2010**, *36*, 798–806.
36. Rousseeuw, P. Silhouettes: A graphical aid to the interpretation and validation of cluster analysis. *J. Comput. Appl. Math.* **1987**, *1*, 53–65.
37. Barbaresco, F. Geometric Theory of Heat from Souriau Lie Groups Thermodynamics and Koszul Hessian Geometry: Applications in Information Geometry for Exponential Families. *Entropy* **2016**, *18*, 386.
38. Wanga, Y.; Huang, X.; Wua, L. Clustering via geometric median shift over Riemannian manifolds. *Inf. Sci.* **2013**, *220*, 292–305.
39. Munkres, J.R. Elementary Differential Topology. In *Annals of Mathematics Studies-54*; Princeton University Press: Princeton, NJ, USA, 1967.

5

A Sequence of Escort Distributions and Generalizations of Expectations on q-Exponential Family

Hiroshi Matsuzoe

Department of Computer Science and Engineering, Nagoya Institute of Technology, Nagoya 466-8555, Japan; matsuzoel@nitech.ac.jp.

Academic Editors: Frédéric Barbaresco and Frank Nielsen

Abstract: In the theory of complex systems, long tailed probability distributions are often discussed. For such a probability distribution, a deformed expectation with respect to an escort distribution is more useful than the standard expectation. In this paper, by generalizing such escort distributions, a sequence of escort distributions is introduced. As a consequence, it is shown that deformed expectations with respect to sequential escort distributions effectively work for anomalous statistics. In particular, it is shown that a Fisher metric on a q-exponential family can be obtained from the escort expectation with respect to the second escort distribution, and a cubic form (or an Amari–Chentsov tensor field, equivalently) is obtained from the escort expectation with respect to the third escort distribution.

Keywords: escort distribution; escort expectation; statistical manifold; deformed exponential family; Tsallis statistics; information geometry

MSC: 53A15; 53B50; 62F99; 94A14

---

1. Introduction

Long tailed probability distributions and their related probability distributions are important objects in anomalous statistical physics (cf. [1–3]). For such long tailed probability distributions, the standard expectation does not exist in general. Therefore, the notion of escort distribution has been introduced [4]. Since an escort distribution gives a suitable weight for tail probability, the escort expectation which is the expectation with respect to an escort distribution is more useful than the standard one.

In anomalous statistics, a deformed exponential function and a deformed logarithm function play essential roles. In fact, a deformed exponential family is an important statistical model in anomalous statistics. Such a statistical model is described by such a deformed exponential function. In particular, the set of all q-normal distributions (or Student's t-distributions, equivalently) is a q-exponential family, which is described by a q-deformed exponential function [5] (see also [6,7]).

On the other hand, a generalized score function is defined from a deformed logarithm function. In the previous works, the author showed that a deformed score function is unbiased with respect to the escort expectation [8,9]. This implies that a deformed score function is regarded as an estimating function on a deformed exponential family. In addition, in information geometry, it is known that a deformed exponential family has a statistical manifold structure. Then a deformed score function is regarded as a tangent vector on this statistical manifold [6,10]. Therefore, properties of escort expectations are closely related to geometric structures on a deformed exponential family.

In this paper, we introduce a sequence of escort distributions, then we consider a sequential structure of escort expectations. It is known that a deformed exponential family naturally has at least three kind of different statistical manifold structures [6,11]. Then we show that such statistical manifold structures can be obtained from a sequential structure of escort expectations. In particular, we show that a Fisher metric on a q-exponential family can be obtained from the deformed expectations with respect to the second escort distribution, and a cubic form (or an Amari–Chentsov tensor field, equivalently) is obtained from the deformed expectations with respect to the third escort distribution.

This paper is written based on the proceeding paper [7]. However, this paper focuses on deformed expectations of a q-exponential family, whereas the previous paper focused on deformed independences. We remark that several authors have been studying deformed expectations recently. See [12,13], for example.

2. Deformed Exponential Families

In this paper, we assume that all objects are smooth for simplicity. Let us review preliminary facts about deformed exponential functions and deformed exponential families. For more details, see [2,6], for example. Historically, Tsallis [14] introduced the notion of q-exponential function and Naudts [5] introduced the notion of q-exponential family together with a further generalization. Such a historical note is provided in [2].

Let $\mathbf{R}_{++}$ be the set of all positive real numbers, $\mathbf{R}_{++} := \{x \in \mathbf{R} | x > 0\}$. Let χ be a strictly increasing function from $\mathbf{R}_{++}$ to $\mathbf{R}_{++}$. We define a *χ-logarithm function* or a *deformed logarithm function* by

$$\ln_\chi s := \int_1^s \frac{1}{\chi(t)} dt.$$

The inverse of $\ln_\chi s$ is called a *χ-exponential function* or a *deformed exponential function*, which is defined by

$$\exp_\chi t := 1 + \int_0^t u(s) ds,$$

where the function $u(s)$ is given by $u(\ln_\chi s) = \chi(s)$.

From now on, we suppose that χ is a power function, that is, $\chi(t) = t^q$. Then the deformed logarithm and the deformed exponential are defined by

$$\ln_q s := \frac{s^{1-q} - 1}{1-q}, \qquad (s > 0),$$

$$\exp_q t := (1 + (1-q)t)^{\frac{1}{1-q}}, \qquad (1 + (1-q)t > 0).$$

We say that $\ln_q s$ is a *q-logarithm function* and $\exp_q t$ is a *q-exponential function*. In this case, the function $u(s)$ is given by

$$u(s) = (1 + (1-q)s)^{\frac{q}{1-q}} = \{\exp_q s\}^q.$$

By taking a limit $q \to 1$, these functions coincide with the standard logarithm $\ln s$ and the standard exponential $\exp t$, respectively.

A statistical model S_q is called a *q-exponential family* if

$$S_q := \left\{ p(x, \theta) \middle| p(x; \theta) = \exp_q \left[\sum_{i=1}^{n} \theta^i F_i(x) - \psi(\theta) \right], \ \theta \in \Theta \subset \mathbf{R}^n \right\}, \tag{1}$$

where $F_1(x), \ldots, F_n(x)$ are functions on a sample space Ω, $\theta = {}^t(\theta^1, \ldots, \theta^n)$ is a parameter, and $\psi(\theta)$ is the normalization with respect to the parameter θ. Under suitable conditions, S_q is regarded as a manifold with a local coordinate system $\{\theta^1, \ldots, \theta^n\}$. In this case, we call $\{\theta^i\}$ a *natural coordinate system*.

In this paper, we focus on the q-exponential case. However, many results for the q-exponential family can be generalized for the χ-exponential family (cf. [6,8]). We remark that a q-exponential family and a χ-exponential family have further generalizations. See [15], for example.

Example 1 (Student's t-distribution (cf. [2,6,7])). *Fix a number q $(1 < q < 1 + 2/d,\ d \in \mathbf{N})$, and set $\nu = -d - 2/(1-q)$. We define a d-dimensional Student's t-distribution with degree of freedom ν or a q-Gaussian distribution by*

$$p_q(x;\mu,\Sigma) := \frac{\Gamma\left(\frac{1}{q-1}\right)}{(\pi\nu)^{\frac{d}{2}}\Gamma\left(\frac{\nu}{2}\right)\sqrt{\det(\Sigma)}}\left[1+\frac{1}{\nu}{}^t(x-\mu)\Sigma^{-1}(x-\mu)\right]^{\frac{1}{1-q}},$$

where $X = {}^t(X_1,\ldots,X_d)$ is a random vector on $\mathbf{R}^d$, $\mu = {}^t(\mu^1,\ldots,\mu^d)$ is a location vector on $\mathbf{R}^d$ and Σ is a scale matrix on $\mathrm{Sym}^+(d)$. For simplicity, we assume that Σ is invertible. Otherwise, we should choose a suitable basis $\{v^\alpha\}$ on $\mathrm{Sym}^+(d)$ such that $\Sigma = \sum_\alpha w_\alpha v^\alpha$. Then, the set of all Student's t-distributions is a q-exponential family. In fact, setting parameters by

$$z_q = \frac{(\pi\nu)^{\frac{d}{2}}\Gamma\left(\frac{\nu}{2}\right)\sqrt{\det(\Sigma)}}{\Gamma\left(\frac{1}{q-1}\right)}, \quad \tilde{R} = \frac{z_q^{q-1}}{(1-q)d+2}\Sigma^{-1}, \quad \text{and} \quad \theta = 2\tilde{R}\mu, \tag{2}$$

we have

$$\begin{aligned} p_q(x;\mu,\Sigma) &= \frac{1}{z_q}\left[1+\frac{1}{\nu}{}^t(x-\mu)\Sigma^{-1}(x-\mu)\right]^{\frac{1}{1-q}} \\ &= \left[\left(\frac{1}{z_q}\right)^{1-q} - \frac{1-q}{(1-q)d+2}\left(\frac{1}{z_q}\right)^{1-q}{}^t(x-\mu)\Sigma^{-1}(x-\mu)\right]^{\frac{1}{1-q}} \\ &= \exp_q\left[-{}^t(x-\mu)\tilde{R}(x-\mu) + \ln_q\frac{1}{z_q}\right] \\ &= \exp_q\left[\sum_{i=1}^d \theta^i x_i - \sum_{i=1}^d \tilde{R}_{ii}x_i^2 - 2\sum_{i<j}\tilde{R}_{ij}x_ix_j - \frac{1}{4}{}^t\theta\tilde{R}^{-1}\theta + \ln_q\frac{1}{z_q}\right]. \end{aligned}$$

Since $\theta \in \mathbf{R}^d$ and $\tilde{R} \in \mathrm{Sym}^+(d)$, the set of all Student's t-distributions is a $d(d+3)/2$-dimensional q-exponential family. The normalization $\psi(\theta)$ is given by

$$\psi(\theta) = \frac{1}{4}{}^t\theta\tilde{R}^{-1}\theta - \ln_q\frac{1}{z_q}.$$

A univariate Student's t-distribution is a well-known probability distribution in elementary statistics. We denote it by

$$t_\nu(x;\mu,\sigma) := \frac{1}{Z_q}\exp_q\left[-\frac{(x-\mu)^2}{(3-q)\sigma^2}\right], \tag{3}$$

where $\mu \in \mathbf{R}$ is a location parameter, $\sigma \in \mathbf{R}_{++}$ is a scale parameter, and Z_q is the normalization defined by

$$Z_q = \sqrt{\frac{3-q}{q-1}}\mathrm{Beta}\left(\frac{3-q}{2(q-1)},\frac{1}{2}\right)\sigma.$$

In this case, the degree of freedom is $\nu = (3-q)/(q-1)$. Conversely, the parameter q is give by

$$q = \frac{\nu+3}{\nu+1}. \tag{4}$$

3. Escort Distributions and Generalizations of Expectations

In anomalous statistics, a generalized expectation, called an escort expectation, is often discussed since the standard expectation does not exist in general (cf. [2,5,6]). In this section, we recall generalizations of expectations and introduce a sequential structure of escort distributions.

Let S_q be a q-exponential family. For a given $p(x;\theta) \in S_q$ we define the *q-escort distribution* $P_q(x;\theta)$ of $p(x;\theta)$ and the *normalized q-escort distribution* $P_q^{esc}(x;\theta)$ by

$$\begin{aligned} P_q(x;\theta) &:= P_{q,(1)}(x;\theta) := \{p(x;\theta)\}^q, \\ P_q^{esc}(x;\theta) &:= \frac{1}{Z_q(p)}\{p(x;\theta)\}^q, \quad \text{where} \quad Z_q(p) = \int_\Omega \{p(x;\theta)\}^q dx, \end{aligned}$$

respectively. For a q-exponential family $S_q = \{p_q(x;\theta)\}$, the set of normalized escort distributions $S_{q'} = \{P_q^{esc}(x;\theta)\}$ is a q'-exponential family with $q' = (2q-1)/q$.

Example 2. *Let $t_\nu(x;\mu,\sigma)$ be a univariate Student's t-distribution with degree of freedom ν. Then its normalized escort distribution is also a univariate Student's t-distribution with degree of freedom $\nu+2$. In fact, from Equation (4), a direct calculation shows that*

$$q' = \frac{2q-1}{q} = \frac{\nu+5}{\nu+3}.$$

This implies that the degree of freedom $\nu' = \nu+2$. Therefore, we obtain a sequence of escort distributions from a given Student's t-distribution t_ν:

$$t_\nu \to t_{\nu+2} \to t_{\nu+4} \to \cdots.$$

This sequence is called a τ-sequence, and the procedure to obtain from a given t-distribution to another t-distribution through an escort distribution is called the τ-transformation [16].

For a given $p_q(x;\theta) \in S_q$, we can define the escort of an escort distribution

$$\widetilde{P}_q(x;\theta) := P_{q,(2)}(x;\theta) := q\{P_q(x;\theta)\}^{q'} = q\{p_q(x;\theta)\}^{2q-1}.$$

We call $\widetilde{P}_q(x;\theta)$ the *second escort distribution* of $p_q(x;\theta)$. The coefficient q before $\{p_q(x;\theta)\}^{2q-1}$ comes from considerations of U-information geometry [17]. We will discuss in the latter part of Section 5.

Similarly, we can define the *n-th escort distribution* $P_{q,(n)}(x;\theta)$ from the sequence of escort distributions:

$$P_{q,(n)}(x;\theta) := \{q(2q-1)\cdots((n-1)q-(n-2))\}\{p_q(x;\theta)\}^{nq-(n-1)}. \tag{5}$$

Let $f(x)$ be a function on Ω. The *q-expectation* $E_{q,p}[f(x)]$ and the *normalized q-expectation* $E_{q,p}^{esc}[f(x)]$ with respect to $p(x;\theta) \in S_q$ are defined by

$$\begin{aligned} E_{q,p}[f(x)] &:= \int_\Omega f(x)P_q(x;\theta)dx, \\ E_{q,p}^{esc}[f(x)] &:= \int_\Omega f(x)P_q^{esc}(x;\theta)dx, \end{aligned}$$

respectively. We denote by $\widetilde{E_{q,p}}[f(x)]$ the expectation with respect to the second escort distribution $\widetilde{P}_q(x;\theta)$, that is,

$$\widetilde{E_{q,p}}[f(x)] := \int_\Omega f(x)\widetilde{P}_q(x;\theta)dx = q\int_\Omega f(x)\{p_q(x;\theta)\}^{2q-1}dx.$$

Since a differential of a power function is also a power function, we can give a characterization for escort distributions.

Proposition 1. *Suppose that S_q is a q-exponential family defined by (1). Then the n-th escort distribution is given by the n-th differential of q-exponential function. That is, by setting $u(t) = (\exp_q t)'$, we have the following formula:*

$$
\begin{aligned}
p_q(x;\theta) &= \exp_q\left(\sum_{i=1}^{n}\theta^i F_i(x) - \psi(\theta)\right), \\
P_q(x;\theta) = P_{q,(1)}(x;\theta) &= u\left(\sum_{i=1}^{n}\theta^i F_i(x) - \psi(\theta)\right), \\
\widetilde{P}_q(x;\theta) = P_{q,(2)}(x;\theta) &= u'\left(\sum_{i=1}^{n}\theta^i F_i(x) - \psi(\theta)\right), \\
&\vdots \\
P_{q,(n)}(x;\theta) &= u^{(n-1)}\left(\sum_{i=1}^{n}\theta^i F_i(x) - \psi(\theta)\right), \\
&\vdots
\end{aligned}
$$

Proof. Since a q-exponential function is $\exp_q(x) = (1 + (1-q))^{1/(1-q)}$, its differential is given by

$$u(x) = \frac{1-q}{1-q}(1+(1-q)x)^{\frac{1}{1-q}-1} = (1+(1-q)x)^{\frac{q}{1-q}} = \{\exp_q x\}^q.$$

Therefore, we obtain $P_q(x;\theta) = u\left(\sum_{i=1}^{n}\theta^i F_i(x) - \psi(\theta)\right)$.

By induction, the n-th differential of $u(x)$ coincides with the n-th escort distribution $P_{q,(n)}$, which is given by Equation (5). □

4. Statistical Manifolds and Their Generalized Conformal Structures

In this section, we us review the geometry of statistical manifolds. For more details about the geometry of statistical manifolds, see [18,19].

Let (S, g) be a Riemannian manifold and ∇ be a torsion-free affine connection on S. We say that the triplet (S, ∇, g) is a *statistical manifold* if ∇g is totally symmetric. In this case, we can define a totally symmetric $(0,3)$-tensor field by

$$C(X,Y,Z) := (\nabla_X g)(Y,Z) = Xg(Y,Z) - g(\nabla_X Y, Z) - g(Y, \nabla_X Z),$$

where X, Y and Z are arbitrary vector fields on S. The tensor field C is called a *cubic form* or an *Amari–Chentsov tensor field*.

The notion of statistical manifold was introduced by Lauritzen [20]. He called the triplet (S, g, C) a statistical manifold. In this paper, the definition is followed to Kurose [18]. Though these two definitions are different, the other statistical manifold structure can be obtained from a given one, However, the motivation for the notion of conformal equivalence using (S, g, C) is different from that one using (S, ∇, g), which we will discuss in the latter part of this section.

For a given statistical manifold (S, ∇, g), we can define another torsion-free affine connection ∇^* on S by

$$Xg(Y,Z) = g(\nabla_X Y, Z) + g(Y, \nabla^*_X Z).$$

The connection ∇^* is called the *dual connection* of ∇ with respect to g. The triplet (S, ∇^*, g) is also a statistical manifold, which is called the *dual statistical manifold* of (S, ∇, g). The cubic form is given by the difference of two affine connections ∇^* and ∇:

$$C(X, Y, Z) = g(\nabla^*_X Y - \nabla_X Y, Z).$$

We define generalized conformal structures for statistical manifolds followed to Kurose [18]. Two statistical manifolds (S, ∇, g) and $(S, \bar{\nabla}, \bar{g})$ are said to be 1-*conformally equivalent* if there exists a function $\lambda : S \to \mathbf{R}_{++}$ such that

$$\bar{g}(X, Y) = \lambda g(X, Y), \tag{6}$$

$$\bar{\nabla}_X Y = \nabla_X Y - g(X, Y)\mathrm{grad}_g(\ln \lambda), \tag{7}$$

where $\mathrm{grad}_g(\ln \lambda)$ is the gradient vector field of $\ln \lambda$ with respect to g, that is, $g(X, \ln \lambda) = X(\ln \lambda)$. We say that (S, ∇, g) is 1-*conformally flat* if (S, ∇, g) is locally 1-conformally equivalent to a flat statistical manifold.

Two statistical manifolds (S, ∇, g) and $(S, \bar{\nabla}, \bar{g})$ are said to be (-1)-*conformally equivalent* if there exists a function $\lambda : S \to \mathbf{R}_{++}$ such that

$$\bar{g}(X, Y) = \lambda g(X, Y),$$

$$\bar{\nabla}_X Y = \nabla_X Y + d(\ln \lambda)(Y)X + d(\ln \lambda)(X)Y, \tag{8}$$

where $d(\ln \lambda)(X) = X(\ln \lambda)$. If two statistical manifolds (S, ∇, g) and $(S, \bar{\nabla}, \bar{g})$ are 1-conformally equivalent, then their dual statistical manifolds (S, ∇^*, g) and $(S, \bar{\nabla}^*, \bar{g})$ are (-1)-conformally equivalent.

Proposition 2. *If two statistical manifolds (S, ∇, g) and $(S, \bar{\nabla}, \bar{g})$ are 1-conformally equivalent, then their cubic forms have the following relation:*

$$\frac{1}{\lambda}\bar{C}(X, Y, Z) = C(X, Y, Z) + g(Y, Z)d(\ln \lambda)(X) + g(Z, X)d(\ln \lambda)(Y) + g(X, Y)d(\ln \lambda)(Z).$$

Proof. From Equations (7) and (8), we obtain

$$\bar{\nabla}_X Y = \nabla_X Y + d(\ln \lambda)(Y)X + d(\ln \lambda)(X)Y + g(X, Y)\mathrm{grad}_g(\ln \lambda).$$

By taking an inner product with respect to g, we obtain the result. □

5. Statistical Manifold Structures on q-Exponential Families

In this section, we consider statistical manifold structures on a q-exponential family. It is known that a q-exponential family naturally has at least three kinds of statistical manifold structures (cf. [6,8]). We reformulate these structures from the viewpoint of the sequence of escort distributions. In this paper, we omit the details about information geometry. See [21,22] for further details.

Firstly, we review basic facts about q-exponential family. Let S_q be a q-exponential family. The normalization $\psi(\theta)$ on S_q is convex, but may not be strictly convex. In fact, we obtain the following proposition.

Proposition 3. *Let $S_q = \{p(x;\theta)\}$ be a q-exponential family. Then the normalization function $\psi(\theta)$ is convex.*

Proof. Set $u(x) = (\exp_q x)'$ and $\partial_i = \partial/\partial\theta^i$. Then we have

$$\begin{aligned}\partial_i p(x;\theta) &= u\left(\sum \theta^k F_k(x) - \psi(\theta)\right)(F_i(x) - \partial_i\psi(\theta)),\\ \partial_i\partial_j p(x;\theta) &= u'\left(\sum \theta^k F_k(x) - \psi(\theta)\right)(F_i(x) - \partial_i\psi(\theta))(F_j(x) - \partial_j\psi(\theta))\\ &- u\left(\sum \theta^k F_k(x) - \psi(\theta)\right)\partial_i\partial_j\psi(\theta).\end{aligned} \tag{9}$$

Since $\partial_i \int_\Omega p(x;\theta)dx = \int_\Omega \partial_i p(x;\theta)dx = 0$ and $\int_\Omega \partial_i\partial_j p(x;\theta)dx = 0$, we have

$$\begin{aligned}Z_q(p) &= \int_\Omega \{(p(x;\theta)\}^q dx = \int_\Omega u\left(\sum \theta^k F_k(x) - \psi(\theta)\right) dx,\\ \partial_i\partial_j\psi(\theta) &= \frac{1}{Z_q(p)}\int_\Omega u'\left(\sum \theta^k F_k(x) - \psi(\theta)\right)(F_i(x) - \partial_i\psi(\theta))(F_j(x) - \partial_j\psi(\theta))dx.\end{aligned} \tag{10}$$

For an arbitrary vector $c = {}^t(c^1, c^2, \ldots, c^n) \in \mathbf{R}^n$, since $Z_q(p) > 0$ and $u''(x) > 0$, we have

$$\sum_{i,j=1}^n c^i c^j (\partial_i\partial_j\psi(\theta)) = \frac{1}{Z_q(p)}\int_\Omega u''\left(\sum_{k=1}^n \theta^k F_k(x) - \psi(\theta)\right)\left\{\sum_{i=1}^n c^i(F_i(x) - \partial_i\psi(\theta))\right\}^2 dx \geq 0.$$

This implies that the Hessian matrix $(\partial_i\partial_j\psi(\theta))$ is semi-positive definite. □

We assume that ψ is strictly convex in this paper. Under this assumption, we can induce many geometric structures for a q-exponential family.

Since ψ is strictly convex, we can define a Riemannian metric and a cubic form by

$$\begin{aligned}g^q_{ij}(\theta) &:= \partial_i\partial_j\psi(\theta),\\ C^q_{ijk}(\theta) &:= \partial_i\partial_j\partial_k\psi(\theta).\end{aligned}$$

We call g^q and C^q a *q-Fisher metric* and a *q-cubic form*, respectively [23,24]. Since g^q is a Hessian of a function ψ, g^q is a *Hessian metric*, and ψ is the *potential* of g^q with respect to the natural coordinate $\{\theta^i\}$ [25].

For a fixed real number α, set

$$g^q\left(\nabla^{q(\alpha)}_X Y, Z\right) := g^q\left(\nabla^{q(0)}_X Y, Z\right) - \frac{\alpha}{2}C^q(X, Y, Z), \tag{11}$$

where $\nabla^{q(0)}$ is the Levi-Civita connection with respect to g^q. Since g^q is a Hessian metric, from standard arguments in Hessian geometry [25], $\nabla^{q(e)} := \nabla^{q(1)}$ and $\nabla^{q(m)} := \nabla^{q(-1)}$ are flat affine connections and mutually dual with respect to g^q. Therefore, the triplets $(S_q, \nabla^{q(e)}, g^q)$ and $(S_q, \nabla^{q(m)}, g^q)$ are flat statistical manifolds, and the quadruplet $(S_q, g^q, \nabla^{q(e)}, \nabla^{q(m)})$ is a dually flat space.

Under q-expectations, we have the following proposition (cf. [10]).

Proposition 4. *For S_q a q-exponential family,* (1) *Set $\eta_i = E^{esc}_{q,p}[F_i(x)]$. Then $\{\eta_i\}$ is a $\nabla^{q(m)}$-affine coordinate system such that*

$$g^q\left(\frac{\partial}{\partial\theta^i}, \frac{\partial}{\partial\eta_j}\right) = \delta^j_i.$$

(2) *Set $\phi(\eta) = E^{esc}_{q,p}[\log_q p(x;\theta)]$, then $\phi(\eta)$ is the potential of g^q with respect to $\{\eta_i\}$.*

Next, let us consider the standard Fisher metric and the standard cubic form. Suppose that $S := \{p(x;\theta)\}$ is a statistical model. Set $p_\theta := p(x;\theta)$, for simplicity. We define the *(standard) Fisher metric* g^F on S_q by

$$g^F_{ij}(\theta) := \int_\Omega (\partial_i \ln p_\theta)(\partial_j \ln p_\theta) p_\theta dx,$$

and the *(standard) cubic form* or the *Amari–Chentsov vector field* C^F by

$$C^F_{ijk}(\theta) := \int_\Omega (\partial_i \ln p_\theta)(\partial_j \ln p_\theta)(\partial_j \ln p_\theta) p_\theta dx.$$

From similar arguments of (11), we can define an *α-connection* $\nabla^{(\alpha)}$ on S_q, and we can obtain a statistical manifold structure $(S_q, \nabla^{(\alpha)}, g^F)$. In this case, $(S_q, \nabla^{(\alpha)}, g^F)$ is called an *invariant statistical manifold* [21,22].

A Fisher metric and a cubic form have the following representation using a sequence of escort distributions,

Theorem 1. *Let S_q be a q-exponential family. For $p(x;\theta) \in S_q$, suppose that $P_{q,(2)}(x;\theta)$ and $P_{q,(3)}(x;\theta)$ are the second and the third escort distribution of $p(x;\theta)$, respectively. Then the Fisher metric g^F and the cubic form C^F are given as follows:*

$$g^F_{ij}(\theta) = \frac{1}{q}\int_\Omega (\partial_i \ln_q p_\theta)(\partial_j \ln_q p_\theta) P_{q,(2)}(x;\theta) dx, \tag{12}$$

$$C^F_{ijk}(\theta) = \frac{1}{q(2q-1)}\int_\Omega (\partial_i \ln_q p_\theta)(\partial_j \ln_q p_\theta)(\partial_k \ln_q p_\theta) P_{q,(3)}(x;\theta) dx. \tag{13}$$

Proof. Differentiating the q-logarithm, we have

$$\partial_i \ln_q p_\theta = \partial_i \left(\frac{p_\theta^{1-q} - 1}{1-q} \right) = p_\theta^{-q} \partial_i p(\theta) = p_\theta^{1-q} \partial_i \ln p(\theta).$$

Therefore, we obtain

$$\begin{aligned}
\frac{1}{q}\int_\Omega (\partial_i \ln_q p_\theta)(\partial_j \ln_q p_\theta) P_{q,(2)}(x;\theta) dx &= \int_\Omega p_\theta^{1-q} (\partial_i \ln p_\theta) p_\theta^{1-q} (\partial_j \ln p_\theta) p_\theta^{2q-1}(x;\theta) dx \\
&= \int_\Omega (\partial_i \ln p_\theta)(\partial_j \ln p_\theta) p_\theta(x;\theta) dx \\
&= g^F_{ij}(\theta).
\end{aligned}$$

By a similar argument, we obtain the representation for C^F. □

We define an *α-divergence* $D^{(\alpha)}$ with $\alpha = 1 - 2q$ and a *q-relative entropy* (or a *normalized Tsallis relative entropy*) D^T_q by

$$D^{(1-2q)}(p(x), r(x)) = \frac{1}{q} E_{q,p}[\ln_q p(x) - \ln_q r(x)] = \frac{1 - \int_\Omega p(x)^q r(x)^{1-q} dx}{q(1-q)}, \tag{14}$$

$$D^T_q(p(x), r(x)) = E^{esc}_{q,p}[\ln_q p(x) - \ln_q r(x)] = \frac{1 - \int_\Omega p(x)^q r(x)^{1-q} dx}{(1-q) Z_q(p)}, \tag{15}$$

respectively. It is known that the α-divergence $D^{(1-2q)}(r,p)$ induces a statistical manifold structure $(S_q, g^F, \nabla^{(2q-1)})$, where g^F is the Fisher metric on S_q and $\nabla^{(2q-1)}$ is the α-connection with $\alpha = 2q - 1$, and the q-relative entropy $D^T_q(r,p)$ induces $(S_q, g, \nabla^{q(e)})$.

Theorem 2 (cf. [10,24]). *For a q-exponential family S_q, two statistical manifolds $(S_q, g^F, \nabla^{(2q-1)})$ and $(S_q, g, \nabla^{q(e)})$ are 1-conformally equivalent. In particular, an invariant statistical manifold $(S_q, g^F, \nabla^{(2q-1)})$ is 1-conformally flat. Riemannian metrics and cubic forms have the following relations:*

$$g^q_{ij}(\theta) = \frac{q}{Z_q(p)} g^F_{ij}(\theta), \tag{16}$$

$$\begin{aligned} C^q_{ijk}(\theta) &= \frac{q}{Z_q(p)}(2q-1)C^F_{ijk}(\theta) \\ &- \frac{q}{Z_q(p)}\left\{ g^F_{ij}\partial_k \ln Z_q(p) + g^F_{jk}(\theta)\partial_i \ln Z_q(p) + g^F_{ki}(\theta)\partial_j \ln Z_q(p) \right\}. \end{aligned} \tag{17}$$

Proof. The results were essentially obtained in [10]. However, we give a simpler proof for Equations (16) and (17). The key idea is a sequence of escort distributions and the escort representations of g^F and C^F in Theorem 1.

From Equation (10), we directly obtain the conformal equivalence relation (16) using the escort representation of g^F in (12).

By differentiating (9) and taking an integration, we obtain

$$\begin{aligned} 0 &= \int_\Omega u''\left(\sum \theta^l F_l(x) - \psi(\theta)\right)(F_i(x) - \partial_i\psi(\theta))(F_j(x) - \partial_j\psi(\theta))(F_k(x) - \partial_k\psi(\theta))dx \\ &- \int_\Omega u'\left(\sum \theta^l F_l(x) - \psi(\theta)\right)(F_k(x) - \partial_k\psi(\theta))\partial_i\partial_j\psi(\theta)dx \\ &- \int_\Omega u'\left(\sum \theta^l F_l(x) - \psi(\theta)\right)(F_i(x) - \partial_i\psi(\theta))\partial_j\partial_k\psi(\theta)dx \\ &- \int_\Omega u'\left(\sum \theta^l F_l(x) - \psi(\theta)\right)(F_j(x) - \partial_j\psi(\theta))\partial_k\partial_i\psi(\theta)dx \\ &- \int_\Omega u\left(\sum \theta^l F_l(x) - \psi(\theta)\right)\partial_i\partial_j\partial_k\psi(\theta)dx. \end{aligned}$$

Since $Z_q(p) = \int_\Omega P_q(x;\theta)dx$, we have

$$\partial_i Z_q(p) = \partial_i \int_\Omega P_q(x;\theta)dx = \int_\Omega \partial_i P_q(x;\theta)dx = \int_\Omega \tilde{P}_q(x;\theta)(F_i(x) - \partial_i\psi(\theta))dx.$$

From the escort representation of C^F in (13), and Proposition 1, we obtain Equation (17) since $g^q_{ij}(\theta) = \partial_i\partial_j\psi(\theta)$ and $C^q_{ijk}(\theta) = \partial_i\partial_j\partial_k\psi(\theta)$. □

We remark that the cubic form of $(S_q, g^F, \nabla^{(2q-1)})$ is not C^F but $(2q-1)C^F$.

The difference of a α-divergence and a q-relative entropy is only the normalization $q/Z_q(p)$. This implies that a normalization for probability density imposes a generalized conformal change for a statistical model.

In the next part of this section, let us consider another statistical manifold on S_q (cf. [6,17,26]).

Recall that a Fisher metric g^F has the following representation:

$$g^F_{ij}(\theta) = \int_\Omega (\partial_i \ln p_\theta)(\partial_j p_\theta)dx.$$

In information geometry, $\partial_i \ln p_\theta$ is called an *e-representation (exponential representation)* of p_θ, and $\partial_j p_\theta$ is called a *m-representation (mixture representation)*. Intuitively, $\partial_i \ln p_\theta$ and $\partial_j p_\theta$ are regarded as tangent vectors on a statistical model. Hence a Fisher metric is regarded as a L^2-inner product of e- and m-representations.

Let us generalize e- and m-representations for a q-exponential family. For $p_\theta \in S_q$, we call $\partial_i \ln_q p_\theta$ a *q-score function*. Then we define a Riemannian metric g^M by

$$g^M_{ij}(\theta) = \int_\Omega (\partial_i \ln_q p_\theta)(\partial_j p_\theta)dx. \tag{18}$$

By differentiating the above equation, we can define mutually dual torsion-free affine connections $\nabla^{M(e)}$ and $\nabla^{M(m)}$:

$$\begin{aligned}\Gamma^{M(e)}_{ij,k}(\theta) &:= \int_\Omega (\partial_i\partial_j \ln_q p_\theta)(\partial_k p_\theta)dx, \\ \Gamma^{M(m)}_{ij,k}(\theta) &:= \int_\Omega (\partial_k \ln_q p_\theta)(\partial_i\partial_j p_\theta)dx,\end{aligned}$$

where $\Gamma^{M(e)}_{ij,k}$ and $\Gamma^{M(m)}_{ij,k}$ are the Christoffel symbols of $\nabla^{M(e)}$ and $\nabla^{M(m)}$ of the first kind, respectively. It is known that g^M is a Hessian metric, and the quadruplet $(S_q, g^M, \nabla^{M(e)}, \nabla^{M(m)})$ is a dually flat space. In addition, a natural parameter $\{\theta^i\}$ is a $\nabla^{M(e)}$-affine coordinate sysem. Therefore, the cubic form for $(S_q, \nabla^{M(e)}, g^M)$ is

$$C^M_{ijk}(\theta) = \Gamma^{M(m)}_{ij,k}(\theta). \tag{19}$$

We remark that the statistical manifold structure $(S_q, \nabla^{M(e)}, g^M)$ is induced from a *β-divergence* [17,26] (or a *density power divergence* [27]):

$$D_{1-q}(p,r) := \int_\Omega \left\{ p(x)\frac{p(x)^{1-q} - r(x)^{1-q}}{1-q} - \frac{p(x)^{2-q} - r(x)^{2-q}}{2-q} \right\} dx. \tag{20}$$

Theorem 3. *For the statistical manifold structure $(S_q, \nabla^{M(e)}, g^M)$, the escort representations of the Riemannian metric g^M and the cubic form C^M are given as follows:*

$$g^M_{ij}(\theta) = \int_\Omega (\partial_i \ln_q p_\theta)(\partial_j \ln_q p_\theta)P_q(x;\theta)dx, \tag{21}$$

$$C^M_{ijk}(\theta) = \int_\Omega (\partial_i \ln_q p_\theta)(\partial_j \ln_q p_\theta)(\partial_k \ln_q p_\theta)\tilde{P}_q(x;\theta)dx. \tag{22}$$

Proof. For the Riemannian metric g^M, since $\partial_i p_\theta = (\partial_i \ln_q p_\theta)P_q(x;\theta)$, we immediately obtain Equation (21) from the definition of g^M.

Let us consider the expression for cubic form (22). The q-score function $\partial_i \ln_q p_\theta$ is unbiased under the q-expectation. In fact,

$$E_{q,p}[\partial_i \ln_q p_\theta] = \int_\Omega (\partial_j \ln_q p_\theta)P_q(x;\theta)dx = \int_\Omega \partial_j p_\theta dx = 0.$$

From Equation (19), we obtain

$$\begin{aligned}C^M_{ijk}(\theta) &= \Gamma^{M(m)}_{ij,k}(x;\theta) \\ &= \int_\Omega (\partial_k \ln_q p_\theta)(\partial_i\partial_j p_\theta)dx \\ &= \int_\Omega (\partial_k \ln_q p_\theta)\partial_i \left\{ (\partial_j \ln_q p_\theta)\, P_q(x;\theta) \right\} dx \\ &= -\partial_{ij}\psi(\theta)\int_\Omega (\partial_k \ln_q p_\theta)P_q(x;\theta)dx + \int_\Omega (\partial_k \ln_q p_\theta)(\partial_j \ln_q p_\theta)\{\partial_i P_q(x;\theta)\}dx \\ &= \int_\Omega (\partial_k \ln_q p_\theta)(\partial_j \ln_q p_\theta)(\partial_i \ln_q p_\theta)\tilde{P}_q(x;\theta)dx.\end{aligned}$$

□

We remark that Naudts [5] gave another generalization of Fisher metric g^N, which is defined by

$$g^N_{ij}(\theta) := \int_\Omega \frac{1}{P^{esc}_q(x;\theta)}(\partial_i p_\theta)(\partial_j p_\theta)dx,$$

The metric g^N is conformally equivalent to g^M with conformal factor $Z_q(p_\theta) = \int_\Omega \{p(x;\theta)\}^q dx$. That is, $g^N(\theta) = Z_q(p_\theta) g^M(\theta)$. (See also [6]). Naudts gave a further generalization of Fisher metric and he showed a Cramér–Rao type bound theorem [5].

6. Concluding Remarks

In this paper, we introduced a sequence of escort distributions. Then we gave representations of Riemannian metrics and cubic forms from a viewpoint of the sequence of escort distributions.

In particular, we can define the following $(0,2)$-tensor fields on a q-exponential family. For $p_\theta \in S_q$, set $\eta_i = \partial_i \psi(\theta)$.

(1) From the standard expectation, we obtain

$$\begin{aligned} g^{(0)}_{ij}(\theta) := G_{ij}(\theta) &:= \int_\Omega (\partial_i \ln_q p_\theta)(\partial_j \ln_q p_\theta) p_\theta dx \\ &= E_p[(F_i(x) - \eta_i)(F_j(x) - \eta_j)]. \end{aligned}$$

The tensor G is a covariance matrix. However, G may not be important in anomalous statistics.

(2) From the q-expectation, we obtain

$$\begin{aligned} g^{(1)}_{ij}(\theta) := g^M_{ij}(\theta) &= \int_\Omega (\partial_i \ln_q p_\theta)(\partial_j \ln_q p_\theta)\{p_\theta\}^q dx \\ &= \int_\Omega (\partial_i \ln_q p_\theta)(\partial_j \ln_q p_\theta) P_q(x;\theta) dx \\ &= E_{q,p}[(F_i(x) - \eta_i)(F_j(x) - \eta_j)]. \end{aligned}$$

The Riemannian metric g^M is a Hessian metric, and it is induced from the β-divergence (20).

(3) From the expectation with respect to the second escort distribution, we obtain

$$\begin{aligned} g^{(2)}_{ij}(\theta) := g^F_{ij}(\theta) &= \int_\Omega (\partial_i \ln_q p_\theta)(\partial_j \ln_q p_\theta)\{p_\theta\}^{2q-1} dx \\ &= \frac{1}{q}\int_\Omega (\partial_i \ln_q p_\theta)(\partial_j \ln_q p_\theta) \tilde{P}_q(x;\theta) dx \\ &= \frac{1}{q} E_{q,(2),p}[(F_i(x) - \eta_i)(F_j(x) - \eta_j)]. \\ g^q_{ij}(\theta) &= \frac{Z_q(p)}{q} g^F_{ij}. \end{aligned}$$

The Riemannian metric g^F is a Fisher metric. Hence g^F is invariant to the choice of reference measure on Ω, but it is not a Hessian metric. In addition, g^F is induced from the α-divergence (14). The conformal Riemannian metric g^q is a q-Fisher metric. It is a Hessian metric, and it is induced from a normalized Tsallis relative entropy (15).

We may define a Riemannian metric and a cubic form from higher order escort expectations:

$$\begin{aligned} g^{(n)}_{ij}(\theta) &:= \int_\Omega (\partial_i \ln_q p_\theta)(\partial_j \ln_q p_\theta) P_{q,(n)}(x;\theta) dx, \\ C^{(n)}_{ij}(\theta) &:= \int_\Omega (\partial_i \ln_q p_\theta)(\partial_j \ln_q p_\theta)(\partial_k \ln_q p_\theta) P_{q,(n+1)}(x;\theta) dx. \end{aligned}$$

Then we obtain a sequence of statistical manifold structures.

$$(S_q, g^{(1)}, C^{(1)}) \to (S_q, g^{(2)}, C^{(2)}) \to \cdots \to (S_q, g^{(n)}, C^{(n)}) \to \cdots$$

However, the geometric meaning of this sequence is not clear at this moment. Elucidating geometric properties of this sequence is a future problem.

Acknowledgments: This research was partially supported by JSPS (Japan Society for the Promotion of Science), KAKENHI (Grants-in-Aid for Scientific Research) Grant Numbers JP26108003 and JP15K04842.

References

1. Tsallis, C. *Introduction to Nonextensive Statistical Mechanics: Approaching a Complex World*; Springer: New York, NY, USA, 2009.
2. Naudts, J. *Generalised Thermostatistics*; Springer: London, UK, 2011.
3. Kaniadakis, G. Theoretical foundations and mathematical formalism of the power-law tailed statistical distributions. *Entropy* **2013**, *15*, 3983–4010.
4. Beck, C.; Schlögl, F. *Thermodynamics of Chaotic Systems: An Introduction*; Cambridge University Press: Cambridge, UK, 1993.
5. Naudts, J. Estimators, escort probabilities, and ϕ-exponential families in statistical physics. *J. Inequal. Pure Appl. Math.* **2004**, *5*, 102.
6. Matsuzoe, H.; Henmi, M. Hessian structures and divergence functions on deformed exponential families. In *Geometric Theory of Information, Signals and Communication Technology*; Nielsen, F., Ed.; Springer: Basel, Switzerland, 2014; pp. 57–80.
7. Sakamoto, M.; Matsuzoe, H. A generalization of independence and multivariate Student's t-distributions. In *Geometric Science of Information*, Proceedings of Second International Conference on Geometric Science of Information (GSI 2015), Palaiseau, France, 28–30 October 2015; Volume 9389, pp. 740–749.
8. Matsuzoe, H.; Wada, T. Deformed algebras and generalizations of independence on deformed exponential families. *Entropy* **2015**, *17*, 5729–5751.
9. Wada, T.; Matsuzoe, H.; Scarfone, A.M. Dualistic hessian structures among the thermodynamic potentials in the κ-Thermostatistics. *Entropy* **2015**, *17*, 7213–7229.
10. Matsuzoe, H. Statistical manifolds and geometry of estimating functions. In *Prospects of Differential Geometry and Its Related Fields, Proceedings of the 3rd International Colloquium on Differential Geomentry and Its Related Fields, Veliko Tarnovo, Bulgaria, 3–7 September 2012*; Adachi, T., Hashimoto, H., Hristov, M.J., Eds.; World Scientific: Hackensack, NJ, USA, 2013; pp. 187–202.
11. Matsuzoe, H.; Henmi, M. Hessian structures on deformed exponential families. In *Geometric Science of Information, Proceedings of First International Conference on Geometric Science of Information (GSI 2013), Paris, France, 28–30 August 2013*; Springer: Berlin/Heidelberg, Germany, 2015; Volume 8085, pp. 275–282.
12. Eguchi, S.; Komori, O. Path connectedness on a space of probability density functions. In *Geometric Science of Information, Proceedings of Second International Conference on Geometric Science of Information (GSI 2015), Palaiseau, France, 28–30 October 2015*; Springer: Berlin/Heidelberg, Germany, 2015; Volume 9389, pp. 615–624.
13. Scarfone, A.M.; Matsuzoe, H.; Wada, T. Consistency of the structure of Legendre transform in thermodynamics with the Kolmogorov-Nagumo average. *Phys. Lett. A* **2016**, *380*, 3022–3028.
14. Tsallis, C. What are the numbers that experiments provide? *Quim. Nova* **1994**, *17*, 468–471.
15. Zhang, J. On monotone embedding in information geometry. *Entropy* **2015**, *17*, 4485–4499.
16. Tanaka M. Meaning of an escort distribution and τ-transformation. *J. Phys. Conf. Ser.* **2010**, *201*, 012007.
17. Murata, N.; Takenouchi, T.; Kanamori, T.; Eguchi, S. Information geometry of U-boost and Bregman divergence. *Neural Comput.* **2004**, *16*, 1437–1481.
18. Kurose, T. On the divergences of 1-conformally flat statistical manifolds. *Tôhoku Math. J.* **1994**, *46*, 427–433.
19. Matsuzoe, H. Statistical manifolds and affine differential geometry. *Adv. Stud. Pure Math.* **2010**, *57*, 303–321.
20. Lauritzen, S.L. Statistical manifolds. In *Differential Geometry in Statistical Inferences*; Gupta, S.S., Ed.; IMS Lecture Notes Monograph Series 10; Institute of Mathematical Statistics: Hayward, CA, USA, 1987; pp. 96–163.
21. Amari, S.; Nagaoka, H. *Method of Information Geometry*; Translations of Mathematical Monographs; American Mathematical Society: Providence, RI, USA; Oxford University Press: Oxford, UK, 2000.
22. Amari, S. *Information Geometry and Its Applications*; Springer: Tokyo, Japan, 2016.
23. Amari, S.; Ohara, A.; Matsuzoe, H. Geometry of deformed exponential families: Invariant, dually-flat and conformal geometry. *Physica A* **2012**, *391*, 4308–4319.

24. Matsuzoe, H.; Ohara, A. Geometry for q-exponential families. In *Recent Progress in Differential Geometry and Its Related Fields, Proceedings of the 2nd International Colloquium on Differential Geomentry and Its Related Fields, Veliko Tarnovo, Bulgaria, 6–10 September 2010*; Adachi, T., Hashimoto, H., Hristov, M.J., Eds.; World Scientific: Hackensack, NJ, USA, 2011; pp. 55–71.
25. Shima, H. *The Geometry of Hessian Structures*; World Scientific: Hackensack, NJ, USA, 2007.
26. Ohara, A.; Wada, T. Information geometry of q-Gaussian densities and behaviors of solutions to related diffusion equations. *J. Phys. A* **2010**, *43*, 035002.
27. Basu, A.; Harris, I.R.; Hjort, N.L.; Jones, M.C. Robust and efficient estimation by minimising a density power divergence. *Biometrika* **1998**, *85*, 549–559.

6

Entropy Minimizing Curves with Application to Flight Path Design and Clustering †

Stéphane Puechmorel * and Florence Nicol

Laboratoire de Mathématiques Appliquées, Informatique et Automatique pour l'Aérien (MAIAA), Département Sciences et Ingénierie de la Navigation Aérienne (SINA), École Nationale de l'Aviation Civile (ENAC), 7 avenue Edouard Belin CS 54005, 31055 Toulouse, France; florence.nicol@enac.fr

* Correspondence: stephane.puechmorel@enac.fr.

† This paper is an extended version of our paper published in the 2nd Conference on Geometric Science of Information, Paris, France, 28–30 October 2015.

Academic Editors: Frédéric Barbaresco and Frank Nielsen

Abstract: Air traffic management (ATM) aims at providing companies with a safe and ideally optimal aircraft trajectory planning. Air traffic controllers act on flight paths in such a way that no pair of aircraft come closer than the regulatory separation norms. With the increase of traffic, it is expected that the system will reach its limits in the near future: a paradigm change in ATM is planned with the introduction of trajectory-based operations. In this context, sets of well-separated flight paths are computed in advance, tremendously reducing the number of unsafe situations that must be dealt with by controllers. Unfortunately, automated tools used to generate such planning generally issue trajectories not complying with operational practices or even flight dynamics. In this paper, a means of producing realistic air routes from the output of an automated trajectory design tool is investigated. For that purpose, the entropy of a system of curves is first defined, and a mean of iteratively minimizing it is presented. The resulting curves form a route network that is suitable for use in a semi-automated ATM system with human in the loop. The tool introduced in this work is quite versatile and may be applied also to unsupervised classification of curves: an example is given for French traffic.

Keywords: curve system entropy; curves manifold; curve clustering; probability distribution estimation; air traffic management

1. Introduction

Based on recent studies [1], traffic in Europe is expected to grow by an average yearly rate of 2.6%, yielding a net increase of two million flights per year at the 2020 horizon. The long-term forecast gives a two-fold increase in 2050 over the current traffic, pointing out the need for a paradigm change in the way flights are managed. Two major framework programs, SESAR (Single European Sky Air traffic management Research) in Europe and Nextgen in the U.S. have been launched in order to first investigate potential solutions and to deploy them in a second phase. One of the main changes that the air traffic management (ATM) system will undergo is a switch from airspace-based to trajectory-based operations with a delegation of the separation task to the crews. Within this framework, trajectories become the basic object of ATM, changing the way air traffic controllers will be working. In order to alleviate the workload of controllers, trajectories will be planned several weeks in advance in such a way that close encounters are minimized and ideally removed. For that purpose, several tools are currently being developed; most of them coming from the field of robotics [2]. Unfortunately, flight paths issued by these algorithms are not tractable for a human controller and need to be simplified. The purpose of the present work is to introduce an automated procedure that takes as input a set of trajectories and outputs a simplified one that can be used in an operational context. Please note that the separation norm constraints were not taken into account in this work. In our algorithm, we cannot enforce the regulatory separation norms, just construct clusters with low interactions. According to the applications, the results of our algorithm may be used as an initial solution of a post-processing algorithm based on optimal control in order to keep in line with the regulatory constraints. Using entropy associated with a curves system, a gradient descent is performed in order to reduce it so as to straighten trajectories while avoiding areas with low aircraft density, thus enforcing route-like behavior. This effect is related to the fact that entropy-minimizing distributions favor concentration.

2. Entropy-Minimizing Curves

2.1. Motivation

As previously mentioned, air traffic management of the future will make an intensive use of 4D trajectories as a basic object. Full automation is a far-reaching concept that will probably not be implemented before 2040–2050, and even in such a situation, it will be necessary to keep humans in the loop so as to gain a wide societal acceptance of the concept. Starting from SESAR or Nextgen initial deployment and aiming towards this ultimate objective, a transition phase with human-system cooperation will take place. Since ATC controllers are used to a well-structured network of routes, it is advisable to post-process the 4D trajectories issued by automated systems in order to make them as close as possible to line segments connecting beacons. To perform this task, in an automatic way, flight paths will be deformed so as to minimize an entropy criterion that enforces avoidance of low density areas and at the same time penalizes length. Compared to already available bundling algorithms [3] that tend to move curves to high density areas, this new procedure generates geometrically-correct curves, without excess curvature.

Let a set $\gamma_1, \ldots, \gamma_N$ of smooth curves be given that will be aircraft flight paths for the air traffic application. It will be assumed in the sequel that all curves are smooth mappings from $[0,1]$ to a domain Ω of $\mathbb{R}^q$ with everywhere non-vanishing derivatives in $]0,1[$. This last condition allows one to view them as smooth immersions with boundaries and is sound from the application point of view, as aircraft velocities are bounded below by the efficiency consideration and ultimately by the stall and, therefore, cannot vanish expect at the endpoints. In air traffic applications, the dimension of the state space is generally two and sometimes three when the evolution of the aircraft in the vertical plane is of interest.

The approach taken in this work is first to get a sound definition of spatial density associated with a curve system, then to derive from it an entropy that will be minimized.

2.2. Spatial Density of a System of Curves

Due to the fact that aircraft positions are acquired through radar measurements, a trajectory is known only at discrete sampling times. In the operational context, the sampling period ranges from 4 to 10 s, which corresponds roughly to a 100–250-m traveling distance. Derived from that, a classical performance indicator used in ATM is the aircraft density [4], obtained from the sampled positions $\gamma_i(t_j)$, $j = 1, \ldots, n_i$ on each flight path γ_i, $i = 1, \ldots, N$. It is constructed from a partition U_k, $k = 1, \ldots, P$ of Ω by counting the number of samples occurring in a given U_k, then dividing out by the total number of samples $n = \sum_{i=1}^{N} n_i$. More formally, the density d_k in the subset U_k of Ω is expressed as:

$$d_k = n^{-1} \sum_{i=1}^{N} \sum_{j=1}^{n_i} 1_{U_k}\left(\gamma_i(t_j)\right) \quad (1)$$

with 1_{U_k} the characteristic function of the set U_k. It seems natural to extend the density obtained from samples to another one based on the trajectories themselves using an integral form:

$$d_k = \lambda^{-1} \sum_{i=1}^{N} \int_0^1 1_{U_k}(\gamma_i(t))\, dt \tag{2}$$

where the normalizing constant λ is chosen so that d_k is a discrete probability distribution:

$$\lambda = \sum_{k=1}^{P} \sum_{i=1}^{N} \int_0^1 1_{U_k}(\gamma_i(t))\, dt = \sum_{i=1}^{N} \int_0^1 \sum_{k=1}^{P} 1_{U_k}(\gamma_i(t))\, dt$$

and since $U_k, k = 1, \ldots, P$ is a partition:

$$\sum_{k=1}^{P} 1_{U_k}(\gamma_i(t)) = 1 \tag{3}$$

so that $\lambda = N$.

Density can be viewed as an empirical probability distribution with the U_k considered as bins in an histogram. It is thus natural to extend the above computation so as to give rise to a continuous distribution on Ω. For that purpose, local weighting techniques, such as kernel density estimation methods, are well known in nonparametric statistics, because they are a useful data-driven way to yield continuous density estimation. Many references may be found in the literature as in [5,6]. Given the observations, the resulting estimation will be the sum of weights taking into account the distance between the observations and the location x at which the density has to be estimated; the more an observation is close to x, the greater is the weighting. The weights are defined by selecting a summable function centered on the observations, called a kernel, usually denoted by $K\colon \mathbb{R} \to \mathbb{R}^+$ in the univariate case, and a smoothed version of the Parzen–Rosenblatt density estimator [7,8] is used. Standard choices for the K function are the ones used for nonparametric kernel estimation, like the Epanechnikov function [9]:

$$K\colon x \mapsto \left(1 - x^2\right) 1_{[-1,1]}(x).$$

There exists a large variety of kernel functions, and any density function satisfying the normalization condition can be considered, so that the estimation is a probability density. Moreover, the kernel function is a symmetric positive function, with the first moment equal to zero and a finite second order moment. In the multivariate case, a multivariate kernel function $\mathcal{K}\colon \mathbb{R}^q \to \mathbb{R}^+$ is selected that can be expressed by means of a real kernel K associated with a norm, denoted by $\|.\|$, in $\mathbb{R}^q$ as follows:

$$\mathcal{K}(x) = K(\|x\|), \qquad x \in \mathbb{R}^q.$$

The normalization condition becomes:

$$\int_{\mathbb{R}^q} \mathcal{K}(x)\, dx = \int_{\mathbb{R}^q} K(\|x\|)\, dx = 1.$$

A kernel version of the density is then defined as a mapping d from Ω to $[0,1]$:

$$d\colon x \mapsto \frac{\sum_{i=1}^{N} \int_0^1 K(\|x - \gamma_i(t)\|)\, dt}{\sum_{i=1}^{N} \int_\Omega \int_0^1 K(\|x - \gamma_i(t)\|)\, dtdx}. \tag{4}$$

Normalizing the kernel is not mandatory, as the normalization occurs with the definition of d. It is nevertheless easier to consider these kinds of kernels, as is done in nonparametric density estimation.

Note that when K is compactly supported, which is the case of the Epanechnikov function and all of its relatives, it becomes:

$$\int_{\Omega} K\left(\|x - \gamma_i(t)\|\right) dx = \int_{\mathbb{R}^q} K\left(\|x\|\right) dx$$

provided that Ω contains the set:

$$\{x \in \mathbb{R}^q, \inf_{i=1\ldots N, t\in[0,1]} \|x - \gamma_i(t)\| \leq A\}$$

where the interval $[-A, A]$ contains the support of K. The case of kernels with unbounded support, like Gaussian functions, may be dealt with provided $\Omega = \mathbb{R}^q$. In the application considered, only compactly-supported kernels are used, mainly to allow fast machine implementation of the density computation.

Using the polar coordinates (ρ, θ) and the rotation invariance of the integrand, the relation becomes:

$$Vol\left(\mathbb{S}^{q-1}\right) \int_{\mathbb{R}^+} K(\rho)\rho^{q-1} d\rho = 1$$

which yields a normalizing constant of $2/\pi$ for the Epanechnikov function in dimension two, instead of the usual $3/4$ in the real case. When the normalization condition is fulfilled, the expression of the density simplifies to:

$$d\colon x \mapsto N^{-1} \sum_{i=1}^{N} \int_0^1 K\left(\|x - \gamma_i(t)\|\right) dt. \tag{5}$$

The normalizing constant is the same as in (2).

As an example, one day of traffic over France is considered and pictured in Figure 1 with the corresponding density map, computed on an evenly-spaced grid with a normalized Epanechnikov kernel.

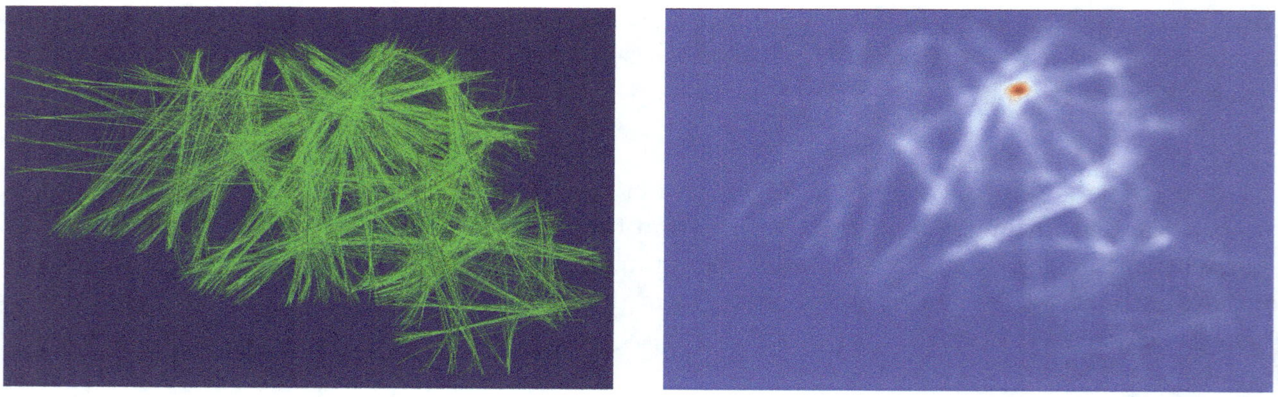

(a) (b)

Figure 1. (**a**) Traffic over France; (**b**) Associated density.

Unfortunately, density computed this way suffers a severe flaw for the ATM application: it is not related only to the shape of trajectories, but also to the time behavior. Formally, it is defined on the set $\text{Imm}\left([0,1], \mathbb{R}^q\right)$ of smooth immersions from $[0,1]$ to $\mathbb{R}^q$ while the space of primary interest will be the quotient by smooth diffeomorphisms of the interval $[0,1]$, $\text{Imm}\left([0,1], \mathbb{R}^q\right) / \text{Diff}([0,1])$. Invariance of the density under the action of $\text{Diff}([0,1])$ is obtained as in [10] by adding a term related to velocity in the integrals. The new definition of d becomes:

$$\tilde{d}\colon x \mapsto \frac{\sum_{i=1}^{N} \int_0^1 K\left(\|x - \gamma_i(t)\|\right) \|\gamma_i'(t)\| dt}{\sum_{i=1}^{N} \int_{\Omega} \int_0^1 K\left(\|x - \gamma_i(t)\|\right) \|\gamma_i'(t)\| dt dx}. \tag{6}$$

Assuming again a normalized kernel and letting l_i be the length of the curve γ_i, the expression of the density simplifies to:

$$\tilde{d}: x \mapsto \frac{\sum_{i=1}^{N} \int_0^1 K\left(\|x-\gamma_i(t)\|\right) \|\gamma_i'(t)\| dt}{\sum_{i=1}^{N} l_i}. \tag{7}$$

The new Diff-invariant density is pictured in Figure 2 along with the standard density. While the overall aspect of the plot is similar, one can observe that routes are more apparent in the right picture and that the density peak located above the Paris area is of less importance and less symmetric due to the fact that near airports, aircraft are slowing down, and this effect exaggerates the density with the non-invariant definition.

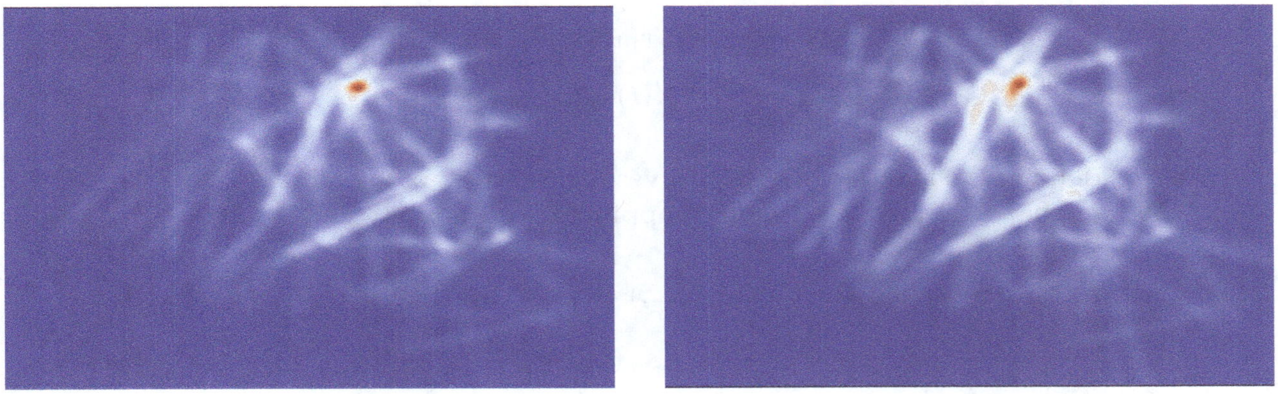

(a) (b)

Figure 2. Density (**a**) and Diff invariant density (**b**) for 12 February 2013 traffic.

The extension of the two-dimensional defined that way to the general case of curves in an arbitrary space $\mathbb{R}^q$ is straightforward.

2.3. Further Properties of the Density

In this section, the curves considered are a smooth mapping from the closed interval $[0,1]$ to $\mathbb{R}^q$, with a non-vanishing derivative in $]0,1[$. All multivariate kernels $\mathcal{K}$ will be assumed smooth, positive, with a unit integral and of the form $x \mapsto K(\|x\|)$. However, it is not required that they are compactly supported unless explicitly stated. All results are presented for the whole space $\mathbb{R}^q$, but apply almost verbatim to an open subset.

Definition 1. *Let f be a smooth summable mapping from $\mathbb{R}$ to $\mathbb{R}$. The scaling f_ν of f is defined, for each $\nu > 0$, to be the mapping:*

$$f_\nu : x \in \mathbb{R} \mapsto \frac{1}{\nu} f\left(\frac{x}{\nu}\right).$$

It is clear that the L^1-norm of the original mapping is preserved by the scaling. Given a summable kernel function K from $\mathbb{R}$ to $\mathbb{R}^+$, it defines a multivariate kernel $\mathcal{K}$ on $\mathbb{R}^q$ that maps x to $K(\|x\|)$. One may derive from it a parametrized family of kernels in $\mathbb{R}$ by mapping each ν in $]0,1]$ to the scaled kernel K_ν. If the original K is of unit integral, so are all of the K_ν.

Proposition 1. *Let $\gamma: [0,1] \to \mathbb{R}^q$ be a smooth path with a non-vanishing derivative in $]0,1[$. Let $K_\nu, \nu > 0$ be a parametrized family of unit integral kernels. The family of Borel measures μ_ν defined for any Borel set A by:*

$$\mu_\nu(A) = \int_A \int_0^1 K_\nu\left(\|x-\gamma(t)\|\right) \|\gamma'(t)\| dt dx$$

is tight and converges narrowly to the Lebesgue measure on $\gamma([0,1])$.

Proof. Let $\epsilon > 0$ be given. By the summability of K, there exists a positive real number r, such that:

$$\int_{\mathbb{R}^q - B(0,r)} K(\|x\|)dx < \epsilon$$

with $B(0,r)$ the open ball of radius r centered at the origin. Since $B(0,r) \subset B(0, r\nu^{-1})$ for $\nu > 0$, the same holds for all of the family K_ν. Let $B(0,M)$ be an open ball containing $\gamma([0,1])$. Then:

$$\begin{aligned} \mu_\nu\left(\mathbb{R}^q - \overline{B(0,M+r)}\right) &= \int_{\mathbb{R}^q - B(0,M+r)} \int_0^1 K_\nu\left(\|x-\gamma(t)\|\right) \|\gamma'(t)\| dt dx \\ &= \int_0^1 \int_{\mathbb{R}^q - B(0,M+r)} K_\nu\left(\|x-\gamma(t)\|\right) \|\gamma'(t)\| dx dt \\ &\leq \epsilon \int_0^1 \|\gamma'(t)\| dt = \epsilon\, l(\gamma) \end{aligned} \tag{8}$$

where $l(\gamma)$ denotes the length of γ. This proves the tightness of the family K_ν.
Let $f\colon \mathbb{R}^q \to \mathbb{R}$ be a bounded continuous mapping. It becomes:

$$\begin{aligned} I_\nu(f) = \int_{\mathbb{R}^q} \int_0^1 K_\nu\left(\|x-\gamma(t)\|\right) f(x) \|\gamma'(t)\| dt dx &= \int_0^1 \int_{\mathbb{R}^q} K_\nu\left(\|x-\gamma(t)\|\right) f(x) \|\gamma'(t)\| dx dt \\ &= \int_0^1 \int_{\mathbb{R}^q} K\left(\|x\|\right) f\left(x\nu + \gamma(t)\right) \|\gamma'(t)\| dx dt \end{aligned} \tag{9}$$

and since f is bounded, the dominated convergence theorem shows that:

$$\lim_{\nu \to 0} I_\nu(f) = \int_0^1 f\left(\gamma(t)\right) \|\gamma'(t)\| dt$$

proving the second part of the claim. □

The density in (7) is for a single curve of the form $d(x) = l(\gamma)^{-1} \int_0^1 K\left(\|x-\gamma(t)\|\right) \|\gamma'(t)\| dt$ with $l(\gamma)$ the length of the curve γ. It is invariant under the change of the parameter and can be written in a more concise way as:

$$\int_0^1 K\left(\|x-\gamma(\eta)\|\right) \| d\eta \tag{10}$$

where η is the arclength times $l(\gamma)^{-1}$.

This form allows a simple probabilistic interpretation of the density d: if a point u is drawn on the curve γ according to a uniform distribution and independently a vector v in $\mathbb{R}^q$ with a density $\mathcal{K}$ (the multivariate kernel corresponding to K), then the density of $x = u + v$ is given by Equation (10).

Proposition 2. *If the multivariate kernel $\mathcal{K}$ has a finite second moment, that is the univariate kernel K is such that:*

$$M = \int_{\mathbb{R}^+} r^{q+1} K(r)\, dr < +\infty$$

then the Wasserstein distance between the densities d_1, d_2 associated with smooth curves γ_1, γ_2 is bounded by:

$$2Vol(\mathbb{S}^{q-1})M + D(\gamma_1, \gamma_2)$$

with :

$$D(\gamma_1, \gamma_2) = \int_0^1 \|\gamma_1(\eta) - \gamma_2(\eta)\|^2 d\eta$$

where each curve is parametrized by the scaled arclength as in (10).

Proof. Let us consider the plan [11] given by the density:

$$d\colon (x,y) \mapsto \int_0^1 K(\|x-\gamma_1(\eta)\|)K(\|y-\gamma_2(\eta)\|)d\eta$$

where each curve is parametrized by the scaled arclength. The associated transport cost is given by:

$$C = \int_{\mathbb{R}^q \times \mathbb{R}^q} \|x-y\|^2 \int_0^1 K\left(\|x-\gamma_1(\eta)\|\right) K\left(\|y-\gamma_2(\eta)\|\right) d\eta dx dy$$

letting $u = y - x$ and using Fubini gives:

$$C = \int_0^1 \int_{\mathbb{R}^q} K\left(\|x-\gamma_1(\eta)\|\right) \int_{R^q} \|u\|^2 K\left(\|u+x-\gamma_2(\eta)\|\right) du dx d\eta.$$

The inner term can be written as:

$$\begin{aligned} \int_{R^q} \|u\|^2 K\left(\|u+x-\gamma_2(\eta)\|\right) du &= \int_{R^q} \|u+\gamma_2(\eta)-x\|^2 K\left(\|u\|\right) du \\ &= \int_{\mathbb{R}^q} \|u\|^2 K(\|u\|) du + 2\langle \gamma_2(\eta)-x, \int_{\mathbb{R}^q} uK(\|u\|)du\rangle + \|\gamma_2(\eta)-x\|^2. \end{aligned} \tag{11}$$

The integral:

$$\int_{\mathbb{R}^q} uK(\|u\|)du$$

is zero and, using spherical coordinates:

$$\int_{\mathbb{R}^q} \|u\|^2 K(\|u\|)du = \int_{\mathbb{R}^+} r^{q+1}K(r) \int_{\mathbb{S}^{q-1}} d\sigma\, dr = Vol(\mathbb{S}^{q-1})M$$

with $M = \int_{\mathbb{R}^+} r^{q+1}K(r)$. Putting back this value in the expression of the cost gives:

$$\begin{aligned} C &= Vol(\mathbb{S}^{q-1})M \int_0^1 \int_{\mathbb{R}^q} K\left(\|x-\gamma_1(\eta)\|\right) dx d\eta + \int_0^1 \int_{\mathbb{R}^q} K\left(\|x-\gamma_1(\eta)\|\right) \|\gamma_2(\eta)-x\|^2 dx d\eta \\ &= Vol(\mathbb{S}^{q-1})M + \int_0^1 \int_{\mathbb{R}^q} K\left(\|x-\gamma_1(\eta)\|\right) \|\gamma_2(\eta)-x\|^2 dx d\eta \\ &= Vol(\mathbb{S}^{q-1})M + \int_0^1 \int_{\mathbb{R}^q} K\left(\|x\|\right) \|\gamma_2(\eta)-\gamma_1(\eta)+x\|^2 dx d\eta. \end{aligned} \tag{12}$$

Finally:

$$\begin{aligned} \int_{\mathbb{R}^q} K\left(\|x\|\right) \|\gamma_2(\eta)-\gamma_1(\eta)+x\|^2 dx = \int_{\mathbb{R}^q} & K\left(\|x\|\right) \|\gamma_2(\eta)-\gamma_1(\eta)\|^2 dx d\eta \\ & + 2\langle \gamma_2(\eta)-\gamma_1(\eta), \int_{\mathbb{R}^q} xK\left(\|x\|\right) dx \\ & + Vol(\mathbb{S}^{q-1})M. \end{aligned} \tag{13}$$

As before, the middle term vanishes, and the first one integrates to:

$$\int_0^1 \|\gamma_2(\eta)-\gamma_1(\eta)\|^2 d\eta$$

so that:

$$C = 2Vol(\mathbb{S}^{q-1})M + \int_0^1 \|\gamma_2(\eta)-\gamma_1(\eta)\|^2 d\eta.$$

□

This result indicates that the densities associated with curves γ_1, γ_2 using the smoothing process described above cannot be too far (with respect to the Wasserstein distance) from each other if the geometric L^2 distance $D(\gamma_1, \gamma_2)$ is small. In fact, the upper bound in Proposition 2 can be interpreted as the cost of moving the smoothed density around γ_1 to the uniform distribution on the curve, then moving γ_1 to γ_2, keeping points with equal scaled arclength in correspondence, and finally, moving the uniform distribution on γ_2 to the smoothed density.

Having the density at hand, the entropy of the system of curves $\gamma_1, \ldots, \gamma_N$ is defined the usual way as:

$$E(\gamma_1, \ldots, \gamma_N) = -\int_\Omega \tilde{d}(x) \log\left(\tilde{d}(x)\right) dx.$$

The entropy is dependent on the particular choice of the kernel K. As mentioned before, it is a common practice in the field of non-parametric statistics to introduce a tuning parameter $\nu > 0$ in the kernel, called bandwidth, so that it is expressed as a scaled version $K = f_\nu$ of a given function $f \colon \mathbb{R}^+ \to \mathbb{R}^+$. The value of ν is the most influential parameter in the estimation of the density and must be selected carefully. For curve clustering applications, it is defined by the desired interaction length: if ν tends to zero, the curves will behave as independent objects, while on the other end of the scale, very high bandwidth will tend to remove the influence of the curves themselves. For the moment, no automated means of finding an optimal ν was used, although it will be part of a future work.

2.4. Minimizing the Entropy

In order to fulfill the initial requirement of finding bundles of curve segments as straight as possible, one seeks after the system of curves minimizing the entropy $E(\gamma_1, \ldots, \gamma_N)$, or equivalently maximizing:

$$\int_\Omega \tilde{d}(x) \log\left(\tilde{d}(x)\right) dx.$$

The reason why this criterion gives the expected behavior will become more apparent after derivation of its gradient at the end of this part. Nevertheless, when considering a single trajectory, it is intuitive that the most concentrated density distribution is obtained with a straight segment connecting the endpoints: this point will be made rigorous later.

Letting ϵ be a perturbation of the curve γ_j, such that $\epsilon(0) = \epsilon(1) = 0$, the first order expansion of $-E(\gamma_1, \ldots, \gamma_N)$ will be computed in order to get a maximizing displacement field, analogous to a gradient ascent (the choice has been made to maximize the opposite of the entropy, so that the algorithm will be a gradient ascent one) in the finite dimensional setting. The notation:

$$\frac{\partial F}{\partial \gamma_j}$$

will be used in the sequel to denote the derivative of a function F of the curve γ_j in the sense that for a perturbation ϵ:

$$F(\gamma_j + \epsilon) = F(\gamma_j) + \frac{\partial F}{\partial \gamma_j}(\epsilon) + o(\|\epsilon\|_2).$$

First of all, please note that since $\tilde{d}$ has integral one over the domain Ω:

$$\int_\Omega \frac{\partial \tilde{d}(x)}{\partial \gamma_j}(\epsilon) dx = 0$$

so that:

$$-\frac{\partial}{\partial \gamma_j} E(\gamma_1, \ldots, \gamma_N)(\epsilon) = \int_\Omega \frac{\partial \tilde{d}(x)}{\partial \gamma_j}(\epsilon) \log\left(\tilde{d}(x)\right) dx. \tag{14}$$

Starting from the expression of $\tilde{d}$ given in Equation (7), the first order expansion of $\tilde{d}$ with respect to the perturbation ϵ of γ_j is obtained as a sum of a term coming from the numerator:

$$\int_0^1 K\left(\|x - \gamma_j(t)\|\right) \|\gamma_j'(t)\| dt. \tag{15}$$

and a second one coming from the length of γ_j in the denominator. This last term is obtained from the usual first order variation formula of a curve length:

$$\int_{[0,1]} \left\|\gamma_j'(t) + \epsilon'(t)\right\| dt =$$
$$\int_{[0,1]} \left\|\gamma_j'(t)\right\| dt + \int_{[0,1]} \left\langle \frac{\gamma_j'(t)}{\|\gamma_j'(t)\|}, \epsilon'(t) \right\rangle dt + o(\|\epsilon\|_2).$$

Using an integration by parts, the first order term can be written as:

$$\begin{aligned} &\int_{[0,1]} \left\langle \frac{\gamma_j'(t)}{\|\gamma_j'(t)\|}, \epsilon'(t) \right\rangle dt = \\ &-\int_{[0,1]} \left\langle \left(\frac{\gamma_j''(t)}{\|\gamma_j'(t)\|}\right)_{\mathcal{N}}, \epsilon(t) \right\rangle dt \end{aligned} \tag{16}$$

with:

$$\left(\frac{\gamma_j''(t)}{\|\gamma_j'(t)\|}\right)_{\mathcal{N}} = \frac{\gamma_j''(t)}{\|\gamma_j'(t)\|} - \frac{\gamma_j'(t)}{\|\gamma_j'(t)\|} \left\langle \frac{\gamma_j'(t)}{\|\gamma_j'(t)\|}, \frac{\gamma_j''(t)}{\|\gamma_j'(t)\|} \right\rangle$$

the normal component of:

$$\frac{\gamma_j''(t)}{\|\gamma_j'(t)\|}.$$

Please note that when dealing with planar curves (i.e., with values in $\mathbb{R}^2$), it is $\kappa_j(t)N_j(t)$ with κ_j (resp. N_j) the curvature (resp. the unit normal vector) of γ_j.

The integral in (15) can be expanded in a similar fashion. Using as above the notation $()_{\mathcal{N}}$ for normal components, the first order term is obtained as:

$$\begin{aligned} &\int_{[0,1]} \left\langle \left(\frac{\gamma_j(t) - x}{\|\gamma_j(t) - x\|}\right)_{\mathcal{N}}, \epsilon(t) \right\rangle K'\left(\|\gamma_j(t) - x\|\right) \|\gamma_j'(t)\| dt \\ &-\int_{[0,1]} \left\langle \left(\frac{\gamma_j''(t)}{\|\gamma_j'(t)\|}\right)_{\mathcal{N}}, \epsilon(t) \right\rangle K\left(\|\gamma_j(t) - x\|\right) dt. \end{aligned} \tag{17}$$

From the expressions in (16) and (17), the first order variation of the entropy is:

$$\begin{aligned} &\frac{1}{\sum_{i=1}^N l_i} \left(\int_{[0,1]} \left\langle \int_\Omega \left(\frac{\gamma_j(t) - x}{\|\gamma_j(t) - x\|}\right)_{\mathcal{N}} K'\left(\|\gamma_j(t) - x\|\right) \log(\tilde{d}(x)) dx, \epsilon(t) \right\rangle \|\gamma_j'(t)\| dt \right. \\ &- \int_{[0,1]} \left(\int_\Omega K\left(\|\gamma_j(t) - x\|\right) \log(\tilde{d}(x)) dx \right) \left\langle \left(\frac{\gamma_j''(t)}{\|\gamma_j'(t)\|}\right)_{\mathcal{N}}, \epsilon(t) \right\rangle dt \\ &\left. + \left(\int_\Omega \tilde{d}(x) \log(\tilde{d}(x)) dx \right) \int_{[0,1]} \left\langle \left(\frac{\gamma_j''(t)}{\|\gamma_j'(t)\|}\right)_{\mathcal{N}}, \epsilon(t) \right\rangle dt \right). \end{aligned} \tag{18}$$

As expected, only moves normal to the trajectory will change at first order the value of the criterion: the displacement of the curve γ_j will thus be performed at t in the normal bundle to γ_j and is given, up to the $(\sum_{i=1}^{N} l_i)^{-1}$ term, by:

$$\begin{aligned}
&\int_\Omega \left(\frac{\gamma_j(t)-x}{\|\gamma_j(t)-x\|}\right)_{\mathcal{N}} K'\left(\|\gamma_j(t)-x\|\right)\log(\tilde{d}(x))dx\|\gamma_j'(t)\| \\
&-\left(\int_\Omega K\left(\|\gamma_j(t)-x\|\right)\log(\tilde{d}(x))dx\right)\left(\frac{\gamma_j''(t)}{\|\gamma_j'(t)\|}\right)_{\mathcal{N}} \\
&+\left(\int_\Omega \tilde{d}(x)\log(\tilde{d}(x))dx\right)\left(\frac{\gamma_j''(t)}{\|\gamma_j'(t)\|}\right)_{\mathcal{N}}.
\end{aligned} \tag{19}$$

The first term in the expression will favor moves towards areas of high density, while the second and third ones are moving along normal vector and will straighten the trajectory. This last point can be enlightened by considering the case of a single planar curve with fixed endpoints.

Proposition 3. *Let a, b be fixed points in $\mathbb{R}^2$ and K be a kernel as in (7). The segment $[a, b]$ is a critical point for the entropy associated with the curve system in $\mathbb{R}^2$ consisting of single smooth paths with endpoints a, b.*

Proof. Let the segment $[a, b]$ be parametrized as $\gamma\colon t \in [0,1] \mapsto a + tv$ with v the vector $(b-a)$. Starting with the expression (19), it is clear that the second and third terms occurring in the formula will vanish as the second derivative of γ is zero. Let u be the unit normal vector to γ. Any point x in $\mathbb{R}^2$ can be written as $x = a + \theta v + \xi u$, $\theta, \xi \in \mathbb{R}$, so that $\gamma(t) - x = (t-\theta)v - \xi u$ and $\|\gamma(t) - x\| = \sqrt{(t-\theta)^2\|b-a\|^2 + \xi^2}$. The change of variables $x \to (\theta, \xi)$ has Jacobian $\|v\| = \|b-a\|$. For a fixed $t \in [0,1]$, it becomes:

$$\begin{aligned}
&\int_{\mathbb{R}^2} \left(\frac{\gamma(t)-x}{\|\gamma(t)-x\|}\right)_{\mathcal{N}} K'\left(\|\gamma(t)-x\|\right)\log(\tilde{d}(x))dx\|\gamma'(t)\| = \\
&\|b-a\|^2 \int_{\mathbb{R}}\int_{\mathbb{R}} \frac{-\xi}{\sqrt{(t-\theta)^2\|b-a\|^2+\xi^2}} K'\left(\sqrt{(t-\theta)^2\|b-a\|^2+\xi^2}\right)\log(\tilde{d}(\theta,\xi))d\xi d\theta.
\end{aligned} \tag{20}$$

The density $\tilde{d}$ for the γ curve is expressed in ξ, θ coordinates as:

$$\int_{[0,1]} K\left(\sqrt{(t-\theta)^2\|b-a\|^2+\xi^2}\right) dt$$

and is an even function in ξ. The same is true for $K'\left(\|\gamma(t)-x\|\right)$. Finally, the mapping:

$$\xi \mapsto \frac{-\xi}{\sqrt{(t-\theta)^2\|b-a\|^2+\xi^2}}$$

is odd for a fixed θ, so that the whole integrand is odd as a function of ξ. By the Fubini theorem, integrating first in ξ will therefore yield a vanishing integral, proving the assertion. □

The result still holds in $\mathbb{R}^q$, the only different aspect being that x is now expanded as $x = a + \theta v + \sum_{i=1}^{q-1} \xi_i u_i$ with $u_i, i = 1, \ldots, q-1$ an orthonormal basis of the orthogonal complement of $\mathbb{R}v$ in $\mathbb{R}^q$. Rewriting $\gamma(t) - x = (t-\theta)v - \sum_{i=1}^{q-1} \xi_i u_i$ and $\|\gamma(t) - x\| = \sqrt{(t-\theta)^2\|b-a\|^2 + \sum_{i=1}^{q-1} \xi_i^2}$, the same parity argument can be applied on any of the components $\xi_i, i = 1, \ldots, q-1$, showing that the integral is vanishing.

The effect of curve straightening is present when minimizing the entropy of a whole curve system, but is counterbalanced by the gathering effect. Depending on the choice of the kernel bandwidth, one or the other effect is dominant: straightening is preeminent for low values, being the only remaining effect in the limit, while gathering dominates at high bandwidths. For the air traffic application, a rule of the thumb is to take 2–3-times the separation norm as an effective support for the kernel. Using an adaptive bandwidth may be of some interest also: starting with medium to high values favors curve gathering; then, gradually reducing it will straighten the trajectories.

Using the scaled arclength in the entropy gives an equivalent, but somewhat easier to interpret result. Starting with the expression (7) that takes in this case the form:

$$\tilde{d}\colon x \mapsto \frac{\sum_{i=1}^{N} l_i \int_0^1 K\left(\|x - \gamma_i(\eta)\|\right) d\eta}{\sum_{i=1}^{N} l_i}. \tag{21}$$

Let $i \in \{1, \ldots, N\}$ be fixed. An admissible variation of the curve γ_i is a smooth mapping from $]-a, a[\times[0,1]$ to $\mathbb{R}^q$, with $a > 0$ satisfying the following properties:

(a) $\forall \eta \in [0,1], \phi(0,\eta) = \gamma_i(\eta)$.
(b) $\forall (t,\eta) \in]-a,a[\times]0,1[, \|\partial_\eta \phi(t,\eta)\| = l_\phi(t)$ with $l_\phi(t)$ the length of the curve $\eta \mapsto \phi(t,\eta)$.
(c) $\forall t \in]-a,a[, \phi(t,0) = \gamma_i(0)$, $\phi(t,1) = \gamma_i(1)$.

Taking the derivative with respect to tat zero of Equation (b) yields:

$$\langle \partial_t \partial_\eta \phi(0,\eta), \partial_\eta \phi(0,\eta) \rangle = \partial_t l_\phi(0) l_i.$$

Letting $T(\eta)$ be the unit tangent vector to γ_i at η and noting that $\partial_\eta \phi(0,\eta) = l_i T(\eta)$, it becomes:

$$\langle \partial_t \partial_\eta \phi(0,\eta), T(\eta) \rangle = \partial_t l_\phi(t). \tag{22}$$

This relation puts a constraint on the variation of the tangential component of the curve derivative and shows that it has to be constant in η.

Proposition 4. *Let D be the mapping from $]-a,a[\times\mathbb{R}^q$ to $\mathbb{R}^+$ defined by:*

$$D\colon (t,x) \mapsto \frac{\sum_{j=1, j\neq i}^{N} l_j \int_0^1 K\left(\|x - \gamma_j(\eta)\|\right) d\eta + \int_0^1 K\left(\|x - \phi(t,\eta)\|\right) d\eta}{\sum_{j=1}^{N} l_j}.$$

where η refers collectively to the scaled arclength parameter for each curve. The partial derivative $\partial_t D(0,x)$ is given by:

$$\partial_t D(0,x) = \frac{l_i}{\sum_{j=1}^{N} l_j} \int_0^1 \left\langle \frac{\gamma_i(\eta) - x}{\|\gamma_i(\eta) - x\|}, \partial_t \phi(0,\eta) \right\rangle K'\left(\|\gamma_i(\eta) - x\|\right) d\eta.$$

The proof is straightforward and is omitted. From Proposition 4, the variation of the entropy is derived:

$$\partial_t E = -\int_{\mathbb{R}^q} \frac{l_i}{\sum_{j=1}^{N} l_j} \int_0^1 \left\langle \frac{\gamma_i(\eta) - x}{\|\gamma_i(\eta) - x\|}, \partial_t \phi(0,\eta) \right\rangle K'\left(\|\gamma_i(\eta) - x\|\right) d\eta dx. \tag{23}$$

This relation is equivalent to (18): it can be seen by splitting the terms into a normal and a tangential component. The first one yields:

$$-\int_{\mathbb{R}^q} \frac{l_i}{\sum_{j=1}^{N} l_j} \int_0^1 \left\langle \left(\frac{\gamma_i(\eta) - x}{\|\gamma_i(\eta) - x\|} \right)_{\mathcal{N}}, (\partial_t \phi(0,\eta))_{\mathcal{N}} \right\rangle K'\left(\|\gamma_i(\eta) - x\|\right) d\eta dx.$$

For the tangential part, the starting point is the relation:

$$\begin{aligned}\partial_\eta \left(K\left(\|\phi(0,\eta)-x\|\right)T(\eta)\right) =& l_i K'\left(\|\phi(0,\eta)-x\|\right)\left\langle \frac{\phi(0,\eta)-x}{\|\phi(0,\eta)-x\|}, T(\eta)\right\rangle T(\eta) \\ &+ K\left(\|\phi(0,\eta)-x\|\right)\partial_\eta T(\eta).\end{aligned} \tag{24}$$

where the subscript $\mathcal{T}$ stands for tangential component. It becomes:

$$\begin{aligned}&\frac{l_i}{\sum_{j=1}^N l_j}\int_0^1 \left\langle \left(\frac{\gamma_i(\eta)-x}{\|\gamma_i(\eta)-x\|}\right)_{\mathcal{T}}, \left(\partial_t\phi(0,\eta)\right)_{\mathcal{T}}\right\rangle K'\left(\|\gamma_i(\eta)-x\|\right)d\eta dx = \\ &\frac{l_i}{\sum_{j=1}^N l_j}\int_0^1 \left\langle \partial_\eta\left(K\left(\|\phi(0,\eta)-x\|\right)T(\eta)\right), \left(\phi(0,\eta)\right)_{\mathcal{T}}\right\rangle d\eta dx \\ &-\frac{l_i}{\sum_{j=1}^N l_j}\int_0^1 \left\langle K\left(\|\phi(0,\eta)-x\|\right)\partial_\eta T(\eta), \left(\phi(0,\eta)\right)_{\mathcal{T}}\right\rangle d\eta dx.\end{aligned} \tag{25}$$

With an integration by parts, the first integral in the right-hand side becomes:

$$\begin{aligned}&-\frac{l_i}{\sum_{j=1}^N l_j}\int_0^1 \left\langle K\left(\|\phi(0,\eta)-x\|\right)T(\eta), \partial_\eta\left(\partial_t\phi(0,\eta)\right)_{\mathcal{T}}\right\rangle d\eta dx = \\ &-\frac{l_i}{\sum_{j=1}^N l_j}\int_0^1 K\left(\|\phi(0,\eta)-x\|\right)\partial_t l_\phi(0) d\eta dx.\end{aligned} \tag{26}$$

Gathering terms, the expression (18) is recovered. As expected, only the normal components enter the relation, but it has to be noted that the tangential component of $\partial_t\phi(0,\eta)$ is not arbitrary and can be deduced from (22). The gradient with respect to the i-th curve is obtained from the expression of the entropy variation and can be written in its simplest form as:

$$\frac{l_i}{\sum_{j=1}^N l_j}\int_{\mathbb{R}^q}\int_0^1 \frac{\gamma_i(\eta)-x}{\|\gamma_i(\eta)-x\|}K'\left(\|\gamma_i(\eta)-x\|\right)d\eta \log\tilde{d}(x)dx. \tag{27}$$

where $\tilde{d}$ is the estimated spatial density. One must keep in mind the constraint on $\partial_t\phi(0,\eta)$ that is hidden within the apparent simplicity of the expression.

3. Numerical Implementation

The two formulations (19) and (27) of the gradient may be used. The first one is more complicated, but does not require any additional constraint to be taken into account. The second one cannot be applied readily as the tangential component must comply with Relation (22). In both cases, it is needed to evaluate a spatial integral, which may yield to prohibitive computational time, especially in high dimensions. In the air traffic application, only planar 3D curves are considered, greatly simplifying the problem. Nevertheless, the performance of the algorithms is still a concern, and the choice made was to replace the spatial integral by a discrete sum over an evenly-spaced grid. From now, it is assumed that all curves are planar, so that the ambient space for the spatial density $\tilde{d}$ is $\mathbb{R}^2$. Going back to the expression of $\tilde{d}$ given by (7), a first step is to replace the integral over t by a discrete sum. In practice, curves are described by a sequence of sampled points $\gamma_i(t_{ij})$ where the sampling times t_{ij} will be assumed to be identical for all curves. This assumption is not satisfied in the air traffic application, so that a pre-processing step must be taken before the actual entropy minimization stage. It will not be described here, as any standard interpolation procedure can be applied with negligible differences on the final result. To obtain the results presented here, a cubic spline smoother was used. Since the sampling times are assumed to be the same for all trajectories, the double subscript will be dropped, so that the samples on each trajectory will be denoted as $\gamma_{ij} = \gamma_i(t_j)$. It is further assumed that the derivative $\gamma'_{ij} = \gamma'_i(t_i)$ is available, most of the time through a numerical approximation. Given a quadrature formula on $[0,1]$ with points $t_j, j = 1, \ldots, m$ and weights $w_j, j = 1, \ldots, m$, the density may be approximated at $x \in \mathbb{R}^2$ by:

$$\tilde{d}(x) = \frac{1}{\sum_{i=1}^{N} l_i} \sum_{i=1}^{N} \sum_{j=1}^{m} w_j K\left(\|x - \gamma_{ij}\|\right) \|\gamma'_{ij}\|. \tag{28}$$

where the lengths $l_i, i = 1, \ldots, N$ are also obtained with the same quadrature rule:

$$l_i = \sum_{j=1}^{m} w_j \|\gamma'_{ij}\|.$$

When γ'_{ij} is computed in a numerical way, it may be expressed as:

$$\gamma'_{ij} = \sum_{k=1}^{m} \tilde{w}_{jk} \gamma_{i,k}.$$

where the weights $\tilde{w}_{jk}$ are often obtained through the application of the Lagrange interpolation formula to ensure exactness on polynomials up to a given degree. In a more compact form, it can be written in matrix form as:

$$\begin{pmatrix} \gamma'_{i1} \\ \vdots \\ \gamma'_{iq} \end{pmatrix} = \tilde{W} \begin{pmatrix} \gamma_{i1} \\ \vdots \\ \gamma_{iq} \end{pmatrix}$$

where the matrix $\tilde{W}$ has as entries the weights $\tilde{w}_{jk}$. The cost of evaluating $\tilde{d}$ at a single point is in $o(Nm)$, with the kernel evaluation being dominant. When dealing with points in $\mathbb{R}^2$ or $\mathbb{R}^3$ and compactly-supported kernels, a simple trick greatly reduces the time needed to compute $\tilde{d}$. First of all, the domain of interest is discretized on an evenly-spaced grid, so that the points of evaluation of the density $\tilde{d}$ are its vertices $x_{ij}, i = 1, \ldots, n_x, j = 1, \ldots, n_y$. The grid step δ_x (resp. δ_y) in the first (resp. second) coordinate is the difference between any two adjacent vertices $\delta_x = x_{i+1,j} - x_{i,j}$ (resp. $\delta_y = x_{i,j+1} - x_{i,j}$ (most of the time, $\delta_x = \delta_y$). Since the expression (28) is linear, the computation can be performed by accumulating values $K(\|x_{kl} - \gamma_{ij}\|)\|\gamma'_{ij}\|$ for a fixed couple (i,j), where only the points x_{kl} close enough to γ_{ij} are considered. In fact, the evaluation can be written as a 2D discrete convolution:

$$\tilde{d}(x_{kl}) = \sum_{i=1,\ldots,N, j=1,\ldots,m} w_j K(\|x_{kl} - \gamma_{ij}\|)\|\gamma'_{ij}\|. \tag{29}$$

When the support of K is small compared to the overall spatial domain, much computation is saved using this procedure. Furthermore, it can be thought of as 2D filtering, so that highly efficient algorithms coming from the field of image processing can be applied: in particular, computing the density on a graphics processing unit (GPU) is straightforward and allows one to decrease the computational time by at least a factor of ten. When dealing with the scaled arclength, the derivative term is not present, and a factor of l_i appears in from of the integral. The discrete version becomes:

$$\tilde{d}(x_{kl}) = \sum_{i=1,\ldots,N, j=1,\ldots,m} l_i w_j K(\|x_{kl} - \gamma_{ij}\|) \tag{30}$$

where $\gamma_{ij} = \gamma_i(\eta_j)$, η_j being in correspondence with t_j. Please note that the quadrature weights must be adapted to the abscissa $\eta_j, j = 1, \ldots, m$ and not to the $t_j, j = 1, \ldots, m$. Therefore, it is advisable to resample the curves so that the points $\eta_j, j = 1, \ldots, m$ are, for example, evenly spaced or of the Gauss–Lobatto form. The former was chosen for the experiments due to its ease of implementation, although the second form is probably more efficient from a numerical point of view and will be investigated in a second stage.

Having the density at hand, the gradient of the entropy with respect to the points $\gamma_{ij}, i = 1, \ldots, N, j = 1, \ldots, m$ can be easily computed using a straightforward application of the formula (19). When dealing with planar curves, a simplification occurs for the second derivative term since for a smooth curve γ_j:

$$\left(\frac{\gamma_j''(t)}{\|\gamma_j'(t)\|} \right)_{\mathcal{N}} = \kappa(t) N(t).$$

where κ is the curvature and N the unit normal vector. These quantities may be computed using numerical differentiation, but a coarse approximation based on the rotation rate of the vectors $\gamma_{i,j+1} - \gamma_{i,j}, \gamma_{i,j+2} - \gamma_{i,j+1}$ works well in many cases.

The case of scaled arclength parametrization needs some extra attention, due to the condition on the tangential component. The simplest approach is to move the points γ_{ij} according to an unconstrained gradient, then to re-sample the obtained curve so as to get adjusted γ_{ij} that correspond to the abscissa $\eta_j, j = 1, \ldots, m$.

In a numerical implementation, the scaling factor in front of the whole expression may be dropped due to the fact that all gradient-based algorithms will use an automatically-tuned step length. As usual with gradient algorithms, one must carefully select the step taking in the maximizing direction in order to avoid divergence. A simple fixed step strategy was first applied and gives satisfactory results on small datasets. A safer approach is to adapt the step size so as to ensure a sufficient decrease of the entropy. Due to the potentially huge dimension of the search space, this procedure has to be simple enough. An approximate quadratic search [12] was used in the final implementation.

The procedure applied to one day of traffic over France yields the picture of Figure 3. As expected, a route-like network emerges. In such a case, since the traffic comes from an already organized situation, the recovered network is indeed a subset of the route network in the french airspace. Please note that there is a trade-off between the density concentration and the minimal curvature of the recovered trajectories, as already mentioned. The kernel bandwidth was chosen empirically in the example presented, with the aid of visual interaction.

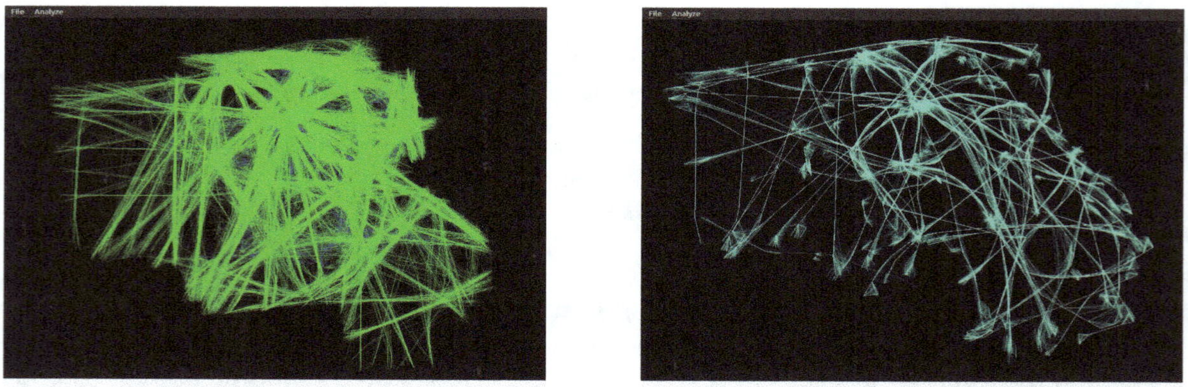

(a) (b)

Figure 3. Traffic of 24 February 2013: (**a**) Initial traffic; (**b**) Bundled traffic.

In the second example of Figure 4, the problem of automatic conflict solving is addressed. In the initial situation, aircraft are converging to a single point, which is unsafe. Air traffic controllers will proceed in such a case by diverting aircraft from their initial flight path so as to avoid each other, but only using very simple maneuvers. An automated tool will make full use of the available airspace, and the resulting set of trajectories may fail to be manageable by a human: in the event of a system failure, no backup can be provided by controllers. The entropy minimization procedure was added to an automated conflict solver in order to end up with flight paths still tractable by humans. The final result is shown in the right part of Figure 4, where encounters no longer exists, but aircraft are bound to simple trajectories, with a merging and a splitting point. Note that since the automated planner acts on velocity, all aircraft are separated in time on the inner part.

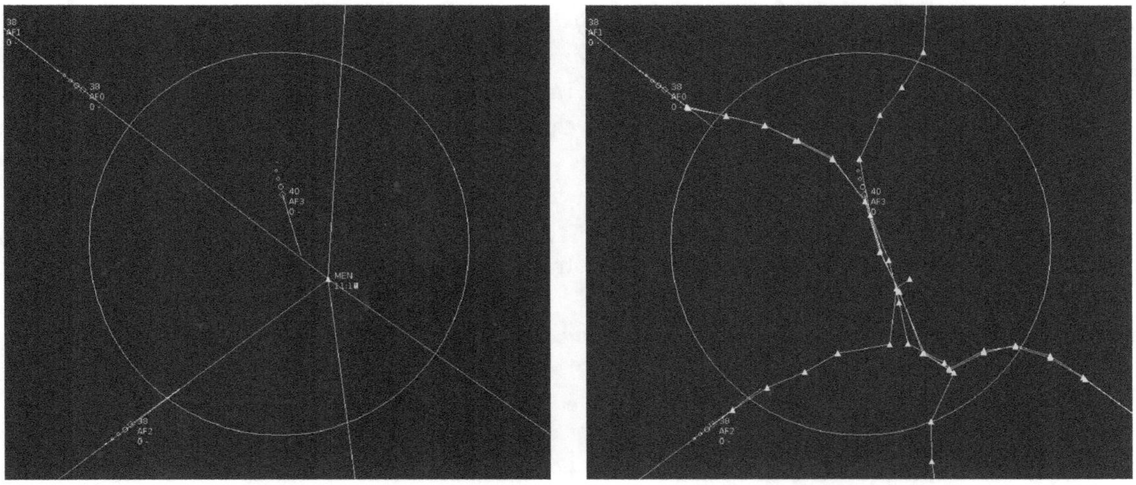

(**a**) (**b**)

Figure 4. (**a**) Initial flight plans; (**b**)Final flight plans.

4. Conclusions and Future Work

Algorithms coming from the field of shape spaces emerge as a valuable tool for applications in ATM. In this work, the foundations of a post-processing procedure that may be applied after an automated flight path planner are presented. Entropy minimization makes straight segment bundles emerge, which fulfills the operational requirements. Computational efficiency has to be improved in order to release a usable building block for future ATM systems. One way to address this issue is to compute kernel density estimators using GPUs, which excel in this kind of task, very similar to texture manipulations. Furthermore, statistical properties, such as the optimal choice of the bandwidth parameter in the kernel estimation, should be explored in more detail in the next step of this work.

Another important point that must be addressed in future works deals with the flight paths that are very similar in shape, but are oriented in opposite directions. As the spatial density is not sensitive to the directional information, the entropy-based procedure presented in this paper will tend to aggregate flight paths that should be sufficiently separated in order to prevent hazardous encounters. In [13], a notion of density based on position and velocity is developed. This work relies on Lie group modeling as a unifying state representation that takes into account the direction and the position of the curves. The curve system entropy has been extended to this setting.

Author Contributions: Stéphane Puechmorel has conceived the theoretical aspects of this work, as well as to the conception and design of the experiments; Florence Nicol has contributed to review the theoretical tools. Both authors have contributed to analyze the data and to write the paper. Both authors have read and approved the final manuscript.

References

1. De Bondt, A.; Leleu, C. *7-Year IFR Flight Movements and Service Units Forecast Update: 2014–2020*; EUROCONTROL: Brussels, Belgium, 2014.
2. Roussos, G.P.; Dimarogonas, D.V.; Kyriakopoulos, K.J. Distributed 3D navigation and collision avoidance for nonholonomic aircraft-like vehicles. In Proceedings of the 2009 European Control Conference, Budapest, Hungary, 23–26 August 2009.

3. Hurter, C.; Ersoy, O.; Telea, A. Smooth bundling of large streaming and sequence graphs. In Proceedings of the 6th PacificVis, Sydney, Australia, 26 February–1 March, 2013; pp. 41–48.
4. Harman, W.H. *Air Traffic Density and Distribution Measurements;* No. ATC-80; Lincoln Laboratory: Lexington, MA, USA, 3 May 1979.
5. Scott, D.W. *Multivariate Density Estimation: Theory, Practice, and Visualization*; Wiley: New York, NY, USA, 1992.
6. Silverman, B.W. *Density Estimation for Statistics and Data Analysis*; CRC: Boca Raton, FL, USA, 1986.
7. Parzen, E. On estimation of a probability density function and mode. *Ann. Math. Stat.* **1962**, *33*, 1065–1076.
8. Rosenblatt, M. Remarks on some nonparametric estimates of a density function. *Ann. Math. Statist.* **1956**, *27*, 832–837.
9. Epanechnikov, V.A. Non-parametric estimation of a multivariate probability density. *Theory Probab. Appl.* **1969**, *14*, 153–158.
10. Michor, P.W.; Mumford, D. Riemannian geometries on spaces of plane curves. *J. Eur. Math. Soc.* **2006**, *8*, 1–48, arXiv:math/0312384.
11. Ambrosio, L.; Gigli, N.; Savaré, G. *Gradient Flows: In Metric Spaces and in the Space of Probability Measures*; Springer: Basel, Switzerland, 2005.
12. Sun, W.; Yuan, Y.X. *Optimization Theory and Methods: Nonlinear Programming;* Springer: New York, NY, USA, 2006.
13. Nicol, F.; Puechmorel, S. Unsupervised aircraft trajectories clustering: A minimum entropy approach. In Proceedings of the Second International Conference on Big Data, Small Data, Linked Data and Open Data, Lisbon, Portugal, 21–25 February 2016.

7

Foliations-Webs-Hessian Geometry-Information Geometry-Entropy and Cohomology †

Michel Nguiffo Boyom

ALEXANDER GROTHENDIECK INSTITUTE, IMAG-UMR CNRS 5149-c.c.051, University of Montpellier, PL. E. Bataillon, F-34095 Montpellier, France; boyom@math.univ-montp2.fr.
† IN MEMORIAM OF ALEXANDER GROTHENDIECK. THE MAN.

Academic Editors: Frédéric Barbaresco and Frank Nielsen

Abstract: Let us begin by considering two book titles: A provocative title, *What Is a Statistical Model? McCullagh (2002)* and an alternative title, *In a Search for Structure. The Fisher Information. Gromov (2012).* It is the richness in open problems and the links with other research domains that make a research topic exciting. Information geometry has both properties. Differential information geometry is the differential geometry of statistical models. The topology of information is the topology of statistical models. This highlights the importance of both questions raised by Peter McCullagh and Misha Gromov. The title of this paper looks like a list of key words. However, the aim is to emphasize the links between those topics. The theory of homology of Koszul-Vinberg algebroids and their modules (KV homology in short) is a useful key for exploring those links. In Part A we overview three constructions of the KV homology. The first construction is based on the pioneering brute formula of the coboundary operator. The second construction is based on the theory of semi-simplicial objects. The third construction is based on the anomaly functions of abstract algebras and their abstract modules. We use the KV homology for investigating links between differential information geometry and differential topology. For instance, "dualistic relation of Amari" and "Riemannian or symplectic Foliations"; "Koszul geometry" and "linearization of webs"; "KV homology" and "complexity of models". Regarding the complexity of a model, the challenge is to measure how far from being an exponential family is a given model. In Part A we deal with the classical theory of models. Part B is devoted to answering both questions raised by McCullagh and B Gromov. A few criticisms and examples are used to support our criticisms and to motivate a new approach. In a given category an outstanding challenge is to find an invariant which encodes the points of a moduli space. In Part B we face four challenges. (1) The introduction of a new theory of statistical models. This re-establishment must answer both questions of McCullagh and Gromov; (2) The search for an characteristic invariant which encodes the points of the moduli space of isomorphism class of models; (3) The introduction of the theory of homological statistical models. This is a pioneering notion. We address its links with Hessian geometry; (4) We emphasize the links between the classical theory of models, the new theory and Vanishing Theorems in the theory of homological statistical models. Subsequently, the differential information geometry has a homological nature. That is another notable feature of our approach. This paper is dedicated to our friend and colleague Alexander Grothendieck.

Keywords: KV cohomoloy; functor of Amari; Riemannian foliation; symplectic foliation; entropy flow; moduli space of statistical models; homological statistical models; geometry of Koszul; localization; vanishing theorem

MSC: 55R10; 55N20; 62B05; 55C12; 55C07

1. Introduction

1.1. The Notation

Throughout the paper we use tha following notation. N is the set of non negative integers, Z is the ring of integers, R is the field of real numbers, $C^{\infty}(M)$ is the associative commutative algebra of real valued smooth functions in a smooth manifold M. Let ∇ be a Koszul connection in a manifold M, R^{∇} is the curvature tensor of ∇. It is defined by

$$R^{\nabla}(X,Y) = \nabla_X \nabla_Y - \nabla_Y \nabla_X - \nabla_{[X,Y]}.$$

T^{∇} is the torsion tensor of ∇. It is defined by

$$T^{\nabla}(X,Y) = \nabla_X Y - \nabla_Y X - [X,Y].$$

Let X be a smooth vector field in M. $L_X\nabla$ is the Lie derivative of ∇ in the direction $X \cdot \iota(X)R^{\nabla}$ is the inner product by X. To a pair of Koszul connections (∇, ∇^*) we assign three differential operators. They are denoted by $D^{\nabla\nabla^*}$, D^{∇} and D_{∇}.

(A.1) $D^{\nabla\nabla^*}$ is a first order differential operator. It is defined in the vector bundle $Hom(TM,TM)$. Its values belong to the vector bundle $Hom(TM^{\otimes 2}, TM)$.

(A.2) D^{∇} and D_{∇} are 2nd order differential operators. They are defined in the vector bundle TM. Their values belong to the vector bundle $Hom(TM^{\otimes 2}, TM)$. Let X be a section of TM and let ψ be a section of $T^*M \otimes TM$. The differential operators just mentioned are defined by

$$D^{\nabla\nabla^*}(\psi) = \nabla^* \circ \psi - \psi \circ \nabla, \tag{1a}$$

$$D^{\nabla}(X) = L_X\nabla - \iota(X)R^{\nabla}, \tag{1b}$$

$$D_{\nabla}(X) = \nabla^2(X). \tag{1c}$$

Part A of this paper is partially devoted to the global analysis of the differential equation

$$FE(\nabla\nabla^*): \quad D^{\nabla\nabla^*}(\psi) = O.$$

The solutions to $FE(\nabla\nabla^*)$ are useful for addressing the links between the KV homology, the differential topology and the information geometry.

The purpose of a forthcoming paper is the study of the differential equations

$$FE^*(\nabla): \quad D^{\nabla}(X) = 0,$$

$$FE^{**}(\nabla): \quad D_{\nabla}(X) = 0.$$

In the Appendix A to this paper we overview the role played by the solutions to $FE^{**}(\nabla)$ in some still open problems.

1.2. Some Explicit Formulas

Let $x = (x_1, ..., x_m)$ be a system of local coordinate functions of M. In those coordinates the Christoffel symbols of both ∇ and ∇^* are denoted by $\Gamma_{ij:k}$ and $\Gamma^*_{ij:k}$ respectively. We use those coordinate functions for presenting an element $\psi \in \mathcal{M}(\nabla\nabla^*)$ as a matrix $[\psi_{ij}]$. Thus by setting $\partial i = \frac{\partial}{\partial x_i}$ one has

$$\nabla_{\partial i}\partial j = \Gamma_{ij:k}\partial k.$$

We focus on $FE(\nabla\nabla^*)$ and of $FE^{**}(\nabla)$. They are equivalent to the following system of partial differential equations

$$[S_{ij:k}]: \quad \frac{\partial\psi_{kj}}{\partial x_i} - \sum_{1\leq\ell\leq m}(\Gamma_{ij:\ell}\psi_{k\ell} - \Gamma^*_{i\ell:k}\psi_{\ell j}) = 0,$$

$$[\Theta^k_{ij}(X)]: \quad \frac{\partial^2 X^k}{\partial x_i \partial x_j} + \sum_{\alpha}[\Gamma^k_{i\alpha}\frac{\partial X^\alpha}{\partial x_j} + \Gamma^k_{j\alpha}\frac{\partial X^\alpha}{\partial x_i} - \Gamma^\alpha_{ij\frac{\partial X^k}{\partial x_\alpha}}] + \sum_{\alpha}[\frac{\partial \Gamma^k_{j\alpha}}{\partial x_i} + \sum_{\beta}[\Gamma^\beta_{j\alpha}\Gamma^k_{i\beta} - \Gamma^\beta_{ij}\Gamma^k_{\beta}\alpha]]X^\alpha = 0.$$

In Part A we address the links between the following topics DTO, HGE, IGE and ENT. Those topics are presented as vertices of a square whose centre is denoted by KVH.

(1) DTO stands for Differential TOpology. In DTO, FWE stands for Foliations and WEbs.
(2) HGE stands for Hessian GEometry. Its sources are the geometry of bounded domains, the topology of bounded domains, the analysis in bounded domains. Among the notable references are [1–3]. Hessian geometry has significant impacts on thermodynamics, see [4,5], About the impacts on other related topics the readers are referred to [6–12].
(3) IGE stands for Information GEometry. That is the geometry of statistical models. More generally its concern is the differential geometry of statistical manifolds. The range of the information geometry is large [13]. Currently, the interest in information geometry is increasing. This comes from the links with many major research domains [14–16]. We address some significant aspects of those links. Non-specialist readers are referred to some fundamental references such as [17,18]. See also [4,19–23]. The information geometry also provides a unifying approach to many problems in differential geometry, see [21,24,25]. The information geometry has a large scope of applications, e.g., physics, chemistry, biology and finance.
(4) ENT stands for ENTropy. The notion of entropy appears in many mathematical topics, in Physics, in thermodynamics and in mechanics. Recent interest in the entropy function arises from its topological nature [14]. In Part B we introduce the entropy flow of a pair of vector fields. The Fisher information is then defined as the Hessian of the entropy flow.
(5) KVH stands for KV Homology. The theory of KV homology was developed in [9]. The motivation was the conjecture of M. Gerstenhaber in the category of locally flat manifolds. In this paper we emphasize other notable roles played by the theory of KV homology. It is also useful for discussing a problem raised by John Milnor in [26].

The conjecture of Gerstenhaber is the following claim.

Every restricted theory of deformation generates its proper cohomology theory [27].

Loosely speaking, in a restricted theory of deformation one has the notion of both infinitesimal deformation and trivial deformation. The challenge is the search for a cochain complex admitting infinitesimal deformations as cocycles. In the present paper, KVH is useful for emphasizing the links between the vertices DTO, HGE, IGE and ENT. That is our reason for devoting a section to KVH.

Warning.

We propose to overview the structure of this paper. The readers are advised to read this paper as through it were a wander around the vertices of the square "DTO-HGE-IGE-ENT". Thus, depending on his interests and his concerns a reader could walk several times across the same vertex. For instance the information geometry appears in many sections, depending on the purpose and on the aims.

1.3. The content of the Paper

This paper is divided into Part A and Part B.

Part A: Sections 1–7.

Section 1 is the Introduction. Section 2 is devoted to algebroids, modules of algebroids and the theory of KV homology of the Koszul-Vinberg algebroids. To introduce the KV cohomology we have adopted three approaches. Each approach is based on its specific machinery. However, the readers will face three cochain complexes which are pairwise quasi isomorphic. The KV cohomology is present throughout this paper. At the end of Part B the reader will see that the theory of statistical models is but a vanishing theorem in the theory of KV hohomology. The first approach is based on the pioneering fundamental brute formula of the coboundary operator. Historically, the brute formula is the first to have been constructed [9].

This first approach is used in many sections of this paper. Regarding the theory of deformation of the Koszul Geometry, the KV cohomology is the solution to the conjecture of Gerstenhaber. In the theory of modules of KV algebroids the role played by the KV cohomology is practically SIMILAR to the role played by the Hocshild cohomology in the category of associative alggebroids and their modules. This last remark holds for the role played by the Chevalley-Eilenberg cohomology in the category of Lie algebroids and their modules. Nevertheless, our comparison fails in the theory of Extension of modules over algebroids. In both categories of extensions of modules over associative algebroids and Lie algebroids the moduli space of equivalence class is encoded by cohomology classes of degree one. In the category of extensions of KV modules the moduli space is encoded by a spectral sequence. That was a unexpected feature in [9]. The pioneering coboundary operator of Nijenhuis [28] may be derived from the total brute coboundary operator introduced in [29].

The second approach is based on the notion of simplecial objects.

The third approach is based on the theory of anomaly functions for abstract algebras and their abstract modules. The idea has emerged from recent correspondences with one of my former teachers. The KV anomaly function of a Koszul connection ∇ may be expressed in terms of the ∇-Hessian operators ∇^2, namely

$$KV_{\nabla}(X,Y,Z) =< \nabla^2(Z),(X,Y) > - < \nabla^2(Z),(Y,X) > .$$

This approach is a powerful for addressing the relationships between the global analysis, the differential topology and the information geometry. The approach by the anomaly functions suggests many conjectures. Among those conjectures is the following.

Conjecture. Every anomaly function of algebras and of modules yields a theory of cohomology of algebras and modules.

Section 3. This section is devoted to the theory of KV (co)homology of Koszul-Vinberg algebroids. We focus on cohomological data which are used in the paper.

Section 4. This section is devoted the KV algebroids which are defined by structures of locally flat manifold. The KV cohomology theory is used for highlighting the impacts on the differential topology of the information geometry and its methods. We make the most of some relationships between the KV cohomology and the global analysis of the differential equation $FE^*(\nabla\nabla^*)$. We also sketch the global analysis of the differential equation

$$FE^{**}(\nabla).$$

This leads to the function

$$\mathcal{LC} \ni \nabla \to r^b(\nabla) \in \mathbb{Z}.$$

We explain how to interpret r^b as a distance. (See the Appendix A to this paper). For instance, the function r^b gives rise to an numerical invariant $r^b(\mathbb{M})$ which measures how far from being an exponential family is a statistical model $\mathbb{M}$. This result is a significant contribution to the information geometry, see [18,22,24].

Section 5. We are interested in how interact the information geometry, the KV cohomology and the geometry and Koszul. In particular we relate the notion of hyperbolicity and vanishing theorems in the KV cohomology.

Section 6. This section is devoted to the homological version of the geometry of Koszul. Our approach involves the dualistic relation of Amari. The KV cohomology links the dualistic relation with the geometry of Koszul.

Section 7. In this section summarize the highlighting features of Part A.

Part B: Sections 8–14.

Section 8. This is the starting section of the second part B. This Part B is devoted to new insights in the theory of statistical models. On 2002 Peter McCullar raised the provocative question.

What Is a Statistical Model

Across the world (Australia, Canada, Europe, US) the $McCullagh^s$ paper became the object of many criticisms and questions by eminent theoretical and applied statisticians [30].

Part B is aimed at supplying some deficiencies in the current theory of statistical models. We address some criticisms which support the need of re-establishing the theory of statistical model for measurable sets. Those criticisms are used for highlight the lack of both Structure and Relations. Those criticisms also highlight the search of M. Gromov [15]. The need for structures and relations was the intuition of Peter McCullagh. Loosely speaking there is a lack of Intrinsic Geometry in the sense of Erlangen. Subsequently the lack of intrinsic geometry yields other things that are lacking. The problem of the moduli space of models is not studied, although this would be crucial for applied information geometry, and for applied statistics. That might be a key in reading some the controversy about [30].

Section 9. In this section we address the problem of moduli space of statistical models. The problem of moduli space in a category is a major question in Mathematic. It is generally a difficult problem that involves finding a characteristic invariant which encodes the point of the moduli space. Such an invariant is a crucial step toward the geometry and the topology of a moduli space. Among other needs, the problem of encoding the moduli space of models has motivated our need of a new approach, that is to say the need of a theory having nice mathematical structure and relations. In this Part B the problem of the moduli space is solved. To summarize the theorem describing the moduli spaces of statistical models we need the following notation.

A gauge structure in a manifold M is a pair (M, ∇) where ∇ is a Koszul connection in M. The category of gauge structures in M is denoted by $\mathcal{LC}(M)$. We are concerned with the vector bundle $T^{*\otimes 2}M$ of bi-linear forms in the tangent bundle TM. The sheaf of sections of $T^{*\otimes 2}M$ is denoted by $\mathcal{BL}(M)$.

The category of m-dimensional statistical models (to be defined) of a measurable (Ξ, Ω) is denoted by $\mathcal{GM}_m(\Xi, \Omega)$. The category of random functors

$$\mathcal{LC}(M) \times \Xi \to \mathcal{BL}(M)$$

is denoted by $\mathcal{F}(\mathcal{LC}, \mathcal{BL})(M)$. One of the interesting breakthrough in this Part.B is the following solution to the problem of moduli.

Theorem 1. *There exists a functor*

$$\mathcal{GM}_m(\Xi, \Omega) \ni \mathbb{M} \to q_{\mathbb{M}} \in \mathcal{BL}(M) \tag{2}$$

which determines a model $\mathbb{M}$ up to isomorphism.
Let p be the probability density of a model $\mathbb{M}$. The mathematical expectation of $q_{\mathbb{M}}(\nabla)$ is defined by

$$E(q_{\mathbb{M}}(\nabla)) = \int_{\Xi} p q_{\mathbb{M}}(\nabla). \tag{3}$$

The quantity $E(q_{\mathbb{M}})(\nabla)$ does not depend on the Koszul connection ∇. It is called the Fisher information of $\mathbb{M}$.

This theorem emphasizes the Search for structure [16].

Section 10. This section is devoted to introduce the category of homological statistical models. This may be interpreted as a variant of the topology of the information. Another approach is to be found in Baudot-Bennequin [31].

The current theory (as in [17]) is called the classical (or local) theory. This means that a statistical model as in [17,18] is derived from the localization of a homological model. Loosely speaking such a model expresses a local vanishing theorem in the theory of homological statistical models.

Section 11. This section is devoted to discussing the links between the geometry of Koszul and the theory of homological statistical models. Those investigations lead to this notable feature.

The Geometry of Koszul, the homological statistical models and the classical information geometry locally look alike.

Section 12. Through Section 9 the framework is the category of equivariant locally trivial fibration. This assumption is weakened in Section 12. We recall the relationships between the Cech cohomology and the theory of locally trivial fiber bundle. We extend the scope of applications of the methods of the information geometry. Those extensions produce some interesting results. Here is an instance.

Theorem 2. *Let M be an oriented compact real analytic manifold and let $C^\omega(M^2)$ be the space of real valued analytic functions in M^2. There exists a non trivial map of $C^\omega(M^2)$ in the family of (positive) stratified Riemannian foliation in M.*

Sections 13. This Section 13 is a variant of Section 7.

Section 14 is an appendix we have mentioned. It is devoted to overview a few new significant results. Those results are derived from the global analysis of the differential operators

$$\left\{D^\nabla, D_\nabla, \nabla \in \mathcal{LC}(M)\right\}.$$

The solutions to a few open problems are announced.

2. Algebroids, Moduls of Algebroids, Anomaly Functions

The purpose of this section is to introduce basic notions in the algebraic topology of locally flat manifolds.

2.1. The Algebroids and Modules

Given a smooth fiber bundle

$$B \to M$$

the set of smooth sections of B is denoted by $\Gamma(B)$.

Definition 1. *An (abstract) real algebra is a real vector space $\mathcal{A}$ endowed with a bilinear map*

$$\mathcal{A} \times \mathcal{A} \to \mathcal{A}$$

Definition 2. *An (abstract) real two-sided module of an (abstract) algebra $\mathcal{A}$ is a real vector space W with two bilinear mappings*

$$\mathcal{A} \times W \to W,$$

$$W \times \mathcal{A} \to W$$

Warning.

Here algebra means a multiplication $a \cdot b$ without any rule of calculations. So the product $a \cdot b \cdot c$ is meaningless.

Throughout this paper, the smooth manifolds we deal with are connected and paracompact. In a smooth manifold M all geometrical objects we are interested in are smooth as well.

The vector space of smooth vector fields in a manifold M is denoted by $\mathcal{X}(M)$. It is a left module of the associative commutative algebra $C^\infty(M)$.

Consider a real vector bundle

$$\mathcal{E} \to M.$$

The real vector space of sections of $\mathcal{E}$ is denoted by $\Gamma(\mathcal{E})$.

Definition 3. *A real algebroid over a smooth manifold M is a real vector bundle whose vector space of sections is a real algebra.*

So the vector space of sections of a real algebroid $\mathcal{E}$ is endowed with a $\mathbb{R}$-bilinear map

$$\Gamma(\mathcal{E}) \times \Gamma(\mathcal{E}) \ni (s, s^*) \to s \cdot s^* \in \Gamma(\mathcal{E})$$

To simplify the multiplication of two sections is denoted $s \cdot s*$.

Definition 4. *A two-sided module of an algebroid $\mathcal{E}$ is a vector bundle*

$$\mathcal{V} \to M$$

whose vector space of sections is a two-sided module of the algebra $\Gamma(\mathcal{E})$.

Let s be section $\mathcal{E}$ and let v be a section of $\mathcal{V}$. Both left action s on v and the right action of s on v are denoted by $s \cdot v$ and $v \cdot s$.

Definition 5. *An anchored vector bundle over M is a pair*

$$(\mathcal{E}, b)$$

formed by a real vector bundle $\mathcal{E}$ and a vector bundle homomorphism

$$\mathcal{E} \ni e \to b(e) \in TM.$$

The homomorphism b is called the anchor map.

2.2. Anomaly Functions of Algebroids and of Modules

Let $\mathcal{V}$ be a two-sided module of an algebroid $(\mathcal{E}, b)$.

Definition 6. *An anomaly function of an algebroid $\mathcal{E}$ is a 3-linear map $A_{\mathcal{E}}$ of $\Gamma(\mathcal{E})^3$ in $\Gamma(\mathcal{E})$ whose values $A_{\mathcal{E}}(s_1, s_2, s_3)$ belong to $span_{\mathbb{R}}[(s_i{\cdot}s_j){\cdot}s_k, s_i{\cdot}(s_j{\cdot}s_k); i, j, k \in [1, 2, 3]]$. An anomaly function of an $\mathcal{E}$-module $\mathcal{V}$ is a 3-linear map $A_{\mathcal{E}\mathcal{V}}$ of $\Gamma(\mathcal{E})^2 \times \Gamma(\mathcal{V})$ in $\Gamma(\mathcal{V})$ whose values $A_{\mathcal{E}\mathcal{V}}(s, s^*, v)$ belong to $span_{\mathbb{R}}[(s{\cdot}s^*){\cdot}v, s{\cdot}(s^*{\cdot}v) \forall s, s^* \in \Gamma(\mathcal{E}), \forall v \in \Gamma(\mathcal{V})]$.*

In this paper we are interested in some anomaly functions which have strong geometrical impacts. They are defined below.

Definition 7. *Let $\mathcal{E}$ be an algebroid and let $s, s^*, s^{*}* \in \Gamma(\mathcal{E})$.*
(1) The associator anomaly function of $\mathcal{E}$ is defined by

$$Ass(s,s^*,s^{**}) = (s \cdot s^*) \cdot s^{**} - s \cdot (s^* \cdot s^{**}).$$

(2) The Koszul-Vinberg anomaly function of $\mathcal{E}$ is defined by

$$KV(s,s^*,s^{**}) = Ass(s,s^*,s^{**}) - Ass(s^*,s,s^{**}).$$

(3) The Jacobi anomaly functions of $\mathcal{E}$ are defined by

$$J(s,s^*,s^{**}) = (s \cdot s^*) \cdot s^{**} + (s^* \cdot s^{**}) \cdot s + (s^{**} \cdot s) \cdot s^*.$$

Definition 8. *Let v be a section of a two-sided $\mathcal{E}$-module $\mathcal{V}$.*
(1) The associator anomaly function of a left module $\mathcal{V}$ is defined as

$$Ass(s,s^*,v) = (s \cdot s^*) \cdot v - s \cdot (s^* \cdot v).$$

(2) The KV anomaly functions of a two sided module $\mathcal{V}$ are defined as

$$KV(s,s^*,v) = Ass(s,s^*,v) - Ass(s^*,s,v),$$

$$KV(s,v,s^*) = (s \cdot v) \cdot s^* - s \cdot (v \cdot s^*) - (v \cdot s) \cdot s^* + v \cdot (s \cdot s^*).$$

Definition 9. *We keep the notation used above. Let s, s^* be sections of $\mathcal{E}$, let v be a section of $\mathcal{V}$ and $f \in C^\infty(M)$.*
(1) The Leibniz anomaly function of an anchored algebroid $\mathcal{E}$ is defined by

$$L(s,f,s^*) = s \cdot (fs^*) - df(b(s))s^* - fs \cdot s^*.$$

(2) The Leibniz anomaly function of the $\mathcal{E}$-module $\mathcal{V}$ is defined by

$$L(s,f,v) = s \cdot (fv) - df(b(s))v - fs \cdot v.$$

A category of algebroids and modules of algebroids is defined by its anomaly functions. The anomaly functions are also used for introducing theories of homology of algebroids.

Some categories of anchored algebroids play important roles in the differential geometry.

Definition 10. *(A1): A Lie algebroid is an anchored algebroid $(\mathcal{E}, b)$ satisfying the identities*

$$s \cdot s* = 0,$$

$$L(s,f,s*) = 0.$$

(B1): A KV algebroid is an anchored algebroid $(\mathcal{E}, b)$ satisfying the identities

$$KV(s,s*,s**) = 0,$$

$$L(s,f,s*) = 0.$$

(B2): A vector bundle $\mathcal{V}$ is a module of Lie algebroid $(\mathcal{E}, b)$ if it satisfies the identities

$$L(s,f,v) = 0,$$

$$(s \cdot s*) \cdot v - s \cdot (s * \cdot v) + s * \cdot (s \cdot v) = 0.$$

A vector bundle $\mathcal{V}$ is a two-sided KV module of a Koszul-Vinberg algebroid $(\mathcal{E}, b)$ if it satisfies the identities

$$L(s, f, v) = 0,$$

$$KV(s, s^*, v) = 0,$$
$$KV(s, v, s^*) = 0.$$

Warning.

Consider a vector V space as the trivial vector bundle

$$V \times O \to 0.$$

Then we get

$$\Gamma(V \times 0) = V.$$

Therefore an algebra is an anchored algebroid over a point; its anchor map of is the zero map. Therefore, the Leibniz anomaly of an algebra is nothing but the bilinearity of the multiplication. So the notion of KV algebra and KV module is clear.

3. The Theory of Cohomology of KV Algebroids and Their Modules

This section is devoted to the cohomology of KV algebroids and KV modules of KV algebroids. KV stands for Koszul-Vinberg. We shall introduce three approaches to the theory of KV cohomology. Each approach has its particular advantage. So, depending on the needs or on the concerns one or other approach may be convenient. The three approaches are called "Version brute formula", "Version semi simplicial objects", "Version anomaly functions". The same graded vector space is common to the three constructions. They differ in their coboundary operators. However, three constructions lead to cohomology complexes which are pairwise quasi isomorphic.

Each construction leads to two cochain complexes. Those complexes are called the KV complex and total KV complex. They are denoted by C^*_{KV} and C^*_τ. In final we obtain six cohomological complexes.

3.1. The Theory of KV Cohomology—Version the Brute Formula of the Coboundary Operator

The geometric framework is the category of real KV algebraoids and their two sided modules. However our machineries only make use of $\mathbb{R}$-multi-linear calculations in the vector spaces of sections of vector bundles. Without any damage we replace the categories of KV algebroids and modules of KV algebroids by the categories of KV algebras and abstract modules of KV algebras.

3.1.1. The Cochain Complex C_{KV}.

Let W be a two-sided module of a KV algebra $\mathcal{A}$.

Definition 11. *The vector subspace $J(W) \subset W$ is defined by*

$$(a \cdot b) \cdot w - a \cdot (b \cdot w) = 0 \quad \forall a, b \in \mathcal{A}$$

We consider the $\mathbb{Z}$-graded vector space

$$C_{KV}(\mathcal{A}, W) = \sum_q C^q_{KV}(\mathcal{A}, W).$$

The homogeneous vector sub-spaces are defined by

$$C^q_{KV}(\mathcal{A}, W) = 0 \quad \forall q < 0,$$
$$C^0_{KV}(\mathcal{A}, W) = J(W),$$
$$C^q_{KV}(\mathcal{A}, W) = Hom_{\mathbb{R}}(\mathcal{A}^{\otimes q}, W) \quad \forall q > 0.$$

Before pursuing we fix the following notation.
Let

$$\xi = a_1 \otimes ... \otimes a_{q+1} \in \mathcal{A}^{\otimes q+1}$$

and let $a \in \mathcal{A}$,

$$\partial_i \xi = a_1 \otimes ...\hat{a}_i... \otimes a_{q+1},$$
$$\partial^2_{i,k+1}\xi = \partial_i(\partial_{k+1}\xi),$$
$$a.\xi = \sum_1^{q+1} a_1 \otimes ...a_{j-1} \otimes a.a_j \otimes a_{j+1}...a_{q+1}.$$

We are going to define the coboundary operator

$$\delta_{KV} : C^q(\mathcal{A}, W) \to C^{q+1}(\mathcal{A}, W).$$

The coboundary operator is a linear map. It is defined by

$$[\delta_{KV}(w)](a) = -a \cdot w + w \cdot a \quad \forall w \in J(W), \tag{4a}$$

$$[\delta_{KV} f](\xi) = \sum_1^q (-1)^i [a_i \cdot f(\partial_i \xi) - f(a_i \cdot \partial_i \xi) + (f(\partial^2_{i,q+1}\xi \otimes a_i)) \cdot a_{q+1}] \forall f \in C^q_{KV}(\mathcal{A}, W),$$
$$\forall \xi \in \mathcal{A}^{\otimes q+1}. \tag{4b}$$

The operator δ_{KV} satisfies the identity

$$\delta^2_{KV} f = 0 \quad \forall f \in C_{KV}(\mathcal{A}, W).$$

Therefore the pair $(C^*_{KV}(\mathcal{A}, W), \delta_{KV})$ is a cochain complex. Its cohomology space is denoted by

$$H_{KV}(\mathcal{A}, W) = \sum_q H^q_{KV}(\mathcal{A}, W).$$

The conjecture of Gerstenhaber: Comments.

A KV algebra $\mathcal{A}$ is a two-sided module of itself. An infinitesimal deformations of $\mathcal{A}$ is a 1-cocycle of $C_{KV}(\mathcal{A}, \mathcal{A})$ [9]. By the conjecture of Gerstenhaber the cohomology complex $C_{KV}(\mathcal{A}, \mathcal{A})$ is generated by the theory of deformations in the category of KV algebras.

The theory of deformation of KV algebras is the algebraic version of the theory of deformation of locally flat manifolds [2]. Therefore, the complex $C_{KV}(\mathcal{A}, \dashv)$ is the solution to the conjecture of Muray Gerstenhaber in the category of locally flat manifolds [27].

Features.

(1) The 2nd cohomology space $H^2_{KV}(\mathcal{A}, \mathcal{A})$ is the space of non trivial deformations of $\mathcal{A}$.

The definition of KV algebra of a locally flat manifold will be given in the next section.
Following [2] every hyperbolic locally flat manifold has non trivial deformations. Thus, if $\mathcal{A}$ is the KV algebra of a hyperbolic locally flat manifold then

$$H^2_{KV}(\mathcal{A}, \mathcal{A}) \neq 0.$$

(2) Let W be a two-sided module of a KV algebra $\mathcal{A}$. We consider W as a trivial KV algebra, viz

$$w \cdot w* = 0 \quad \forall w, w* \in W.$$

Let $EXT_{KV}(\mathcal{A}, W)$ be the set of equivalence classes of short exact sequences of KV algebras

$$0 \to W \to \mathcal{B} \to \mathcal{A} \to 0.$$

An interpretation of the 2nd cohomology space of $C_{KV}(\mathcal{A}, W)$ is the identification

$$H^2_{KV}(\mathcal{A}, W) = EXT_{KV}(\mathcal{A}, W).$$

Let $W, W*$ be two-sided modules of $\mathcal{A}$. Let $EXT_{\mathcal{A}}(W*, W)$ be the set of equivalence classes of exact short sequences of two-sided $\mathcal{A}$-modules

$$0 \to W \to T \to W* \to 0.$$

In both the category of associative algebras and the category of Lie algebras we have

$$HH^1(A, Hom_{\mathbb{R}}(W*, W)) = EXT_A(W*, W),$$

$$H^1_{CE}(A, Hom_{\mathbb{R}}(W*, W)) = EXT_A(W*, W).$$

Here $HH(A, -)$ stands for Hochschild cohomology of an associative algebra A and $H_{CE}(A, -)$ stands for cohomology of Chevalley-Eilenberg of a Lie algebra A.

Unfortunately in the category of KV modules of KV algebras this interpretation of the first cohomology space fails. Loosely speaking in the category of KV algebras the set $H^1(\mathcal{A}, Hom(W*, W))$ is not canonically isomorphic to set $EXT_{\mathcal{A}}(W*, W)$ [9].

3.1.2. The Total Cochain Complex C_τ.

The purpose is the total complex

$$C_\tau(\mathcal{A}, W) = \sum_q C^q_\tau(\mathcal{A}, W).$$

Its homogeneous vector subspaces are defined by

$$C^q_\tau(\mathcal{A}, W) = 0 \quad \forall q < 0,$$

$$C^0_\tau(\mathcal{A}, W) = W,$$

$$C^q_\tau(\mathcal{A}, W) = Hom_{\mathbb{R}}(\mathcal{A}^{\otimes q}, W) \quad \forall q > 0.$$

The total coboundary operator is a linear map

$$C^q_\tau(\mathcal{A}, W) \to C^{q+1}_\tau(\mathcal{A}, W).$$

That operator is defined by

$$(1): \quad [\delta_\tau w](a) = -a \cdot w + wa \quad \forall (a, w) \in \mathcal{A} \times W,$$

$$(2): \quad [\delta_\tau f](\xi) = \sum_{1}^{q+1}(-1)^i[a_i \cdot f(\partial_i\xi) - f(a_i \cdot \partial_i\xi) + (f(\partial^2_{i,q+1}\xi \otimes a_i)) \cdot a_{q+1}] \quad \forall f \in C^q_\tau(\mathcal{A}, W).$$

The pair

$$(C^*_\tau(\mathcal{A}, W), \delta_\tau)$$

is a cochain complex, viz

$$\delta^2_\tau = 0.$$

The derived cohomology space is denoted by

$$H_\tau(\mathcal{A}, W) = \sum_q H^q_\tau(\mathcal{A}, W).$$

It is called the W-valued total KV cohomology of $\mathcal{A}$.

3.2. *The Theory of KV Cohomology—Version: the Semi-Simplicial Objects*

Let V be a two-sided module of a KV algebra $\mathcal{A}$. Our aim is the construction of semi simplicial $\mathcal{A}$-modules whose derived cochain complex is quasi isomorphic to the KV cochin complex $C_{KV}(\mathcal{A}, V)$.

3.2.1. Extension

We start by considering the vector space

$$\mathcal{B} = \mathcal{A} \oplus \mathbb{R}.$$

Its elements are denoted by $(s + \lambda)$. We endow $\mathcal{B}$ with the multiplication which is defined by

$$(s + \lambda) \cdot (s^* + \lambda^*) = s \cdot s^* + \lambda s^* + \lambda^* s + \lambda\lambda^*.$$

With the multiplication we just defined, $\mathcal{B}$ is a real KV algebra. In other words we have

$$KV(X_1, X_2, X_3) = 0.$$

Here

$$X_j = s_j + \lambda_j.$$

In the $\mathcal{A}$-module V we have a structure of left $\mathcal{B}$-module which is defined by

$$(s + \lambda)\cdot v = s\cdot v + \lambda v \quad \forall (s + \lambda) \in \mathcal{B}, \quad \forall v \in V.$$

3.2.2. Construction

Let $\tilde{\mathcal{B}}$ be the vector space spanned by $\mathcal{A} \times \mathbb{R}$. Its elements are finite linear combinations of $(s, \lambda), s \in \mathcal{A} \times \mathbb{R}$.

The tensor algebra of $\tilde{\mathcal{B}}$ is denoted by $T(\tilde{\mathcal{B}})$. It has a $\mathbb{Z}$-grading. its homogeneous vector sub-spaces are defined by

$$T_q(\tilde{\mathcal{B}}) = \tilde{\mathcal{B}}^{\otimes q}.$$

A monomial element is denoted by

$$\xi = x_1 \otimes x_2 \otimes ... \otimes x_q.$$

Here

$$x_j = (s_j, \lambda_j) \in \mathcal{A} \times \mathbb{R}.$$

The KV algebra $\mathcal{A}$ is a two-sided ideal of the KV algebra $\mathcal{B}$. Thereby, the vector space $\tilde{\mathcal{B}}$ is canonically a left module of $\mathcal{A}$.

We define the natural two-sided action of $\mathbb{R}$ in $\tilde{\mathcal{B}}$ by setting

$$\lambda \cdot (s^*, \lambda^*) = (\lambda s, \lambda\lambda^*),$$

$$(s^*, \lambda^*) \cdot \lambda = (\lambda s^*, \lambda^* \lambda).$$

Thereby every vector subspace $T_q(\tilde{\mathcal{B}})$ is a left KV module of $\mathcal{B}$. Here the left action of $\mathcal{B}$ in $T_q(\tilde{B})$ is defined

$$(s + \lambda) \cdot \xi = s \cdot \xi + \lambda \xi.$$

Before continuing we recall the (extended) action of $\mathcal{A}$ in tensor space $T_q(\tilde{\mathcal{B}})$,

$$s \cdot (x_1 \otimes x_2 \otimes ... \otimes x_q) = \sum_{j=1}^{q} x_1 \otimes x_2 ... \otimes s \cdot x_j \otimes ... \otimes x_q.$$

We recall a notation which has been used in the last subsections,

$$\partial_j \xi = x_1 \otimes x_2 \otimes ... \hat{x}_j ... \otimes x_q.$$

The symbol $\hat{x}_j$ means that x_j is missing. Let $1 \in \mathbb{R}$ be the unit element, then $\tilde{1}$ stands for $(0, 1) \in \tilde{\mathcal{B}}$. We are going to construct semi simplicial modules of $\mathcal{B}$.

3.2.3. Notation-Definitions

Implicitly we use set isomorphism

$$\tilde{\mathcal{B}} \ni x = (s, \lambda) \to X^* = s + \lambda \in \mathcal{B}.$$

Then $\forall \xi \in T_q(\tilde{\mathcal{B}})$ one has

$$\tilde{1}^* \cdot \xi = \xi \cdot$$

We go back to the $\mathbb{Z}$-graded $\mathcal{B}$-module

$$T_*(\tilde{\mathcal{B}}) = \sum_q T_q(\tilde{\mathcal{B}}).$$

Definition 12. *Let j, q be two positive integers with $j < q$, let*

$$\xi = x_1 \otimes x_2 ... \otimes x_q.$$

The linear maps

$$d_j : T_q(\tilde{\mathcal{B}}) \to T_{q-1}(\tilde{\mathcal{B}})$$

and

$$S_j : T_q(\tilde{\mathcal{B}}) \to T_{q+1}(\tilde{\mathcal{B}})$$

are defined by

$$d_j \xi = X_j^* \cdot \partial_j \xi,$$

$$S_j \xi = e_j(\tilde{1}) \xi$$

The right member of the last equality has the following meaning

$$e_j(\tilde{1})\xi = x_1 \otimes x_2 ... \otimes x_{j-1} \otimes \tilde{1} \otimes x_j ... \otimes x_q$$

Structure. *The maps d_j and S_j satisfy the following identities*

$$d_i d_j = d_{j-1} d_i \quad if \quad i \leq j, \tag{5a}$$

$$S_i S_j = S_{j+1} S_i \quad if \quad i < j, \tag{5b}$$

$$(S_{j-1} d_i - d_i S_j)(\xi) = e_{j-1}(x_i)\partial_i \xi \quad if \quad 1 < i < j, \tag{5c}$$

$$(d_{i+1} S_j - S_j d_i)(\xi) = e_j(x_i)\partial_i \xi \quad if \quad j+1 < i \leq q, \tag{5d}$$

$$d_i(S_i(\xi)) = \xi \quad if \quad i = j. \tag{5e}$$

Definition 13. *The system $\{T_q(\mathcal{B}), d_i, S_i\}$ is called the canonical semi simplicial module of $\mathcal{B}$.*

3.2.4. The KV Chain Complex

From the canonical simplicial $\mathcal{B}$-module we derive the chain complex $C_*(\mathcal{B})$. it has a $\mathbb{Z}$-grading which is defined by

$$C_q(\mathcal{B}) = 0 \quad if \quad q < 0, \tag{6a}$$

$$C_0(\mathcal{B}) = \mathcal{R}, \tag{6b}$$

$$C_q(\mathcal{B}) = T_q(\tilde{\mathcal{B}}) \qquad if \quad q > 0. \tag{6c}$$

Now one defines the (linear) boundary operator

$$d : C_q(\mathcal{B}) \to C_{q-1}(\mathcal{B})$$

by setting

$$d(C_0(\mathcal{B})) = 0,$$

$$d(C_1(\mathcal{B})) = 0,$$

$$d\xi = \sum_1^q (-1)^j d_j \xi \quad if \quad q > 1.$$

By the virtue of (5a) we have

$$d^2 = 0.$$

3.2.5. The V-Valued KV Homology

We keep the notation used in the preceding sub-subsection. So the vector spaces $\mathcal{A}$, $\mathcal{B}$ and V are the same as in the preceding subsubsection.

We consider the $\mathbb{Z}$-graded vector space

$$C_*(\mathcal{B}, V) = \oplus_q C_q(\mathcal{B}, V).$$

Its homogeneous sub-spaces are defined by

$$C_q(\mathcal{B}, V) = 0 \quad if \quad q < 0,$$

$$C_0(\mathcal{B}, V) = V,$$

$$C_q(\mathcal{B}, V) = T_q(\tilde{\mathcal{B}}) \otimes V \quad if \quad q > 0.$$

Every homogeneous vector subspace $C_q(\mathcal{B}, V)$ is a left module of the KV algebra $\mathcal{B}$. The left action is defined by

$$s \cdot (\xi \otimes v) = s \cdot \xi \otimes v + \xi \otimes s \cdot v.$$

Let j and q be two positive integers such that $j < q$.
Let $\xi = x_1 \otimes x_2 ... \otimes x_q$. To define the linear map

$$d_j : C_q(\mathcal{B}, V) \to C_{q-1}(\mathcal{B}, V)$$

we put

$$d_j(\xi \otimes v) = X_j^* \cdot (\partial_j \xi \otimes v).$$

Henceforth one defines the boundary operator

$$d : C_q(\mathcal{B}, V) \to C_{q-1}(\mathcal{B}, V)$$

by setting

$$d = \sum_1^q (-1)^j d_j.$$

So we obtain a chain complex whose homology space of degree q is denoted by $H_q(\mathcal{B}, V)$.

Definition 14. *The graded vector space*

$$H_*(\mathcal{B}, V) = \sum_q H_q(\mathcal{B}, V)$$

is called the total homology of $\mathcal{B}$ with coefficients in V.

3.2.6. Two Cochain Complexes

We are going to define two cochain complexes. They are denoted by $C_{KV}(\mathcal{B}, V)$ and by $C_\tau(\mathcal{B}, V)$ respectively.
We recall that the vector subspace $J(V) \subset V$ is defined by

$$(s \cdot s*) \cdot v - s \cdot (s * \cdot v) = 0 \quad \forall s \quad s* \in \mathcal{B}.$$

Let us set

$$C^0_{KV}(\mathcal{B}, V) = J(V),$$
$$C^0_\tau(\mathcal{B}, V) = V,$$
$$C^q(\mathcal{B}, V) = Hom_{\mathbb{R}}(T_q(\tilde{\mathcal{B}}) \quad \forall q \geq 1.$$

Let (j, q) be a pair of non negative integers such that $j < q$. We are going to define the linear map

$$d_j : C^q(\mathcal{B}, V) \to C^{q+1}(\mathcal{B}, V).$$

Given $f \in C_q(\mathcal{B}, V)$ and

$$\xi = x_1 \otimes ... \otimes x_{q+1}$$

we put

$$d_j f(\xi) = X_j^* \cdot f(\partial_j \xi) - f(d_j \xi).$$

The family of linear mappings d_j has property $S \cdot 1$, viz

$$d_j d_i = d_i d_{j-1} \quad \forall i, j \quad with \quad i < j.$$

We use these data for constructing two cochain complexes. They are denoted by (C^*_{KV}, d_{KV}) and by (C^*_τ, d_τ) respectively. The underlying graded vector spaces are defined by

$$C_{KV} = J(V) \oplus \sum_{q>0} C^q(\mathcal{B}, V),$$

$$C_\tau = V \oplus \sum_{q>0} C^q(\mathcal{B}, V).$$

Their coboundary operators are defined by

$$(d_{KV} v)(s) = -s \cdot v,$$

$$(d_\tau w)(s) = -sw,$$

$$d_{KV}(f) = \sum_1^q (-1)^j d_j(f) \quad if \quad q > 0,$$

$$d_\tau(f) = \sum_1^{q+1} (-1)^j d_j(f) \quad if \quad q > 0.$$

The simplicial formula (5a) yields the identities

$$d^2_{KV} = 0,$$

$$d^2_\tau = 0.$$

The cohomology space

$$H_{KV}(\mathcal{B}, V) = \sum_q H^q_{KV}(\mathcal{B}, V)$$

is called the V-valued KV cohomology of $\mathcal{B}$.

The cohomology space

$$H_\tau(\mathcal{B}, V) = \sum_q H^q_\tau(\mathcal{B}, V)$$

is called the V-valued total KV cohomology of $\mathcal{B}$.

The algebra $\mathcal{A}$ is a two-sided ideal of the KV algebra $\mathcal{B}$. Mutatis mutandis our construction gives the cohomology spaces $H_{KV}(\mathcal{A}, V)$ and $H_\tau(\mathcal{A}, V)$. They are called the V-valued KV cohomology and the V-valued total KV cohomology of $\mathcal{A}$.

Comments.

Though the spectral sequences are not the purpose of this paper we recall that the pair $(\mathcal{A} \subset \mathcal{B})$ gives rise to a spectral sequences E_r^{ij} [32–34]. The term E_0^{ij} is nothing other than $H_{KV}(\mathcal{A}, V)$ [29]. In other words one has

$$H^q_{KV}(\mathcal{A}, V) = \sum_{0 \le j \le q} E_0^{j, q-j}.$$

3.2.7. Residual Cohomology

Before pursuing we introduce the notion of residual cohomology. It will be used in the section be devoted the homological statistical models.

The machinery we are going to introduce is similar to the machinery of Eilenberg [35]. In particular we introduce the residual cohomology. Our construction leads to an exact cohomology sequence which links the residual cohomology with the equivariant cohomology. We restrict the attention to the category of left modules of KV algebroids. We keep our previous notation.

We recall that for every positive integer $q > 0$ the vector space $C^q(\mathcal{B}, V)$ is a left module of $\mathcal{B}$. The left action of $s \in \mathcal{B}$ is defined by

$$(s \cdot f)(\xi) = s \cdot f(\xi) - f(s \cdot \xi).$$

Definition 15. *A cochain $f \in C^q(\mathcal{B}, V)$ is called a left invariant cochain if*

$$s \cdot f = 0 \quad \forall s \in \mathcal{B} \quad \forall s.$$

A straightforward consequence of this definition is that a left invariant cochain is a cocycle of both C^*_{KV} and C^*_τ. The vector subspace of left invariant q-cochains of $\mathcal{B}$ is denoted by $H^q_e(\mathcal{B}, V)$. It is easy to see that

$$Z^q_\tau(\mathcal{B}, V) \cap Z^q_{KV}(\mathcal{B}, V) = H^q_e(\mathcal{B}, V),$$
$$Z^q_\tau(\mathcal{A}, V) \cap Z^q_{KV}(\mathcal{A}, V) = H^q_e(\mathcal{A}, V).$$

Definition 16. *A KV cochain of degree q whose coboundary is left invariant invariant is called a residual KV cocycles.*

(1) *The vector subspace of residual KV cocycles of degree q is denoted by Z^q_{KVres}.*
(2) *The vector subspace of residual coboundaries of degree q is defined by $B^q_{KVres} = H^q_e(\mathcal{B}, V) + d_{KV}(C^{q-1}_{KV}(\mathcal{B}, V))$. The residual KV cohomology space of degree q is the quotient vector space.*
(3) $H^q_{KVres}(\mathcal{B}, V) = \frac{Z^q_{KVres}}{B^q_{KVres}}$.
(4) *By replacing the KV complex by the total KV complex one defines the vector space of residual total cocycles $Z^q_{\tau res}$ and the space of residual total coboundaries $B^q_{\tau res}$. Therefore we get the residual total KV cohomology space*

$$H^q_{\tau res}(\mathcal{A}, V) = \frac{Z^q_{\tau res}}{B^q_{\tau,res}}$$

The definitions above lead to the cohomological exact sequences which is similar to those constructed by Eilenberg machinery [35]. We are going to pay a special attention to two cohomology exact sequences.

(1) At one side the operator d_{KV} yields a canonical linear map

$$H^q_{KVres}(\mathcal{B}, V) \to H^{q+1}_e(\mathcal{B}, V).$$

(2) At another side every KV cocycle is a residual cocycle and every KV coboundary is a residual coboundary as well. Then one has a canonical linear map

$$H^q_{KV}(\mathcal{B}, V) \to H^q_{KVres}(\mathcal{A}, V).$$

Those canonical linear mappings yield the following exact sequences

$$\to H^{q-1}_{KVres}(\mathcal{B}, V) \to H^q_e(\mathcal{B}, V) \to H^q_{KV}(\mathcal{B}, V) \to H^q_{KVres}(\mathcal{B}, V) \to$$

$$\to H^{q-1}_{\tau res}(\mathcal{B}, V) \to H^q_e(\mathcal{B}, V) \to H^q_\tau(\mathcal{B}, V) \to H^q_{\tau res}(\mathcal{B}, V) \to$$

Some Comments.

(c.1): *We replace the KV $\mathcal{B}$ by $\mathcal{A}$. Then we obtain the exact sequences*

$$\rightarrow H^{q-1}_{KVres}(\mathcal{A},V) \rightarrow H^q_e(\mathcal{A},V) \rightarrow H^q_{KV}(\mathcal{A},V) \rightarrow H^q_{KVres}(\mathcal{A},V) \rightarrow$$

$$\rightarrow H^{q-1}_{\tau res}(\mathcal{A},V) \rightarrow H^q_e(\mathcal{A},V) \rightarrow H^q_\tau(\mathcal{A},V) \rightarrow H^q_{\tau res}(\mathcal{A},V) \rightarrow$$

(c.2): *The KV cohomology difers from the total cohomology. Loosely speaking their intersecttion is the equivariant cohomology $H^*_e(\mathcal{B},V)$ their difference is the residual cohomology. The domain of their efficiency are different as well. Here are two illustrations.*

Example 1.

In the introduction we have stated a conjecture of M. Gerstenhaber, namely *Every Restricted Theory of Deformation Generates Its Proper Theory of Cohomology.*

From the viewpoint of this conjecture, the KV cohomology is the completion a long history [2,9,28]. Besides Koszul and Nijenhuis, other pioneering authors are Vinberg, Richardson, Gerstenhaber, Matsushima, Vey.

The challenge was the search for a theory of cohomology which might be generated by the theory of deformation of locally flat manifolds [8]. The expected theory is the now known KV theory of KV cohomolgy [9].

Example 2.

The total cohomology is close to both the pioneering Nijenhuis work [28,36]. In [29] we have constructed a spectral sequence which relates to [28,36].

From another viewpoint, the total KV cohomology is useful for exploring the relationships between the information geometry and the theory of Riemannian foliations. This purpose will be addressed in the next sections.

3.3. The Theory of KV Cohomology—Version the Anomaly Functions

This subsection is devoted to use the KV anomaly functions for introducing the theory of cohomology of KV algebroids and their modules.

This viewpoint leads to an unifying framework for introducing the theory of cohomology of abstract algebras and their abstract two-sided modules. Here are a few examples of cohomology theory which are based on the anomaly functions.

Example 1. The theory of Hochschild cohomology of associative algebras is based on the associator anomaly function.

Example 2. The theory of Chevalley-Eilenberg-Koszul cohomology of Lie algebras is based on the Jacobi anomaly function.

Example 3. The theory of cohomology of Leibniz algebras is based on the Jacobi anomaly function as well.

3.3.1. The General Challenge $CH(\mathbb{D})$

We consider data

$$\mathbb{D} = [(\mathcal{A}, A_{\mathcal{A}}), (V, A_{\mathcal{A}V}), Hom(T(\mathcal{A}), V)].$$

Here

(1) V is an (abstract) two sided module of an (abstract) algebra $\mathcal{A}$.

(2) $A_{\mathcal{A}}$ and $A_{\mathcal{A}V}$ are fixed anomaly functions of $\mathcal{A}$ and of V respectively.

(3) $Hom(T(\mathcal{A}), V)$ stands for the $\mathbb{Z}$-graded vector space

$$Hom(T(\mathcal{A}), V) = \oplus_q Hom_{\mathbb{R}}(\mathcal{A}^{\otimes q}, V).$$

Let $A_{\mathbb{D}}$ be the category of (abstract) algebras and (abstract) modules whose structures are defined by the pair $(A_{\mathcal{A}}, A_{\mathcal{AV}})$. So the rules of calculations in the category $\mathbb{A}$ are defined by the identities

$$A_{\mathcal{A}}(a, b, c) = 0,$$

$$A_{\mathcal{AV}}(a, b, v) = 0.$$

The challenge is the search of a particular family of linear maps

$$Hom(\mathcal{A}^{\otimes q}, V) \ni f \to d_q(f) \in Hom(\mathcal{A}^{\otimes q+1}, V).$$

Such a particular family d_q must satisfy a condition that we call the property Δ.

Property Δ

$\forall \xi = a_1 \otimes a_2 ... \otimes a_{q+2} \in \mathcal{A}^{q+2}$, $\forall f \in Hom(\mathcal{A}^{\otimes q}, V)$ the quantity $[d_{q+1}(d_q(f))](\xi)$ depends linearly on the values of the anomaly functions

$$\{A_{\mathcal{A}}(a_i, a_j, a_k), \quad A_{AV}(a_i, a_j, v)\}$$

Let us assume that a family d_q is a solution to $CH(\mathbb{D})$. Then the category $A_{\mathbb{D}}$ admits a theory of cohomology with coefficients in modules.

The next is devoted to this challenge in the category of KV algebras and KV modules. The geometry version is the category of KV algebroids and KV modules of KV algebroids.

3.3.2. Challenge $CH(\mathcal{D})$ for KV Algebras

Let W be a two-sided module of an abstract algebra $\mathcal{A}$. We assume that the following bilinear mappings are non trivial applications

$$\mathcal{A} \times W \ni (X, w) \to X \cdot w \in W,$$

$$W \times \mathcal{A} \ni (w, X) \to w \cdot X \in W.$$

Let $f \in Hom(\mathcal{A}^{\otimes q}, W)$. We consider a monomial $\xi \in \mathcal{A}^{\otimes q+1}$, so

$$\xi = X_1 \otimes ... \otimes X_{q+1} \in \mathcal{A}^{\otimes q+1}.$$

Our construction is divided into many STEPS.

Step 1.

Let $(i < j)$ be a pair of positive integers with $1 \leq i < j \leq q$. The linear the map

$$S_{[i,j]}(f) \in Hom(\mathcal{A}^{q+1}, V).$$

$S_{[ij]}$ is defined by

$$S_{[i,j]}(f)(X_1 \otimes ... \otimes X_{q+1}) = (-1)^j [X_j \cdot f(X_1 \otimes ... \otimes X_i \otimes ... \hat{X}_j \otimes X_{j+1} ... \otimes X_{q+1})$$
$$+ (f(X_1 \otimes ... \otimes X_i \otimes ... \hat{X}_j \otimes ... \hat{X_{q+1}} \otimes X_j) \cdot X_{q+1}$$
$$- \omega(f) f(X_1 \otimes ... \otimes X_j \cdot X_i \otimes ... \hat{X}_j \otimes ... \otimes X_{q+1})]$$

$$+(-1)^i[X_i\cdot f(X_1\otimes ...\hat{X}_i\otimes ...\otimes X_j\otimes .\otimes X_{q+1})$$
$$+(f(X_1\otimes ...\hat{X}_i\otimes ...\otimes X_j\otimes ...\hat{X_{q+1}}\otimes X_i))\cdot X_{q+1}$$
$$-\omega(f)f(X_1\otimes ...\hat{X}_i\otimes ...\otimes X_i\cdot X_j\otimes ...\otimes X_{q+1})].$$

In the right side member of $S_{[i,j]}(f)(\xi)$ the coefficient $\omega(f)$ is the degree of f, viz $\omega(f)=q$ for all $f\in Hom(\mathcal{A}^q,W)$.

Step 2.

For every pair $(i,q+1)$ with $1\le i\le q$ we define the map $S_{[i,q+1]}(f)$ by

$$S_{[i,q+1]}(f)(X_1\otimes ...\otimes X_{q+1})=(-1)^i[X_i\cdot f(X_1\otimes ...\hat{X}_i\otimes ...\otimes X_{q+1})$$
$$+(f(X_1\otimes ...\hat{X}_i\otimes ...\hat{X_{q+1}}\otimes X_i))\cdot X_{q+1}$$
$$-\omega(f)f(X_1\otimes ...\hat{X}_i\otimes ...X_i\cdot X_{q+1})].$$

Step 3.

Let $g\in Hom(\mathcal{A}^{\otimes q+1},W)$ and let

$$\xi=X_1\otimes ...\otimes X_{q+2}\in \mathcal{A}^{\otimes q+2}.$$

Let i,j,k be three positive integers such that $i<j<q+2;\quad k\le q+2$. We have already introduced the notation

$$\partial_k\xi=X_1\otimes ...\hat{X}_k\otimes ...\otimes X_{q+2},$$
$$\partial^2_{k,q+2}\xi=X_1\otimes ...\hat{X}_k\otimes ...\otimes ...\hat{X}_{q+2}.$$

We define $S^k_{[i,j]}(g)\in Hom(\mathcal{A}^{\otimes q+2},W)$ by setting

$$S^k_{[i,j]}(g)(\xi)=(-1)^{i+k}[X_k\cdot g(\partial_k\xi)+(g(\partial^2_{k,q+2}\xi\otimes X_k))\cdot X_{q+2}$$
$$+\omega(g)g(X_1\otimes ...\otimes X_k\cdot X_i\otimes ...\hat{X}_k\otimes ...\otimes X_{q+2})]$$
$$+(-1)^{j+k}[X_k\cdot g(\partial_k\xi)+(g(\partial^2_{k,q+2}\xi)\otimes X_k)\cdot X_{q+2}$$
$$+\omega(g)g(X_1\otimes ...\otimes X_k\cdot X_j\otimes ...\hat{X}_k\otimes ...\otimes X_{q+2})].$$

Given a triple (i,j,k) with $i<j<k<q+2$ we put

$$S_{[i,j,k]}(g)(\xi)=S^k_{[i,j]}(g)(\xi)+S^j_{[i,k]}(g)(\xi)+S^i_{[j,k]}(g)(\xi).$$

The proof of the following statement is based on direct calculations.

Lemma 1.

$$(****):\quad \sum_{[i<j]}S_{[i,j]}(g)(\xi)=\sum_{[i<j<k]}S_{[i,j,k]}(g)(\xi)$$

Let $f\in Hom(\mathcal{A}^q,W)$. In both the left side and the right side of the equality $(****)$ we replace g by $\sum_{i<j}S_{[i,j]}(f)$. Then we obtain a linear mapping

$$Hom(\mathcal{A}^{II},W)\ni f\rightarrow E^{****}(f)\in Hom(\mathcal{A}W)^{+2},$$

Our aim is to evaluate ξ of $E^{****}(f)$ at $\xi \in \mathcal{A}^{\otimes q+2}$. Here

$$\xi = X_1 \otimes ... \otimes X_{q+2}.$$

To calculate $[E^{****}(f)](\xi)$ we take into account both STEP1 and STEP2. Then we obtain

$$[E^{****}(f)](\xi) = \sum_{[i<j<q+2;1\leq k\leq q+2]} [E^{****}_{[ijk]}(f)](\xi).$$

At the right side member

$$[E^{****}_{[ijk]}(f)](\xi) = (-1)^{i+j}[KV(X_i, X_j, f(X_1 \otimes ...\hat{X}_i \otimes ..\hat{X}_j \otimes ...X_k \otimes ... \otimes X_{q+2}))$$
$$+KV(X_i, f(X_1 \otimes ...\hat{X}_i...\hat{X}_j \otimes ... \otimes X_{q+1} \otimes X_j), X_{q+2})$$
$$+KV(X_j, f(X_1 \otimes ...\hat{X}_i...\hat{X}_j \otimes ... \otimes X_{q+1} \otimes X_i), X_{q+2})$$
$$+\omega(f)(\omega(f)+1)f(X_1 \otimes ...\hat{X}_i...\hat{X}_j \otimes ... \otimes KV(X_i, X_j, X_k) \otimes ... \otimes X_{q+2})].$$

Step 4.

We are in position to face $CH(\mathbb{D})$.

Definition 17. *Let $f \in Hom(\mathcal{A}^{\otimes q}, W)$ and $\xi = X_1 \otimes ... \otimes X_{q+1} \in \mathcal{A}^{\otimes q+1}$. We take into account Step 1, Step 2 and Step 3. Therefore, we define the linear map*

$$Hom(\mathcal{A}^{\otimes q}, W) \ni f \rightarrow \partial f \in Hom(\mathcal{A}^{\otimes q+1}, W)$$

by putting

$$[\partial f](\xi) = \sum_{1\leq i<j\leq q+1} S_{[i,j]}(f)(\xi)$$

The following lemma is a straightforward consequence of the machinery in STEP3.

Lemma 2.

$$\partial^2 f(\xi) = \sum_{[i<j<q+2];1\leq k\leq q+2} [E^{****}_{[ijk]}(f)](\xi)$$

Lemma 2 tells us that $\partial^2 f(\xi)$ depends linearly on the values of the KV anomaly functions.

The challenge $CH(\mathbb{D})$ is won in the category of KV algebras and their two-sided KV modules.

We replace the category of KV algebras and their two-sided modules by the category of KV algebroids and their bi-modules. Then we win the geometry version of $CH(\mathbb{D}$.

We use Lemma 2 for introducing a theory of KV homology of KV algebras and their two-sided modules.

3.3.3. The KV Cohomology

Let W be a two sided KV module of a KV algebra $\mathcal{A}$. We consider the graded vector space

$$\mathcal{C}_{KV} = \oplus_q C^q_{KV}.$$

The homogeneous subspaces are defined by $C^q_{KV} = 0$ if q is a negative integer, $C^0_{KV} = J(W)$, $\mathcal{C}^q_{KV} = Hom(\mathcal{A}^{\otimes q}, W)$ if q is a positive integer.

We define the linear map

$$C^q_{KV} \ni f \rightarrow \partial_{KV} f \in C^q + 1_{KV}$$

by setting

$$\partial_{KV}(w)(X) = -X \cdot w + w \cdot X \quad if \quad w \in J(W), \tag{7a}$$

$$\partial_{KV} f = \sum_{[i<j]} S_{[i,j]}(f) \quad if \quad q > 0. \tag{7b}$$

By Lemma 2 we obtain the following statement

Theorem 3. *For every two sided KV module W of a KV algebra $\mathcal{A}$ the pair $(C^*_{KV}, \partial_{KV})$ is a cochain complex.*

3.3.4. The Total Cohomology

Let W be a two-sided module of a KV algebra $\mathcal{A}$. Our concern is the $\mathbb{Z}$-graded vector space

$$C_\tau = W + \oplus_{q>0} C^q(\mathcal{A}, W).$$

For our present purpose the maps S_{ij} are not subject the requirement as in Step 2.
We define the coboundary operator ∂_τ by setting

$$\partial_\tau w(X) = -X \cdot w + w \cdot X \quad \forall w in W,$$

$$\partial_\tau f(\xi) = \sum_{1 \le i < j \le q+1} S_{[i,j]}(f)(\xi) \quad \forall q > 0.$$

The quantity $(\partial^2_\tau f(\xi)$ depends linearly on the KV anomaly functions of the pair $(\mathcal{A}, W)$. Thus the pair $(C^*_\tau, \partial_\tau)$ is a cochain complex. Its cohomology is called the W-valued total KV cohomology of $\mathcal{A}$. We denote it by $H^*_\tau(\mathcal{A}, W)$.

3.3.5. The Residual Cohomology, Some Exact Sequences, Related Topics, DTO-HEG-IGE-ENT

In the next sections we will see that the links between the information geometry and the differential topology involve the real valued total KV cohomology of KV algebroids. Many relevant relationships are based on the exact sequences

$$\to H^{q-1}_{KVres}(\mathcal{A}, \mathbb{R}) \to H^q_e(\mathcal{A}, \mathbb{R}) \to H^q_{KV}(\mathcal{A}, \mathbb{R}) \to H^q_{KVres}(\mathcal{A}, \mathbb{R}) \to$$

$$\to H^{q-1}_{\tau res}(\mathcal{A}, \mathbb{R}) \to H^q_e(\mathcal{A}, \mathbb{R}) \to H^q_\tau(\mathcal{A}, \mathbb{R}) \to H^q_{\tau res}(\mathcal{A}, \mathbb{R}) \to$$

Now we are provided with cohomological tools which will be used in the next sections.
We plan to perform KV cohomological methods for studying some links between the vertices of the square "DTO, IGE, ENT HGE" as in Figure 1. We recall basic notions.

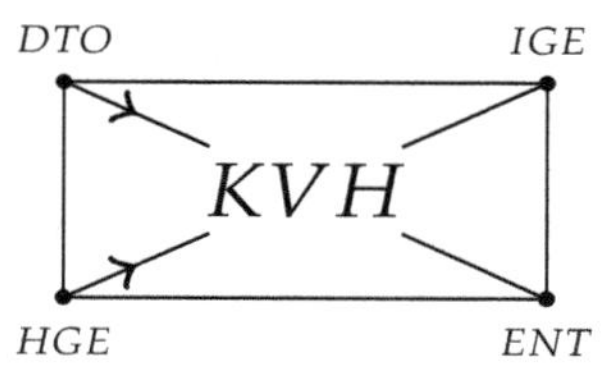

Figure 1. Federation.

DTO stands for Differential TOpology.

The purposes: Riemannian foliations and Riemannian webs. Symplectic foliations and symplectic webs. Linearization of webs.

Our aims: We use cohomological methods for constructing Riemannian foliations, Riemannian webs, linearizable webs.

Nowadays, there does not exist any criterion for deciding whether a manifold supports those differential topological objects. Our aim is to discuss sufficient conditions for a manifold admitting those structures. Our approach leads to notable results. The key tools are the KV cohomology and the dualistic relation of Amari. Both the KV cohomology and the dualistic relation product remarkable split exact sequences. Notable results are based on those exact sequences. *HGE stands for Hessian GEometry.*

The purposes: Hessian structures, geometry of Koszul, hyperbolicity, cohomological vanishing theorems. Our aims: The geometry of Koszul is a cohomological vanishing theorem. Statistical geometry and vanishing theorem, the solution to a hold question of Alexander K Guts (announced).

Theorem 3 as in [2] may be rephrased in the framework of the theory of KV homology. For a compact locally flat manifold (M, ∇) being hyperbolic it is necessary and sufficient that $C^2_{KV}(\mathcal{A}, C^\infty(M))$ contains a positive definite EXACT cocycle. To be hyperbolic is a geometrical-topological property of the developing map of locally flat manifolds. To be hyperbolicitic means that the image of the developing is a convex domain not containing any straight line. This formulation is far from being a homological statement. So the Hessian GEOmetry is a link between the theory of KV homology and the Riemannian Riemannian geometry.

The geometry of Koszul, the geometry of homogeneous bounded domains and related topics have been studied by Vinberg, Piatecci-Shapiro and many other mathematicians [3]. The geometry of Siegel domains belongs to that galaxy [7,12]. Almost all of those studies are closely related to the Hessian geometry.

Among the open problems in the Hessian geometry are two questions we are concerned with. The first is to know whether the metric tensor g of a Riemannian manifold is a Hessian metric. Alexander K. Guts raised this question in a mail (to me) forty years ago. The second question is to know whether a locally flat manifold admits a Hessian tensor metric. The solutions to those two problems are announced in the Appendix A to this paper.

IGE stands for Information GEometry.

The purposes: The differential geometry of statistical models, the complexity of statistical models, ramifications of the information geometry.

Our aims: We revisit the classical theory of statistical models, requests of McCullagh and Gromov. A search of a characteristic invariant. The moduli space of models. The homological nature of the information geometry.

The information geometry is the differential geometry in statistical models for measurable sets. In both the theoretical statistics and the applied statistics the exponential families and their generalizations are optimal statistical models. There are many references, e.g., [17,18,22,37]. Here Murray-Rice 1.15 means Murray-Rice Chapter 1, Section 15. A major problem is to know whether a given statistical model is isomorphic to an exponential model. That is what we call the complexity problem of statistical models. This challenge is a still open problem. It explicitly arises from the purposes which are discussed in [22] here, see also [30]. In the appendix to this paper we present a recently discovered invariant which measures how far from being an exponential family is a given model. That invariant is useful for exploring the differential topology of statistical models. That is particularly important when models are singular, viz models whose Fisher information is not inversible.

ENT stands for ENTropy.

Pierre Baudot and Daniel Bennequin recently discovered that the entropy function has a homological nature [31]. We recall that in 2002 Peter McCullagh raised a fundamental geometric-topological question in the theory of information: What Is a Statistical Model? [30] A few years after Misha Gromov raised a similar request: The Search of Structure. Fisher Information [15,16].
Those two titles are two formulations of the same need.

The paper of McCullagh became the subject of controversy. It gave rise to questions, discussions, criticisms, see [30].

In Part B of this paper we will be addressing this fundamental problem. A reading of the McCullagh paper would be useful for drawing a comparison between our approach and [15,16,30].

4. The KV Topology of Locally Flat Manifolds

4.1. The Total Cohomology and Riemannian Foliations

In this section we focus on the KV algebroids which are defined by structures of locally flat manifolds. To facilitate a continuous reading of this paper we recall fundamental notions which are needed.

Definition 18. *A locally flat manifold is a pair (M, D). Here D is a torsion free Koszul connection whose curvature tensor R^D vanishes identically.*

The pair (M, D) defines a Koszul-Vinberg algebroid

$$A = (TM, D, 1)$$

The anchor map is the identity map of TM. The multiplication of sections is defined by D, viz

$$X \cdot Y = D_X Y$$

forall $X, Y \in \mathcal{X}(M)$.

The KV algebra of (M, D) is the algebra

$$\mathcal{A} := (\mathcal{X}(M), D).$$

The cotangent bundle T^*M is a left module of the KV algebroid $(TM, D, 1)$. For every $(X, Y, \theta) \in \mathcal{X}(M) \times \mathcal{X}(M) \times \Gamma(T^*M)$ the differential 1-form $X \cdot \theta$ is defined by

$$[X \cdot \theta](Y) = [d(\theta(Y))](X) - \theta(X \cdot Y).$$

In the right hand member of the equality above $d(\theta(Y))$ is the exterior derivative of the real valued function $\theta(Y)$.

Let $S^2(T^*M)$ be the vector bundle of symmetric bi-linear forms in M.

The vector space of sections of $S^2(T^*M)$ is denoted by $\mathcal{S}_2(M)$, viz

$$\mathcal{S}_2(M) = \Gamma(S^2(T^*M)).$$

The vector space $\mathcal{S}_2(M)$ is a left module of the KV algebra $\mathcal{A}$. The left action of $\mathcal{A}$ in $\mathcal{S}_2(M)$ is defined by

$$(X \cdot g)(Y, Z) = [dg(Y, Z)](X) - g(X \cdot Y, Z) - g(Y, X \cdot Z).$$

We put

$$\Omega^1(M) = \Gamma(T^*M).$$

The T^*M-valued cohomology of the KV algebroid $(TM, D, 1)$ is but the cohomology of $\mathcal{A}$ with coefficients in $\Omega^1(M)$. The KV cohomology and the total cohomology are denoted by

$$H^*_{KV}(\mathcal{A}, \Omega^1(M)),$$

$$H^*_\tau(\mathcal{A}, \Omega^1(M)).$$

Warning.

We observe that elements of $\mathcal{S}_2(M)$ may be regarded as 1-cochains of $\mathcal{A}$ with coefficients in its left module $\Omega^1(M)$. By [29] we have

$$Z^2_\tau(\mathcal{A}, C^\infty(M)) = \mathcal{S}_2^{\mathcal{A}}(M). \tag{8}$$

At another side we have the cohomolgy exact sequence

$$\to H^1_{KVres}(\mathcal{A}, V) \to H^2_{KVe}(\mathcal{A}, V) \to H^2_{KV}(\mathcal{A}, V) \to H^2_{KVres}(\mathcal{A}, V) \to \tag{9}$$

By Equations (8) and (9) we obtain the inclusion maps

$$\mathcal{S}_2^{\mathcal{A}}(M) \subset Z^1_{KV}(\mathcal{A}, \Omega^1(M)) \subset Z^2_{KV}(\mathcal{A}, \mathbb{R}).$$

Mutatis mutandis one also has

$$\mathcal{S}_2^{\mathcal{A}}(M) \subset Z^1_\tau(\mathcal{A}, \Omega^1(M) \cap Z^2_\tau(\mathcal{A}, \mathbb{R}).$$

Remark 1 (Important Remarks). *We give some subtle consequences of (1).*

(R.1) Every exact total 2-cocycle $\omega \in C^2_\tau(\mathcal{A}, \mathbb{R})$ is a skew symmetric bilinear form. Viz one has the identity

$$\omega(X, X) = 0 \quad \forall X \in \mathcal{A}.$$

(R.2) Every symmetric KV 2-cocycle $g \in Z^2_{KV}(\mathcal{A}, \mathbb{R})$ is locally an exact KV cocycle, viz in a neighbourhood of every point there exists a local section $\theta \in \Omega^1(M)$ such that

$$g = \delta_{KV}\theta.$$

(R.3) Every symmetric total 2-cocycle is a left invariant cochain, viz

$$Z^2_\tau(\mathcal{A}, \mathbb{R}) \cap \mathcal{S}_2(M) = \mathcal{S}_2^{\mathcal{A}}(M).$$

By (R.1) and (R.3) we obtain the inclusion map

$$\mathcal{S}_2^{\mathcal{A}}(M) \subset H^2_\tau(\mathcal{A}, \mathbb{R}).$$

Let $H^2_{dR}(M)$ be the second cohomology space of the de Rham complex of M. The following theorem is useful for relating the total KV cohomology and the differential topology.

Theorem 4. *[29] There exists a canonical linear injection of $H^2_{dR}(M)$ in $H^2_\tau(\mathcal{A}, \mathbb{R})$ such that*

$$H^2_\tau(\mathcal{A}, \mathbb{R}) = H^2_{dR}(M) \oplus \mathcal{S}_2^{\mathcal{A}}(M)$$

The theorem above highlights a fruitful link between the total KV cohomology and the differential topology. We are particularly interested in D-geodesic Riemannian foliations in a locally flat manifold (M, D).

Warning.

Throughout this paper a Riemannian metric tensor in a manifold M is a non-degenerate symmetric bilinear form in M.

A positive metric tensor is a positive definite metric tensor.

In the next we use the following definition of Riemannian foliation and symplectic foliation.

Definition 19. *A Riemannian foliation is an element $g \in \mathcal{S}_2(M)$ which has the following properties*

(1.1) $rank(g) = constant,$

(1.2) $L_X g = 0 \quad \forall X \in \Gamma(Ker(g))$. *A symplectic foliation is a (de Rham) closed differential 2-form ω which satisfies*

(2.1) $rank(\omega) = constant,$

(2.2) $L_X \omega = 0 \quad \forall X \in \Gamma(Ker(\omega))$.

Warning.

When g is positive semi-definite our definition is equivalent to the classical definition of Riemannian foliation [38–40].

The complete integrability of Ker(g) and the conditions to be satisfied by the holonomy of leaves are equivalent to the Property (2.2).

The set of Riemannian foliations in a manifold M is denoted by $\mathcal{RF}(M)$. The last theorem above yields the inclusion map

$$\frac{H^2_\tau(\mathcal{A}, \mathbb{R})}{H^2_{dR}(M)} \subset \mathcal{RF}(M).$$

We often use the notion of affine coordinates functions in a locally flat manifold. For non specialists we recall two definitions and the link between them.

Definition 20. *An m-dimensional affinely flat manifold is an m-dimensional smooth manifold M admitting a complete atlas $\{(U_j, \phi_j)\}$ whose local coordinate changes coincide with affine transformations of the affine space $\mathbb{R}^m$.*

We denoted an affine atlas by

$$\mathbb{A} = \{(U_j, \phi_j)\}.$$

Definition 21. *An affinely flat structure $(M, \mathbb{A})$ and a locally flat structure (M, ∇) are compatible if local coordinate functions of $(M, \mathbb{A})$ are solutions to the Hessian equation*

$$\nabla^2 x_j = O$$

Theorem 5. *For every positive integer m the relation to be compatible with a locally flat manifold is an equivalence between the category of m-dimensional affinely flat manifolds and the category of m-dimensional locally flat manifolds.*

4.2. The General Linearization Problem of Webs

In the framework $\mathcal{RF}(M)$ the inclusion

$$\frac{H^2_\tau(\mathcal{A}, \mathbb{R})}{H^2_{dR}(M)} \subset \mathcal{RF}(M)$$

may be rewritten as the exact sequence

$$O \to H^2_{dR}(M) \to H^2_\tau(\mathcal{A},\mathbb{R}) \to \mathcal{RF}(M).$$

Let (M,∇) be a locally flat manifold whose KV algebra is denoted by $\mathcal{A}$. Every finite family in $H^2_\tau(\mathcal{A},\mathbb{R})$ is a family of ∇-geodesic Riemannian foliations.

There does not exist any criterion to know whether a manifold supports Riemannian foliations. The exact cohomology sequences we have been performing provide us with a cohomological method for constructing Riemannian foliations in the category of locally flat manifolds. This is an impact of the theory of KV homology on DTO.

In the next section we will introduce other new ingredients which highlight the impacts on DTO of the information geometry.

Further we will see that those new machineries from the information geometry have a homological nature.

Another major problems in the differential topology is the linearization of webs. Among references are [41–43].

Definition 22. *Consider a finite family of distributions $\mathcal{D}_j \subset TM, j := 1,2,...,k$. Those distributions are in general position at a point $x \in M$ if for every subset $J \subset \{1,2,...,k\}$ one has*

$$dim(\sum_{j\in J}\mathcal{D}_j(x)) = \min\left\{dim(M), \sum_{j\in J} dim(\mathcal{D}_j(x))\right\}.$$

Definition 23. *A k-web in M is a family of completely integrable distributions which are in general position everywhere in M.*

A Comment.

The distributions belonging to a web may have different dimensions. An example of problem is the symplectic linearization of lagrangian 2-webs.

Let $(\mathcal{D}_j, j := 1,2)$ be a lagrangian 2-web in a 2n-dimensional symplectic manifold (M,ω). The challenge is the search of special local Darboux coordinate functions

$$(x,y) = (x_1,...,x_n,y_1,...,y_n).$$

Those functions must have three properties
(1): $\omega(x,y) = \Sigma_j dx_j \wedge dy_j$; (2) : *The leaves of $\mathcal{D}_1$ are defined by $x =$ constant;* (3): *The leaves of $\mathcal{D}_2$ are defined by $y =$ constant.*

Definition 24. *An affine web in an affine space is a web whose leaves are affine subspaces.*

Definition 25. *A web in a m-dimensional manifold is linearizable if it is locally diffeomorphic to an affine web in a m-dimensional affine space.*

Example 1. In the symplectic manifold $(\mathbb{R}^2, e^{xy}dx\wedge dy)$ one considers the lagrangian 2-web which is defined by

$$\mathcal{L}_1 = \{(x,y)|x = constant\},$$

$$\mathcal{L}_2 = \{(x,y)|y = constant\}.$$

This lagrangian 2-web is not symplectic linearizable.

Example 2. We keep $(\mathcal{L}_1,\mathcal{L}_2)$ as in example.1. It is symplectic linearizable in $(\mathbb{R}^2,(e^x+e^y)dx\wedge dy)$. The linearization problem for lagrangian 2-webs is closely related to the locally flat geometry [10,44,45].

Example 3. What about the linearization of the 3-web defined by $L_1 := \{(x = constant, y)\}$, $L_2 := \{(x, y = constant)\}$, $L_3 := \{e^{-x}(x+y) = constant\}$, $(x,y) \in \mathbb{R}^2$.

Up to today the question as to whether it is linearizable is subject to controversies, see [42] and references therein.

4.3. The Total KV Cohomology and the Differential Topology Continued

We implement the KV cohomology to address some open problems in the differential topology. For our purpose we recall a few classical notions which are needed.

Definition 26. *A metric vector bundle over a manifold M is a vector bundle $\mathcal{V}$ endowed with a non-degenerate inner product $< v, v^* >$.*

A Koszul connection in a vector bundle $\mathcal{V}$ is a bilinear map

$$\Gamma(TM) \times \Gamma(\mathcal{V}) \ni (X, v) \rightarrow \nabla_X v \in \Gamma(\mathcal{V})$$

which has the properties

$$\nabla_{fX} v = f\nabla_X v \forall v, \forall f \in C\infty(M), \tag{10a}$$

$$\nabla_X f v = df(X)v + f\nabla_X v \forall v, \forall f \in C\infty(M). \tag{10b}$$

Definition 27. *A metric connection in $(\mathcal{V}, < -, >)$ is a Koszul connection ∇ which satisfies*

$$d(< v, v^* >)(X) - < \nabla_X v, v^* > - < v, \nabla_X v^* > = 0.$$

Definition 28. *Let $(M, \mathcal{D})$ be a foliation in the usual sense, viz $\mathcal{D}$ has constant rank and is in involution.*
(1): $(M, \mathcal{D})$ is transversally Riemannian if there exists a $g \in \mathcal{S}_2(M)$ such that

$$\mathcal{D} = \mathcal{K}er(g).$$

(2): $(M, \mathcal{D})$ is transversally symplectic if there exists a (de Rham) closed differential 2-form ω such that

$$\mathcal{D} = Ker(\omega)$$

A transversally Riemannian foliation and a transversally symplectic foliation are denoted by

$$(\mathcal{D}, g),$$
$$(\mathcal{D}, \omega).$$

Definition 29. *Given a Koszul connection ∇, a transversally Riemannian foliation $(\mathcal{D}, g)$ (respectively a transversally symplectic foliation $(\mathcal{D}, \omega)$) is called ∇-geodesic if*

$$\nabla g = 0,$$

$$\nabla \omega = 0$$

The notions of transversally Riemannian foliation and transversally symplectic foliation are weaker than the notion of Riemannian foliation and symplectic foliation. However if ∇ a torsion free Koszul connection every ∇geodesic transversally Riemannian foliation is a Riemannian foliation. Every ∇-geodesic transversally symplectic foliation is a symplectic foliation.

For the general theory of Riemannian foliations the readers are referred to [39,40,46], see also the monograph [38] and the references therein.

We have pointed out that criterions for deciding whether a smooth manifold admits Riemannian foliations (respectively symplectic foliations) are missing. Our purpose is to address this existence problem in the category $\mathcal{SLC}$ whose objects are symmetric gauge structures. Such an object is a pair (M, ∇) where ∇ is a torsion free Koszul connection in M. The category of locally flat structure $\mathcal{LF}$ is a subcategory of $\mathcal{SLC}$. The theory of KV homology is useful for discussing geodesic Riemannian foliations in the category $\mathcal{LF}$. In a locally flat manifold (M, D) we have been dealing with the decomposition

$$H^2_\tau(\mathcal{A}, \mathbb{R}) = H^2_{dR}(M) \oplus \mathcal{S}^D_2(M).$$

Here $\mathcal{A}$ is the KV algebra of (M, D).

Let $b_2(M)$ be the second Betti number of M. We define the numerical geometric invariant $r(D)$ by

$$r(D) = dim(H^2_\tau(\mathcal{A}, \mathbb{R})) - b_2(M).$$

Formally $r(D)$ is the codimension of $H^2_{dR}(M) \subset H^2_\tau(\mathcal{A}, \mathbb{R})$, viz

$$r(D) = dim(\frac{H^2_\tau(\mathcal{A}, \mathbb{R})}{H^2_{dR}(M)}).$$

We consider the exact sequences

$$O \to H^2_{dR}(M) \to H^2_\tau(\mathcal{A}, \mathbb{R}) \to \mathcal{S}^{\mathcal{A}}_2(M) \to 0$$

and

$$\to H^2_{\tau,e}(\mathcal{A}, \mathbb{R}) \to H^2_\tau(\mathcal{A}, \mathbb{R}) \to H^2_{\tau,res}(\mathcal{A}, \mathbb{R}) \to H^3_{\tau,e}(\mathcal{R}, \mathbb{R}) \to$$

From those exact sequences, one deduces the equality

$$\frac{H^2_\tau(\mathcal{A}, \mathbb{R})}{H^2_{dR}(M)} = \frac{H^2_{\tau,e}(\mathcal{A}, \mathbb{R})}{H^2_{dR}(M)}.$$

Thus $r(D)$ is formally the dimension of $\mathcal{S}^{\mathcal{A}}_2(M)$.

The present approach leads to the following statement

Proposition 1. *If $r(D) > 0$ then M admits non trivial D-geodesic Riemannian foliations.*

Proof. Let B be a non zero element of $\mathcal{S}^{\mathcal{A}}_2(M)$ and let $\mathcal{D}$ be the kernel of B.

(1) Suppose that

$$0 < rank(\mathcal{D}) < dim(M)$$

Therefore, (M, B) is a D-geodesic Riemannian foliation.

(2) Suppose that

$$rank(\mathcal{D}) = O.$$

Then (M, B) is a Riemannian manifold the Levi-Civita connection of which is D. Therefore, the proposition holds. □

Before proceeding we define three numerical invariants

$$r(M) = \max \{r(D) | D \in \mathcal{LC}(M)\},$$

$$s(M, \mathcal{A}) = \max \left\{rank(B) | B \in \mathcal{S}^{\mathcal{A}}_2(M)\right\},$$

$$s(M) = \max \{s(M, \mathcal{A}) | D \in \mathcal{LF}(M)\}.$$

The non negative integers $r(M)$ and $s(M)$ are global geometric invariants. They connect the total KV cohomology to geodesic Riemannian foliations. By this viewpoint the proposition has an interesting corollary.

Corollary 1. *In an m-dimensional manifold M suppose that the following inequalities are satisfied*

$$0 < s(M) < m.$$

Then the manifold M admits a locally flat structure (M, D^*) which supports a non trivial D^*-geodesic Riemannian foliation.

The integer $s(M)$ is a local characteristic invariant of some class of 2-webs in Hessian manifolds. Let (M, D) be a locally flat manifold whose KV algebra is denoted by $\mathcal{A}$. we recall that a Hessian metric tensor in (M, D) is a inversible cocycle $g \in Z^2_{KV}(\mathcal{A}, \mathbb{R})$.

Theorem 6. *Let (M, D, g) and (M^*, D^*, g^*) be m-dimensional Hessian manifolds. We assume that the following inequalities hold*

$$0 < s(M, D) = s(M^*, D^*) = s < m.$$

Then M and M^* admit linearizable 2-webs which are locally isomorphic.

Proof. The proof is based on methods of the information geometry.

Let $\mathcal{A}$ and $\mathcal{A}*$ be the KV algebras of (M, D) and of (M^*, D^*) respectively. By the hypothesis there exists a pair of geosic Riemannian foliations

$$(B, B^*) \in \mathcal{S}_2^{\mathcal{A}} \times \mathcal{S}_2^{\mathcal{A}^*}$$

such that

$$rank(B) = rank(B^*) = s.$$

By the dualistic relation both M and M^* admit locally flat structures $(M, \tilde{D})$ and $(M^*, \tilde{D}^*)$ defined by

$$g(Y, \tilde{D}_X Z) = Xg(Y, Z) - g(D_X Y, Z),$$

$$g^*(Y, \tilde{D}^*_X Z) = Xg^*(y, Z) - g^*(D^*_X Y, Z).$$

Their KV algebras are denoted by $\tilde{\mathcal{A}}$ and $\tilde{A}^*$.

Step a

There exists a 1-cocycle

$$\psi \in Z^1_\tau(\tilde{\mathcal{A}}, \tilde{A})$$

such that

$$B(X, Y) = g(\psi(X), Y),$$

$$Ker(B) = Ker(\psi).$$

By the definition of $\tilde{D}$ we have

$$TM = Ker(\psi) \oplus im(\psi).$$

Further $im(\psi)$ is $\tilde{D}$-geodesic and $Ker(B)$ is D-geodesic. Therefore, the pair

$$(Ker(\psi), im(\psi))$$

is a 2-web in M.

In (M^*, D^*, g^*) we obtain similar 2-web

$$(Ker(\psi^*), im(\psi^*)).$$

By the choice of B and B^* we have

$$rank(Ker(\psi)) = rank(Ker(\psi^*)) = m - s.$$

Now we perform the following arguments.

(a): The foliation B is D-geodesic. In a neighbourhood of every point $p_0 \in (M, D)$ we linearize B by choosing appropriate local affine coordinate functions

$$(x, y) = (x_1, ..., x_{m-s}, y_1, ..., y_s).$$

The leaves of $Ker(\psi)$ are defined by

$$y = constant.$$

Thereby those leaves are locally isomorphic to affine sub-spaces.

Step b

The distribution $im(\psi)$ is $\tilde{D}$-geodesic. Therefore, near the same point $p_0 \in (M, \tilde{D})$ we linearize $im(\psi$ by choosing appropriate local affine coordiante functions

$$(x^*, y^*) = (x_1^*, ..., y_1^*, ...).$$

The leaves of $im(\psi)$ are defined by

$$x^* = constant.$$

Thus near p_0 the foliation defined by $m(\psi)$ is isomorphic to an linear foliation.

Step c

By both step a and step b we choose a neighbourhood of p_0 which is the domain of systems of appropriate local coordinate functions (x, y) and (x^*, y^*). From those data we pick the local coordinate functions

$$(x, y^*) = (x_1, ..., x_{m-s}, y_1^*, ..., y_s^*).$$

So we linearize the 2-web $(Ker(\psi), im(\psi))$ with the local coordinate functions (x, y^*).

$$(Ker(\psi), im(\psi)).$$

Thus near the p_0 the 2-web $(Ker(\psi), im(\psi))$ is isomorphic to the linear 2-web (L_1, L_2) which is defined in $\mathbb{R}^m$ by

$$\mathbb{R}^m = \mathbb{R}^{m-s} \times \mathbb{R}^s.$$

Step d

At a point p_0^* in M^* we perform the construction as in step a and in steps b and c, then we linearize $(Ker(\psi^*), im(\psi*))$ by choosing appropriate local coordinate functions

$$(x^0, y^{0*}) = (x_1^0, ..., x_{m-s}^0, y_1^{0*}, ..., y_s^{0*}).$$

In final, near the $p_O^* \in M^*$ the web $(Ker(\psi^*), im(\psi^*))$ is diffeomorphic to the affine web whose leaves are parallel to a decomposition

$$\mathbb{R}^m = V^{m-s} \times V^s.$$

Here V^{m-s} and V^s are vector subspaces of $\mathbb{R}^m$. Their dimensions are $m-s$ and s.

Conclusion.

There exists a unique linear transformation ϕ of $\mathbb{R}^m$ such that

$$\phi(\mathbb{R}^{m-s} \times 0) = V^{m-s},$$

$$\phi(0 \times \mathbb{R}^s) = V^s.$$

Thereby there is a local diffeomorphism Φ of M in M^* subject to the requirements

$$\Phi(p_O) = p_O^*,$$

$$(x^0, y^{0*}) \circ \Phi = (x, y^*).$$

The differential of Φ is denoted by Φ_*. We express the properties above by

$$\Phi(p_O) = p_O^*,$$

$$\Phi_*[Ker(\psi), im(\psi)] = [Ker(\psi^*), im(\psi^*)].$$

This ends the sketch of proof of Theorem. □

In the next we use the following definitions.

Definition 30. *A finite family*

$$\{B_J, \quad J \subset \mathbb{Z}\} \subset \mathcal{S}_2^{\mathcal{A}}(M)$$

is in general position if the distributions $\{Ker(B_j), j \in J\}$ *are in general position.*

The following statement is a straight corollary of the theorem we just demonstrated.

Proposition 2. *In a locally flat manifold* (M, D) *with* $r(D) > 0$ *every finite family in general position define a linearizable Riemannian web.*

4.4. The KV Cohomology and Differential Topology Continued

We have seen how the total cohomology and linearizable Riemannian webs are related. More precisely the theory of KV cohomology provides sufficient conditions for a locally flat manifold admitting linearizable Riemannian webs. That approach is based on the split exact sequence

$$0 \to H^2_{dR}(M) \to H^2_\tau(\mathcal{A}, C^\infty(M)) \to \mathcal{S}_2^{\mathcal{A}}(M) \to 0.$$

4.4.1. Kernels of 2-Cocycles and Foliations

Not all locally flat manifolds admit locally flat foliations. The existence of locally flat foliations is related to the linear holomnomy representation, viz the linear component of the affine holonomy representation of the fundamental group. Via the developing map the affine holonomy representation is conjugate to the natural action of the fundamental group in the universal covering. The KV homology is useful for investigating the existence of locally flat foliations. To simplify we work in the analytic category. So our purposes include singular foliations.

For those purposes we focus on an elementary item which has a notable impacts on our request.

Let (M,D) be a locally flat manifold whose KV algebra is denoted by $\mathcal{A}$. Let $g \in C^2(\mathcal{A}, C^\infty(M))$. The left kernel and the right kernel of g are denoted by $Ker(g)$ and $K^0er(g)$ respectively.

$Ker(g)$ is defined by

$$g(X,Y) = 0 \quad \forall Y \in \mathcal{A}.$$

$K^0er(g)$ is defined by

$$g(Y,X) = 0 \quad \forall Y \in \mathcal{A}.$$

The scalar KV 2-cocycles have elementary relevant properties

(1) The left kernel of every KV 2-cocycle is closed under the Poisson bracket of vector fields.
(2) The right kernel of every KV 2-cocycle is a KV subalgebra of the KV algebra $\mathcal{A}$.

We translate those elementary properties in term of the differential topology

Theorem 7. *In an analytic locally flat manifold* (M,D)

(1) The arrow

$$Z^2_{KV}(\mathcal{A}, C^\infty(M)) \ni g \to Ker(g)$$

maps the set of analytic 2-cocycles in the category of analytic stratified foliations M,

(2) The arrow

$$Z^2_{KV}(\mathcal{A}, C^\infty(M)) \ni g \to K^0er(g)$$

maps the set of analytic 2-cocycles in the category of stratified locally flat foliations,

(3) If a 2-cocycle g is a symmetric form then Ker(g) is a stratified locally flat transversally Riemannian foliation.

The vector subspace of symmetric 2-cocycles the kernels of which are D-geodesic is denoted by $\tilde{Z}^2_{KV}(\mathcal{A})$. The corresponding cohomology vector subspace is denoted by

$$\tilde{H}^2_{KV}(\mathcal{A}) \subset H^2_{KV}(\mathcal{A}, C^\infty(M)).$$

By the exact sequence

$$O \to H^2_{dR}(M) \to H^2_\tau(\mathcal{A}, C^\infty(M)) \to \mathcal{S}_2^{\mathcal{A}}(M) \to 0$$

we have the inclusion map

$$\frac{H^2_\tau(\mathcal{A}, C^\infty(M))}{H^2_{dR}(M)} \subset \tilde{H}^2_{KV}(\mathcal{A}) \subset \mathcal{RF}(M).$$

5. The Information Geometry, Gauge Homomorphisms and the Differential Topology

We combine the dualistic relation with gauge homomorphisms to relate the total cohomology and two problems.

(i) The first is the existence problem for Riemannian foliations.
(ii) The second is the linearization of webs.

Those relationships highlight other roles played by the total KV cohomology. Through this section we use the brute coboundary operator.

5.1. The Dualistic Relation

We are interest in the foliation counterpart of the reduction in statistical models. The statistical reduction theorem is Theorem 3.5 as in [18]. We recall the notions which are needed.

Definition 31. *A dual pair is a quadruple (M, g, D, D^*) where (M, g) is a Riemannian manifold, D and D^* are Koszul connections in M which are related to the metric tensor g by*

$$Xg(Y,Z) = g(D_X Y, Z) + g(Y, D^*_X Z) \quad \forall X, Y, Z.$$

We recall that a Riemannian tensor is a non degenerate symmetric bilinear 2-form.

The dualistic relation between linear connections plays a central role in the information geometry [17,18,47,48].

Definition 32. *Let (M, g) be a Riemannian manifold.*

(1) *A dual pair (M, g, D, D^*) is called a flat pair if the connection D is flat, viz $R^\nabla = 0$.*
(2) *A flat pair (M, g, D, D^*) is called a dually flat pair if both (M, D) and (M, D^*) are locally flat manifolds.*

Given a dual pair (M, D, D^*) let us set $A = D - D^*$. Here are the relationships between the torsion tensors T^D and T^{D^*} (respectively the relationship between the curvature tensors R^D and R^{D^*})

$$g(R^D(X,Y) \cdot Z, T) + g(Z, R^D(X,Y) \cdot T) = 0,$$

$$g(T^D(X,Y), Z) - g(T^{D^*}(X,Y), Z) = g(Y, A(X,Z)) - g(X, A(Y,Z)).$$

Proposition 3. *Given a flat pair (M, g, D, D^*), the following assertions are equivalent.*

(1) *Both D and D^* are torsion free.*
(2) *D is torsion free and A is symmetric, viz*

$$A(X,Y) = A(Y,X).$$

(3) *D^* is torsion free and the metric tensor g a is KV cocycle of the KV algebra $\mathcal{A}^*$ of the locally flat manifold (M, D^*).*
(4) *The flat pair (M, g, D, D^*) is a dually flat pair.*

Proof. Let us prove that 1 implies (2)
If both T^D and T^{D^*} vanish identically then A is symmetric, viz $A(X,Y) = A(Y,X)$.

Let us prove that (2) implies (3).

Since D is a flat connection, (2) implies that both the torsion tensor and the curvature tensor of D vanish identically. Then (M, D) is a locally flat manifold whose KV complex is denoted by $(C^*(\mathcal{A}, \mathbb{R}), \delta_{KV})$. Using the dualistic relation of the pair (M, g, D, D^*) one obtains the identity

$$\delta_{KV} g(X,Y,Z) = g(A(X,Y) - A(Y,X), Z) = g(T_{D^*}(X,Y), Z),$$

therefore (2) implies (3).

Let us prove that (3) implies (4).

The assertion (3) implies that (M, D^*) is a locally flat manifold. Since g is δ_{KV}-closed D is torsion free. Thereby (M, g, D, D^*) is a dually flat pair.

Let us prove that (4) implies (1).

This implication derives directly from the definition of dually flat pair. □

A Comment.

From the proposition just proved arises a relationship between the dually flatness and the KV cohomology.

Indeed let (M, D^0) *be a fixed locally flat manifold whose KV algebra is denoted by* $\mathcal{A}^0$. *Let* $C^*_{KV}(\mathcal{A}^0, \mathbb{R})$ *be the KV complex of* $\tilde{\mathbb{R}}$*-valued cochains of the KV algebroid* $(TM, D^0, 1)$. *We know that every* $g \in \mathcal{R}ie(M)$ *yields a flat pair* (M, g, D^0, D^g).

Here D^g *is the flat Koszul connection defined by*

$$g(D^g_X Y, Z) = Xg(Y, Z) - g(Y, D^0_X Z).$$

Proposition 4. *The following assertions are equivalent.*

(1) (M, g, D^0, D^g) *is a dually flat pair.*
(2) $\delta^0_{KV}(g) = 0$

The scalar KV cohomology of a fixed locally flat manifold (M, D^0) provides a way of constructing new locally flat structures in M. Indeed let us set

$$\mathcal{H}es(M, D^0) = Z^2_{KV}(\mathcal{A}^0, \mathbb{R}) \cap \mathcal{R}ie(M).$$

For every $g \in \mathcal{H}es(M, D^0)$ there is a unique $D^g \in \mathcal{LF}(M)$ such that (M, g, D^0, D^g) is a dually flat pair.

So the dualistic relation leads to the map

$$\mathcal{H}es(M, D^0) \ni g \to D^g \in \mathcal{LF}(M).$$

We recall that a gauge map in TM is a vector bundle morphism of TM in TM which projects on the identity map of M. The readers interested in others topological studies involving connections and gauge transformations are referred to [49].

Given two symmetric cocycles $g, g^* \in \mathcal{H}es(M, D^0)$ there is a unique gauge transformation

$$\phi^* : TM \to TM$$

such that

$$g^*(X, Y) = g(\phi^*(X), Y).$$

The following properties are equivalent

$$\phi(D^0_X Y) = D^0_X \phi(Y), \tag{11a}$$

$$D^g = D^{g^*}. \tag{11b}$$

We fix a metric tensor $g* \in \mathcal{H}es(M, D^0)$. A gauge transformation ϕ is called g-symmetric if we have

$$g(\phi(X), Y) = g(X, \phi(Y)) \quad \forall (X, Y).$$

Every g-symmetric gauge transformation ϕ defines the metric tensor

$$g_\phi(X, Y) = g(\phi(X), Y).$$

This gives rise to the flat pair

$$(M, g_\phi, D^0, D^{g_\phi}).$$

To simplify we set

$$D^\phi = D^{g_\phi}.$$

We note $Sym(g)$ the subset of g-symmetric gauge transformations ϕ such that the following assertions are equivalent

(1) $\phi \in Sym(g)$.
(2) (M, g_ϕ, D^0, D^ϕ) is a dually flat pair.

The Lie group of D^0-preserving gauge transformations of TM is denoted by G^0. It is easy to see that for every $\phi \in Sym(g)$ the following assertions are equivalent

(1) $\phi \in G^0$,
(2) $g_\phi \in \mathcal{H}es(M, D^0)$.

Henceforth we deal with a fixed $g^* \in \mathcal{H}es(M, D^0$. The triple (M, g^*, D^0) leads to the dually flat pair (M, g, D^0, D^{g^*}). We set

$$D^* = D^{g^*}.$$

The tangent bundle TM is regarded as a left KV module of the KV algebroid $(TM, D^*, 1)$.

The KV algebras of (M, D^0) and of (M, D^*) are denoted by $\mathcal{A}^0$ and by $\mathcal{A}^*$ respectively. Their coboundary operators are noted δ^0 and δ^* respectively.

We focus on the role played by the total KV cohomology of the algebroid $(M, D^*, 1)$.

Let ϕ be a g^*-symmetric gauge transformation. Then ϕ gives rise to the metric tensor g_ϕ which is defined by

$$g_\phi(X, Y) = g^*(\phi(X), Y).$$

Lemma 3. *The following assertions are equivalent,*

(1) $g_\phi \in \mathcal{H}es(M, D^0)$,
(2) $\phi \in Z^1_\tau(\mathcal{A}^*, \mathcal{A}^*)$.

Hint.

Use the following formula

$$\delta^0_{KV} g_\phi(X, Y, Z) = g^*(\delta^*_\tau \phi(X, Y), Z).$$

Following the pioneering definition as in [2] a hyperbolic locally flat manifold is a positive exact Hessian manifold $(M, D, \delta_{KV}\theta)$. *We extend the notion of hyperbolicity by deleting the condition that* $\delta_{KV}\theta$ *is positive. Now denote by* $\mathcal{H}yp(M, D^0)$ *the set of exact Hessian structures in* (M, D^0).

A hyperbolic structure is defined by a triple (M, D, θ) *where* (M, D) *is a locally flat manifold and* θ *is a de Rham closed differential 1-form such that the symmetric bilinear* $\delta_{KV}\theta$ *is definite.*

The following statement is a straightforward consequence of Lemma 3.

Corollary 2. *The following statements are equivalent.*

(1) $g_\phi \in \mathcal{H}yp(M, D^0)$,
(2) $\phi \in B^1\tau(\mathcal{A}^*, \mathcal{A}^*)$

Proof of Corollary. By (1) there exists a (de Rham) closed differential 1-form θ such that

$$g_\phi(X, Y) = X\theta(Y) - \theta(D^0_X Y).$$

Let ξ be the unique vector field such that

$$\theta = \iota_\xi g^*.$$

Therefore one has

$$g^*(\phi(X), Y) = Xg^*(\xi, Y) - g^*(\xi, D^0_X Y).$$

Since the quadruple

$$(M, g^*, D^0, D^*)$$

is a dually flat pair one has the identity

$$g^*(\phi(X), Y) = g^*(D^*_X \xi, Y).$$

Thus we get the expected conclusion, viz

$$\phi(X) = D^*_X \xi.$$

Conversely let us assume that there exists a vector ξ satisfying the identity

$$\phi(X) = D^*_X \xi.$$

That leads to the identity

$$g^*(D^*_X \xi, Y) = Xg^*(\xi, Y) - g(\xi, D^0_X Y).$$

In other words one has

$$g_\phi \in \mathcal{H}yp(M, D^0).$$

This ends the proof of Corollary 2. □

The set of $g*$-symmetric gauge transformation is denoted by $\Sigma(g^*)$.
We have the canonical isomorphism

$$\Sigma(g^*) \ni \phi \to g_\phi \in \mathcal{R}ie(M). \tag{12}$$

Now we define the sets

$$\tilde{Z}^1_\tau(\mathcal{A}^*, \mathcal{A}^*) = \Sigma(g^*) \cap Z^1_\tau(\mathcal{A}^* \mathcal{A}^*),$$
$$\tilde{B}^1_\tau(\mathcal{A}^*, \mathcal{A}^*) = \Sigma(g^*) \cap B^1_\tau(\mathcal{A}^*, \mathcal{A}^*).$$

Combining Lemma 3 and its corollary with the isomorphism Equation (12). Then we obtain the identifications

$$\tilde{Z}^1_\tau(\mathcal{A}^*, \mathcal{A}^*) = \mathcal{H}es(M, D^*),$$
$$\tilde{B}^1_\tau(\mathcal{A}^*, \mathcal{A}^*) = \mathcal{H}yp(M, D^*).$$

Reminder.

We recall that a hyperbolic manifold (or a Koszul manifold) is δ_{KV}-exact Hessian manifold (M, g, D).

It is easily seen that the set of positive hyperbolic structures in a locally flat manifold (M, D) is a convex subset of $\mathcal{H}es(M, D)$.

So show the Koszul geometry is a vanishing theorem in the theory of KV homology of KV algebroids. The theory of homological statistical model (to be introduced in Part B) is another impact on the information geometry of the KV cohomology.

At the present step we have the relations

$$\frac{\mathcal{H}es(M, D^*)}{\mathcal{H}yp(M, D^*)} \subset H^2_{KV}(\mathcal{A}^*, \mathbb{R}),$$

$$\frac{\tilde{Z}^1_\tau(\mathcal{A}^*,\mathcal{A}^*)}{\tilde{B}^1_\tau(\mathcal{A}^*,\mathcal{A}^*)} = \frac{\mathcal{H}es(M,D^*)}{\mathcal{H}yp(M,D^*)}.$$

Another outstanding result of Koszul is the non rigidity of compact positive hyperbolic manifolds [2]. The non rigidity means that every open neighborhood of a positive Hyperbolic locally flat manifold $(M, D, \delta_{KV}\theta)$ contain another positive hyperbolic locally flat structure which is not isomorphic to (M, D). This non rigidity property may be expressed with the Maurer–Cartan polynomial function P^A_{MC} of (M, D) (see the local convexity theorem in [29]. In the next sub-subsection we revisit the notion of dual pair of foliations as in [18].

5.1.1. Statistcal Reductions

The statistical reduction theorem is the following statement.

Theorem 8 ([18])**.** *Let (M, g, D, D^*) be a dually flat pair and let N be a submanifold of M. Assume that N is either D-geodesic or D^*-geodesic. Then N inherits a structure of dually flat pair which is either (N, g_N, D, D^*_N) or (N, g_N, D_N, D^*)).*

The foliation counterpart of the reduction theorem is of great interest in the differential topology of statistical models see [18]. In the preceding sections we have addressed a cohomological aspect of this purpose. The matter will be more extensively studied in a forthcoming paper (See the Appendix A).

In mathematical physics a principal connection 1-form is called a gauge field.

In the differential geometry a principal connection 1-form in a bundle of linear frames is called a linear connection.

In the category of vector bundle Koszul connections are algebroid counterpart of principal connection 1-forms.

In a tangent bundle TM, depending on concerns and needs Koszul connections may called linear connections or linear gauges.

Definition 33. *Let $D, D^* \in \mathcal{LC}(M)$. A vector bundle homomorphism*

$$\psi : TM \to TM$$

is called a gauge homomorphism of (M, D) in (M, D^*) if for all pairs of vector fields (X, Y) one has

$$D^*_X\psi(Y) = \psi(D_X Y).$$

The vector space of gauge homomorphisms of (M, D) in (M, D^*) is denoted by $\mathcal{M}(D, D^*)$. The vector space $\mathcal{M}(D, D^*)$ is not a $C^\infty(M)$-module.

5.1.2. A Uselful Complex

In this subsubsection we fix a dually flat pair (M, g, D, D^*) whose KV algebras are denoted by $\mathcal{A}$ and by $\mathcal{A}^*$. The tangent bundle TM is endowed the structure left module of the anchored KV algebroids $(TM, D, 1)$ and $(TM, D^*, 1)$. This means that each of the KV algebras $\mathcal{A}$ or $\mathcal{A}^*$ is regarded as a left module of itself.

We consider the tensor product

$$C = C^*_\tau(\mathcal{A}^*,\mathcal{A}^*) \otimes C^*_\tau(\mathcal{A},\mathbb{R}).$$

We endow C with the $\mathbb{Z}$ bi-grading.

$$C^{i,0} = C^i_\tau(\mathcal{A}^*,\mathcal{A}^*) \otimes C^\infty(M),$$
$$C^{0,j} = \mathcal{A}^* \otimes C^j_\tau(\mathcal{A},\mathbb{R}),$$

$$C^{i,j} = C^i_\tau(\mathcal{A}^*,\mathcal{A}^*) \otimes C^j_{KV}(\mathcal{A},\mathbb{R}).$$

We recall that $C^*(\mathcal{A},\mathbb{R})$ stands for $C^*(\mathcal{A},C^\infty(M))$.

For every non negative integer q we set

$$C^q = \Sigma_{i+j=q}C^{i,j}.$$

We defines the linear map

$$\delta_{i,j} : C^{i,j} \to C^{i+1,j} \oplus C^{i,j+1}$$

by

$$\delta_{i,j} = \delta_\tau \otimes 1 + (-1)^i \otimes \delta_\tau.$$

So we obtain a linear map

$$C^q \to C^{q+1}$$

Therefore, we consider the bi-graded differential vector space

$$C := (C^{**},\delta_{**}).$$

That is a bi-graded cochain complex whose q^{th}*cohomology* is denoted by $H^q(C)$. The cohomology inherits the bi-grading

$$H^q(C) = \sum_{[i+j=q]} H^{i,j}(C).$$

Here

$$H^{i,j}(C) = \frac{C^{i,j} \cap [Z^i_\tau(\mathcal{A}^*,\mathcal{A}^*) \otimes Z^j_\tau(\mathcal{A},\mathbb{R})]}{im(\delta_{i-1,j}) + im(\delta_i, j-1)}$$

In the next subsubsection we shall discuss the impacts of this cohomology.

Remark 2. *The pair (C^{**},δ_{**}) generates a spectral sequence [34]. That spectral sequence is a useful tool for simultaneously computing both the KV cohomology and the total KV cohomology of KV algebroids. Those matters are not the purpose of this paper.*

5.1.3. The Homological Nature of Gauge Homomorphisms

Giving a dually flat pair (M,g,D,D^*) one considers the linear map

$$C^{1,0}_\tau(\mathcal{A}^*,\mathcal{A}^*) \ni \psi \to \psi \otimes q_\psi \in C^{1,2}.$$

Here the symmetric 2-form q_ψ is defined by

$$q_\psi(X,Y) = \frac{1}{2}[g(\psi(X),Y) + g(X,\psi(Y))].$$

To relate the bi-complex (C^{**},δ^{**}) and the space of gauge homomorphisms we use the following statement.

Theorem 9. *Given a gauge morphism*

$$\psi : TM \to TM$$

the following statements are equivalent

(1) $\psi \in \mathcal{M}(D,D^*)$,

(2) $\delta_{1,2}(\psi \otimes q_\psi) = 0$

Proof. (1) implies (2).
Suppose that $\psi \in \mathcal{M}(D, D^*)$. Then we have

$$D^*_X\psi(Y) = \psi(D_X Y) \quad \forall (X,Y).$$

Since both D and D^* are torsion free one has the identity

$$D^*_X.\psi(Y) - \psi(D^*_X Y) - D^*_Y\psi(X) + \psi(D^*_Y X) = 0.$$

Thus ψ is a (1,0)-cocycle of the total KV complex (C^{**}, δ_{**}).
At another side the relation $D^*_X \circ \psi = \psi \circ D_X$ leads to the identity

$$D_X q_\psi = 0.$$

So q_ψ is a (0,2)-cocycle of complex (C^{**}, δ_{**}). We conclude that

$$\delta_{1,2}(\psi \otimes q_\psi) = 0, \quad QED.$$

(2) implies (1).
We recall the formula

$$\delta_{1,2}(\psi \otimes q_\psi) = (\delta_\tau \psi) \otimes q_\psi - \psi \otimes \delta_\tau q_\psi.$$

By this formula

$$\delta_{1,2}(\psi \otimes q_\psi) \in C^{2,2} \oplus C^{1,3}$$

Thus the statement (2) is equivalent to the system

$$\delta_\tau \psi = 0,$$

$$\delta_\tau q_\psi = 0. \quad \square$$

To continue the proof we perform the following lemma.

Lemma 4 ([29]). *For every symmetric cochain $B \in C^{0,2}$, viz*

$$B(X,Y) = B(Y,X)$$

the following identities are equivalent

$$\delta_\tau B = 0, \tag{13a}$$
$$\nabla B = 0, \tag{13b}$$

By Lemma 4 the bilinear form q_ψ is D-parallel. Thereby we get the identity

$$Xq_\psi(Y,Z) - q_\psi(D_X Y, Z) - q_\psi(Y, D_X Z) = 0.$$

To usefully interpret this identity we involve the dualistic relation

$$Xg(Y,Z) = g(D_X Y, Z) + g(Y, D^*_X Z).$$

This expression leads to the identity

$$g(D^*_X\psi(Y) - \psi(D_X Y), Z) + g(Y, D^*_X\psi(Z) - \psi(D_X Z)) = 0. \tag{14}$$

A highlighting consequence is the identity

$$D^*_X\psi(Y) - \psi(D_XY) = D^*_Y\psi(X) - \psi(D_YX). \tag{15}$$

To every vector field X we assign the linear map

$$Y \to S_X(Y) = D^*_XY - \psi(D_XY).$$

Then we rewrite Equations (14) and (15) as

$$g(S_X(Y), Z) + g(Y, S_X(Z)) = 0,$$

$$S_X(Y) = S_Y(X).$$

We consider the last identities in the framework of the Sternberg geometry [50,51].

Since the application

$$(X, Y) \to S_X(Y)$$

is $C^\infty(M)$-bi-linear it belongs to the first Kuranishi-Spencer prolongation of the orthogonal Lie algebra $so(g)$. Thereby $S_X(Y)$ vanishes identically. In other words we have

$$\psi \in \mathcal{M}(D, D^*).$$

This ends the proof of Theorem

A Comment.

The Sternberg geometry is the algebraic counterpart of the global analysis on manifolds. It has been introduced by Shlomo Sternberg and Victor Guillemin. It is an algebraic model for transitive differential geometry [50]. In that approach the Riemannian geometry is a geometry of type one. All of its Kuranishi-Spencer prolongations are trivial. The unique relevant geometrical invariant of the Riemnnian geometry is the curvature tensor of the Levi-Civita connection. Except the connection of Levi-Civita the other metric connections have been of few interest. Really other metric connections may have outstanding impacts on the differential topology. I shall address this purpose in a forthcoming paper.

5.1.4. The Homological Nature of the Equation $FE^{\nabla\nabla^*}$

Before proceeding we plan to discuss some homological ingredients which are connected to the differential equation

$$FE^{\nabla\nabla^*}: \quad D^{\nabla\nabla^*}(\psi) = 0.$$

Let us consider a dually flat pair (M, g^*, D, D^*) and the KV complex

$$\psi \in C^{1,0} = C^1_\tau(\mathcal{A}^*, \mathcal{A}^*).$$

Lemma 4 yields the following corollary.

Corollary 3. *We keep the notation used the preceding sub-subsection. Given a gauge morphism ψ the following statements are equivalent.*

(1) $\psi \otimes q_\psi$ *is an exact (1,2)-cocyle,*
(2) $\psi \in B^1_\tau(\mathcal{A}^*, \mathcal{A}^*)$.

Proof. Assume that the assertion (2) holds. Then there is $\xi \in \mathcal{A}^*$ satisfying the condition

$$\psi(X) = D^*_X\xi.$$

Thereby one has

$$q_{D^*\xi}\otimes \in Z^2_\tau(\mathcal{A},\mathbb{R}).$$

So one gets

$$D^*\xi \otimes q_{D^*\xi} = \delta_{0,2}[\xi \otimes q_{D^*\xi}].$$

Therefore assumption (2) implies (1).

Conversely assume that (1) holds, viz the (1,2)-cochain $\psi \otimes q_\psi$ is exact.

There exists

$$\xi \otimes \alpha \oplus \psi^* \otimes \beta \in C^{0,2} + C^{1,1}$$

such that

$$\psi \otimes q_\psi = \delta_{0,2}(\xi \otimes \alpha) + \delta 1, 1(\psi^* \otimes \beta).$$

Thus for vector fields Z, X, Y we have

$$\psi(Z) \otimes q_\psi(X,Y) = \delta_\tau\xi \otimes \alpha(X,Y) + \xi \otimes \delta_\tau\alpha(Z,X,Y) + \delta\tau\psi^*(Z,X) \otimes \beta(Y) + \psi^*(Z) \otimes \delta\tau\beta(X,Y).$$

Since

$$\psi \otimes q_\psi \in C^{1,2} = C^1_\tau(\mathcal{A}^*, \mathcal{A}^*) \otimes C^2_\tau(\mathcal{A}, C^\infty(M))$$

the exactness of $\psi \otimes q_\psi$ implies

$$\delta_\tau\alpha = 0,$$

$$\alpha(X,Y) = \alpha(Y,X).$$

Therefore

$$\psi(Z) \otimes q_\psi(X,Y) = \delta\xi(Z) \otimes \alpha(X,Y) + \psi^*(Z) \otimes \delta_\tau\beta(X,Y).$$

Now we observe that

$$\delta\tau\beta(X,Y) + \delta_\tau\beta(Y,X) = 0.$$

In final we get

$$\psi(Z) \otimes q_\psi(X,Y) = \delta\tau\xi(Z) \otimes \alpha(X,Y).$$

So we obtain

$$\psi(Z) = D^*_Z\xi,$$

$$q_\psi(X,Y) = \alpha(X,Y).$$

This end the proof of the corollary. □

From the mapping

$$C^{1,0} \ni \psi \to \psi \otimes q_\psi \in C^{1,2}$$

we deduce the canonical linear map

$$H^1_\tau(\mathcal{A}^*, \mathcal{A}^*) \ni [\psi] \to [\psi \otimes q_\psi] \in H^{1,2}(C).$$

We define another map

$$C^{1,0} \to C^{1,2}$$

by

$$\psi \to \psi \otimes \omega_\psi.$$

Here the differential 2-form ω is defined by

$$\omega_\psi(X,Y) = \frac{1}{2}[g(\psi(X),Y) - g(X,\psi(Y))].$$

This yields a linear map

$$H^1_\tau(\mathcal{A}^*, \mathcal{A}^*) \ni [\psi] \to [\psi \otimes \omega_\psi] \in H^{1,2}(C).$$

Now let (M, g, ∇, ∇^*) be a dually flat pair whose KV algebras are denoted by $\mathcal{A}$ and $\mathcal{A}^*$.
We identify the vector space $\Gamma(Hom(TM, TM))$ with the space $C^1_\tau(\mathcal{A}^*, \mathcal{A}^*)$.
We keep the notation $D^{\nabla\nabla^*}$, C^{ij}, δ_{ij} and q_ψ. Therefore, we can rephrase Lemma 4 as it follows.

Proposition 5. *For every section ψ of $Hom(TM, TM)$ the following assertions are equivalent.*

$$(1): \quad D^{\nabla\nabla^*}(\psi) = 0,$$

$$(2): \quad \delta_{12}(q_\psi) = 0$$

Here is an interesting feature. In a dually flat pair (M, g, ∇, ∇^*) we combine the double complex

$$\left\{C^{ij}, \delta_{ij}\right\}$$

with the correspondence

$$\psi \to q_\psi.$$

That allow us to replace the differential equation

$$FE^{\nabla\nabla^*}: \quad D^{\nabla\nabla^*}(\psi) = 0$$

by the homological equation

$$\delta_{12}(\psi) = 0.$$

That is a relevant impact on the global analysis of combinations of the KV cohomological methods with methods in the information geometry.

5.1.5. Computational Relations. Riemannian Foliations. Symplectic Foliations: Continued

We continue to relate the vector space of gauge homomorphisms and the differential topology. The tools we use are the KV cohomology and the Amari dualistic relation.

Let (M, g, D, D^*) be a dual pair. The vector subspace of g-preserving elements of $\mathcal{M}(D, D^*)$ is denoted by $\mathcal{M}(g, D, D^*)$. Thus every $\psi \in \mathcal{M}(g, D, D^*)$ satisfies the identity

$$g(\psi(X), Y) + g(X, \psi(Y)) = 0.$$

Now we fix a Koszul connection D^0 and we define the map

$$\mathcal{R}ie(M) \ni g \to D^g \in \mathcal{LC}(M).$$

by setting

$$g(D^g_X Y, Z) = Xg(Y, Z) - g(Y, D^0_X Z).$$

We define the non negative integers

$$n_x(D^0) = dim[\frac{\mathcal{M}_x(D^0, D^g)}{\mathcal{M}_x(g, D^0, D^g)}],$$

$$n(D^O) = \min_{x \in M} dim[\frac{\mathcal{M}_x(D^0, D^g)}{\mathcal{M}_x(g, D^0, D^g)}].$$

Lemma 5. *The integer $n(D^0)$ does not depend on the choice of $g \in \mathcal{R}ie(M)$.*

An Idea.

We fix a metric tensor g. For every $g^* \in \mathcal{R}ie(M)$ there is a unique g-symmetric vector bundle morphism $\phi \in \Sigma(g)$ such that

$$g^*(X,Y) = g(\phi(X),Y).$$

Therefore, we have

$$\phi^{-1} \circ \mathcal{M}(D^0, D^g) = \mathcal{M}(g^*, D^0, D^{g^*}),$$
$$\phi^{-1}\mathcal{M}(g, D^0, D^g) = \mathcal{M}(g^*, D^0, D^{g^*}).$$

Now one defines the numerical invariant $n(M)$.

Definition 34.

$$n(M) = \max\{n(D)|D \in \mathcal{SLC}(M)\}.$$

Given a Koszul connection ∇ the vector space of ∇-parallel differential 2-forms is denoted by $\Omega_2^\nabla(M)$.

Every dual pair (M, g, ∇, ∇^*) gives rise to the linear isomorphisms

$$(1): \quad \frac{\mathcal{M}(\nabla,\nabla^*)}{\mathcal{M}(g,\nabla,\nabla^*)} \ni [\psi] \to q_\psi \in \mathcal{S}_2^\nabla(M),$$
$$(2): \quad \mathcal{M}(g,\nabla,\nabla^*) \ni \psi \to \omega_\psi \in \Omega_2^\nabla(M).$$

The isomorphism (1) derives from the linear map

$$(1*): \quad \psi \to q_\psi(X,Y) = \frac{1}{2}[g(\psi(X),Y) + g(X,\psi(Y))].$$

The isomorphism (2) is defined by

$$(2*): \quad \psi \to \omega_\psi(X,Y) = \frac{1}{2}[g(\psi(X),Y) - g(X,\psi(Y))].$$

Proposition 6. *Let (M, g, ∇, ∇^*) be a dual pair. The inclusion map*

$$\mathcal{M}(g,\nabla,\nabla^*) \to \mathcal{M}(\nabla,\nabla^*)$$

induced the split short exact sequence

$$(*****): \quad 0 \to \Omega_2^\nabla(M) \to \mathcal{M}(\nabla,\nabla^*) \to \mathcal{S}_2^\nabla(M) \to 0.$$

Reminder.

According our previous notation elements of $\Omega_2^\nabla(M)$ are ∇-geodesic symplectic foliations. Those of $\mathcal{S}_2^\nabla(M)$ are ∇-geodesic Riemannian foliations. Thus we apply methods of the information geometry to relate the gauge geometry and the differential topolgy.

Digressions.

Our construction may open to new developments. Here are some unexplored perspectives.

(a) *A ∇-geodesic symplectic foliation $\omega \in \Omega^\nabla$ might carry richer structures such as Kahlerian structures.*

(b) *Suppose that the manifold M is compact and suppose that $g \in \mathcal{S}_2^{\nabla}(M)$ is a positive Riemannian foliation, viz*

$$g(X, X) \geq 0 \quad \forall X.$$

Then the theory of Molino may be applied to study g [38]. Therefore, the structure theorem of Molino tells that g gives rise to a Lie foliation whoses leaves are the adherences leaves of $\bar{g}$ [39].

(c) *In the principal bundle of first order linear frames of M the analog of a Koszul connection ∇ is a principal connection 1-form ω whose curvature form is denoted by Ω. The curvature form is involved in constructing characteristic classes of M, (the formalism of Chern-Weill.)*

At another side ∇-geodesic Riemannian foliations and ∇-geodesic symplectic foliations are Lie algebroids. They have their extrinsic algebraic topology. In particular the theory of integrable systems may be performed in every leaf of $\omega \in \Omega_2^{\nabla}(M)$. If one considers the α-connections in a statistical model those new insights may be of interest.

Here is an interpretation of the numerical invaraint $n(\nabla)$.

Theorem 10. *We assume there exists $\nabla \in \mathcal{SLC}(M)$ whose linear holonomy group $H(\nabla)$ is neither an orthogonal subgroup nor a symplectic subgroup. If $n(\nabla) > 0$ then the manifold M admits a couple $(\mathcal{F}_r, \mathcal{F}_s)$ formed by a ∇-geodesic Riemannian foliation $\mathcal{F}_r$ and a ∇-geodesic symplectic foliation $\mathcal{F}_s$.*

Proof. Let g be a Riemannian metric tensor in M. Since

$$n(\nabla) \leq dim(\mathcal{S}_2^{\nabla}(M)(x))$$

for all $x \in M$ there exists $\psi \in \mathcal{M}(\nabla, \nabla(g)$ such that

$$q_\psi \in \mathcal{S}_2^{\nabla}(M) \setminus \{0\},$$

$$\omega_\psi \in \Omega_2^{\nabla}(M) \setminus \{0\}.$$

The assumption that the holonomy group $H(\nabla)$ is neither orthogonal nor symlectic implies

$$Ker(q_\psi) \neq 0,$$

$$Ker(\omega_\psi) \neq 0.$$

Both distributions $Ker(q_\psi)$ and $Ker(\omega_\psi)$ are ∇-geodesic. Since ∇ is torsion free those distributions are completely integrable.

For all $X \in \Gamma(Ker(q_\psi))$ we have

$$L_X q_\psi = 0.$$

Mutatis mutandis for all $X \in \Gamma(Ker(\omega_\psi))$ we have

$$L_X \omega_\psi = 0.$$

From those properties we conclude

$(M, Ker(q_\psi), q_\psi)$ is a ∇-geodesic Riemannian foliation, $(M, Ker(\omega_\psi), \omega_\psi)$ is a ∇-geodesic symplectic foliation.

The theorem is proved. □

A Useful Comment.

Let (M, D) be a locally flat manifold whose KV algebra is denoted by $\mathcal{A}$. To every dual pair (M, g, D, D^g) we assign the short split exact sequence

$$0 \to \Omega_2^{\mathcal{A}}(M) \to \mathcal{M}(D, D^g) \to \mathcal{S}_2^{\mathcal{A}}(M) \to 0$$

which is canonically isomorphic to the short exact sequence

$$0 \to \mathcal{M}(g, D, D^g) \to \mathcal{M}(D, D^g) \to \mathcal{S}_2^{\mathcal{A}}(M) \to 0.$$

We have already defined the geometric invariant

$$r(D) = dim(H_\tau^2(\mathcal{A}, \mathbb{R})) - b_2(M).$$

We observe that the integer $n(D)$ is a byproduct of methods of the information geometry while $r(D)$ is a byproduct of homological methods. However the split short exact sequence (****) leads to the equality

$$n(D) = r(D).$$

Here is a straight consequence of the theorem we just proved.

Proposition 7. *Every odd-dimensional manifold M with $n(M) > 0$ admits a geodesic symplectic foliation.*

The dualistic relation of Amari has another significant impact on the differential topology.

Definition 35. *We consider a dual pair (M, g, ∇, ∇^*). Let $\psi \in \mathcal{M}(\nabla, \nabla^*)$.*

(1) The g-symmetric part of ψ, ψ^+ is defined by

$$g(\psi^+(X), Y) = \frac{1}{2}[g(\psi(X), Y) + g(X, \psi(Y))].$$

(2) The g-skew symmetric part of ψ, ψ^- is defined by

$$g(\psi^-(X), Y) = \frac{1}{2}[g(\psi(X), Y) - g(X, \psi(Y))].$$

Theorem 11. *Let (M, g, ∇, ∇^*) be a dual pair where (M, g) is a positive Riemannian manifold. Let $\psi \in \mathcal{M}(\nabla, \nabla^*)$.*

(1) The g-symmetric part ψ^+ is an element $\mathcal{M}(\nabla, \nabla^*)$ whose rank is constant.
(2) We have the g-orthogonal decomposition

$$TM = Ker(\psi^+) \oplus im(\psi^+).$$

(3) If both ∇ and ∇^* are torsion free then $Ker(\psi^+)$ and $Im(\psi^+)$ are completely integrable.

A Digression.

We recall that a statistical manifold is a torsion free dual pair (M, g, ∇, ∇^*). If the vector space $\mathcal{M}(\nabla, \nabla^*)$ is non-trivial then it plays an outstanding role in the differential topology of M. We define a canonical map of $\mathcal{M}(\nabla, \nabla^*)$ in the category of 2-webs by

$$\mathcal{M}(\nabla, \nabla*) \ni \psi \to Ker(\psi^+) \oplus im(\psi^+).$$

Thus one may regard elements of $\mathcal{M}(\nabla, \nabla^*)$ as orthogonal 2-webs in M.
We keep our previous notation. The we have

$$q_\psi(X,Y) = g(\psi^+(X),Y),$$

$$\omega_\psi(X,Y) = g(\psi^-(X),Y).$$

Now suppose that (M, g, ∇, ∇^*) is a dually flat pair whose KV algebras are noted $\mathcal{A}$ and $\mathcal{A}^*$. We take into account the inclusion

$$\mathcal{M}(\nabla, \nabla^*) \subset Z^1_\tau(\mathcal{A}^*, \mathcal{A}^*).$$

We have a map of $\mathcal{M}(\nabla, \nabla^*)$ in the space of de Rham 2-cocyles which is defined by

$$\mathcal{M}(\nabla, \nabla*) \ni \psi \to \omega_\psi.$$

Assume that the cocycle $\psi \in \mathcal{M}(\nabla, \nabla^*)$ is exact. Then there exists $\xi \in \mathcal{A}^*$ such that

$$\psi(X) = \nabla^*_X \xi \quad \forall X \in \mathcal{A}.$$

By the dualistic relation one easily sees that

$$\omega_\psi = d_{dR}(\iota_\xi g).$$

Therefore one gets a canonical linear map

$$H^1_\tau(\mathcal{A}^*, \mathcal{A}^*) \ni [\psi] \to [\omega_\psi] \in H^2_{dR}(M, \mathbb{R}).$$

The next subsubsection is devoted to a few consequences of items we just discussed.

5.1.6. Riemannian Webs—Symplectic Webs in Statistical Manifolds

We introduce Riemannian webs and symplectic webs and we discuss their impacts on the topology of statistical manifolds. We recall that a Riemannian foliation is a symmetric bilinear form $g \in \mathcal{S}_2(M)$ with the following properties

(a) $rank(g) = constant$,
(b) $L_X g = 0 \forall X \in \Gamma(Ker(g))$.

We put

$$\mathcal{D} = Ker(g).$$

To avoid confusions the pair $(\mathcal{D}, g)$ stands for the Riemannian foliation g.

Definition 36. *A Riemannian k-web is a family of k Riemannian foliations in general position $(\mathcal{D}_j, g_j)$, $j := 1, ..., k$. A symplectic k-web is a family of k symplectic foliations in general position $(\mathcal{D}_j, \omega_j); j := 1, ..., k)$.*

Let (M, g, D, D^*) be a dually flat pair whose KV algebras are denoted by $\mathcal{A}$ and $\mathcal{A}^*$. We recall the inclusion

$$\mathcal{M}(D, D^*) \subset Z^1_\tau(\mathcal{A}^*, \mathcal{A}^*).$$

Consider a statistical manifold (M, g, ∇, ∇^*). By the classical theorem of Frobenius every ∇-parallel differential system in M is completely integrable.

In a statistical manifold (M, g, ∇, ∇^*) we consider a ∇-geodesic Riemannian k-web

$$[\tilde{g}_j \in \mathcal{S}_2^{\nabla}(M); j : 1, ..., k].$$

The distributions

$$\mathcal{D}_j = Ker(\tilde{g}_j)$$

are in general position. We consider the family $\Psi_j^+ \in \Sigma(g)$ defined by

$$g(\Psi_j^+(X), Y) = \tilde{g}_j(X, Y).$$

We get the family of g-orthogonal 2-webs defined by

$$TM = Ker(\Psi_j^+) \oplus im(\Psi_j^+).$$

Mutatis mutandis we can consider a ∇-geodesic symplectic web

$$[\omega_j \in \Omega_2^{\nabla}(M); j := 1, ..., k].$$

There is family of g-skew symmetric gauge morphisms Ψ_j^- defined by

$$g(\Psi_j^-(X), Y) = \omega_j(X, Y).$$

Since ∇ and ∇^* are torsion free $Ker(\Psi_j^-)$ and $im(\Psi_j^-)$ are completely integrable. Since g is positive definite we get the 2-web

$$TM = Ker(\Psi_j^-) \oplus im(\Psi_j^-).$$

Further every leaf of $im(\Psi_j^-)$ is a symplectic manifold.

Definition 37. *Let $\omega \in \Omega_2^{\nabla}(M)$ be a non trivial symplectic foliation in a statistical manifold (M, g, ∇, ∇^*). Consider $\Psi^- \in \mathcal{M}(\nabla, \nabla^*)$ defined by*

$$g(\Psi^-(X), Y) = \omega(X, Y).$$

The differential 2-form ω is called simple if the foliation $Ker(\Psi^-)$ is simple.

In a statistical manifold (M, g, ∇, ∇^*) every non trivial symplectic web

$$[\omega_j; j := 1, ...] \subset \Omega_2^{\nabla}(M)$$

gives rise to a family of g-orthogonal 2-webs. So in this approach the role played by $\mathcal{S}_2^{\nabla}(M)$ is similar to the role played by $\Omega_2^{\nabla}(M)$. Our construction of Riemannian webs and symplectic webs in the category of dually flat pairs holds in the category of statistical manifolds.

At one side, in a dually flat pair (M, g, D, D^*) our approach yields linearizable webs. This property does not hold in all statistical manifolds.

At another side, in a statistical manifold (M, g, ∇, ∇^*) a Riemannian web

$$[\tilde{g}_j, j \in J] \subset \mathcal{S}_2^{\nabla}(M)$$

or symplectic web

$$[\omega_j, j \in J] \subset \Omega_2^{\nabla}(M)$$

gives rise to families of orthogonal 2-webs. This property does not hold in all dually flat pairs.

The considerations we just discussed may have remarkable impacts on the topological-geometrical structure of statistical manifolds.

From our brief discussion we conclude

Theorem 12. *Consider a statistical manifold (M, g, ∇, ∇^*). Every non trivial simple symplectic foliation $\omega \in \Omega_2^\nabla(M)$ is defined by a Riemannian submersion on a symplectic manifold.*

Corollary 4. *Every non trivial simple symplectic web $[\omega_j, j \in J] \subset \Omega_2^\nabla(M)$ is defined by family of Riemannian submersions on symplectic manifols.*

5.2. The Hessian Information Geometry, Continued

In [52] Shima pointed out that the Fisher informations of many classical statistical models are Hessian metric tensors.

At another side the exponential models (or exponential family) may be considered as optimal Statistical models.

As already mentioned there does not exists any criterion for knowing whether a given statistical model is isomorphic to an exponential model [22], [13]

In the category of regular models, viz models whose Fisher information is a Riemannian metric, it is known that the Fisher information of an exponential model is a Hessian Riemannian metric [18,52].

In this subsection we address the general situation. We give a cohomological characterization of exponential models. We also introduce a new numerical invariant r^b which measures how far from being an exponential family is a given statistical model. See the Appendix A to this paper.

We recall that the metric tensor g of a Hessian structure (M, D, g) is a 2-cocycle of the KV complex $[C^*_{KV}(\mathcal{A}, \mathbb{R}), \delta_{KV}]$.

To non specialists we go to recall some constructions in the geometry of Koszul [2,52], see also [53].

Let $((M, x^*), D)$ be a pointed locally flat manifold whose universal covering is denoted by $(\tilde{M}, \tilde{D})$. Here the topological space $\tilde{M}$ is the set of homotopy class of continuous paths

$$\{([0,1], 0) \to (M, x^*)\}.$$

Its topology is the compact-open topology. Let c be a continuous path with

$$c(0) = x^*.$$

For $s \in [0,1]$ the parallel transport of $T_{x^*}M$ in $T_{c(s)}M$ is denoted by τ_s. One defines $Q(c) \in T_{x^*}M$ by

$$Q(c) = \int_0^1 \tau_s^{-1}(\frac{dc(s)}{ds})ds.$$

The tangent vector $Q(c)$ depends only on the homotopy class of the path c. Therefore, Q defines a local homeomorphism

$$\tilde{Q} : \tilde{M} \to T_{x^*}M.$$

Let $\pi_1(x^*)$ be the fundamental group at x^*. Let $[\gamma] \in \pi_1(x^*)$. The natural left action

$$\pi_1(x^*) \times \tilde{M} \to \tilde{M}$$

is given by the composition of paths, viz

$$[\gamma].[c] = [\gamma \circ c].$$

The parallel transport along a loop $\gamma(t)$ yields a linear action of $\pi_1(x^*)$ in $T_{x^*}M$ which is denoted by $f([\gamma])$.

Let $[\gamma], [\gamma^*] \in \pi_1(x^*)$. The composition of paths leads to the formula

$$Q([\gamma].[\gamma^*]) = f([\gamma])Q([\gamma^*]) + Q([\gamma]).$$

The last relation shows that the pair (f, Q) is an affine representation of $\pi_1(x^*)$ in $T_{x^*}M$. This representation is called the holonomy representation of the locally flat manifold (M, D). The group $(f, Q)(\pi_1(x^*))$ is called the affine holonomy group of (M, D). Its linear component $f(\pi_1(x^*))$ is called the linear holonomy group of (M, D).

Definition 38 ([2])**.** *An m-dimensional locally flat structure (M, D) is called hyperbolic if $\tilde{Q}(\tilde{M})$ is a convex domain not containing any straight line in $T_{x^*}M$.*

Definition 39. *A locally flat manifold (M, D) is complete if the map $\tilde{Q}$ is a diffeomorphism onto $T_{x^*}M$.*

Among the major open problems in the theory of space groups is the conjecture of Markus.

Conjecure of Markus: a compact locally flat manifold (M, D) whose linear holonomy group is unimodular is complete.

Before pursuing we recall KV cohomological version of Theorem 3 as in [2].

Theorem 13 ([2])**.** *A necessary condition for a locally flat manifold (M, D) being hyperbolic is that (M, D) carries a positive Hessian structure whose metric tensor is exact in the KV complex of (M, D). This condition is sufficient if M is compact.*

We have already mentioned a notable consequence of this theorem of Koszul. In the category of compact locally flat manifolds the subcategory of hyperbolic locally flat structures is the same thing as the category of positive exact Hessian structures. So The geometry of compact hyperbolic local flat manifolds is an appropriate vanishing theorem.

In the preceding sections the family of Hessian metrics in a locally flat manifold (M, D) is denoted by $\mathcal{H}es(M, D)$. Therefore, $\mathcal{H}^+es(M)$ stands for the sub-family of positive Hessian metric tensors. It is a convex cone in $\mathcal{R}ie(M)$.

We have already used the KV complex for expressing the dually flatness. More precisely let (M, D) be a fixed locally flat manifold whose KV algebra is noted $\mathcal{A}$. A dual pair $(M, g, D, D(g))$ is dually flat if and only if $g \in Z^2_{KV}(\mathcal{A}, \mathbb{R})$. Therefore, every dually flat pair (M, g, D, D^*) yields two cohomology classes

$$[g]_D \in H^2_{KV}(\mathcal{A}, \mathbb{R}),$$

$$[g]_{D^*} \in H^*_{KV}(\mathcal{A}^*, \mathbb{R}).$$

Thereby, we can use methods of the information geometry for rephrasing Theorem 3 as in [2].

Theorem 14. *A necessary condition for (M, D) being hyperbolic is the existence a positive dually flat pair (M, g, D, D^*) such that*

$$[g] = 0 \in H^2_{KV}(\mathcal{A}, \mathbb{R}).$$

If M is compact this (vanishing) condition is sufficient.

About the geometry of Koszul the non specialists are referred to [2,7,8,52] and bibliography therein [12].

About applications of the geometry of Koszul the readers are refereed to [12,13,54,55].

About relationships between the theory of deformation and the theory of cohomology, the readers are referred to [9,27,56].

5.3. The α-Connetions of Chentsov

Still, nowadays, the information geometry deals with models (Θ, P) whose underlying m-dimensional manifold Θ is an open subset of the euclidean space $\mathbb{R}^m$. Further the Fisher information g is assumed to be regular, viz (Θ, g) is a Riemannian manifold. In this paper this classical information geometry is called the local information geometry. This "local nature" will be explained in Part B of this paper.

At the moment we plan to investigate other topological properties of the local statistical models.

Let (Θ, P) be an m-dimensional local statistical model for a measurable set (Ξ, Ω). The manifold Θ is a domain in the Euclidean space $\mathbb{R}^m$. The function P is non negative. It is defined in $\Theta \times \Xi$. We recall the requirements P is subject to.

(1) $\forall \xi \in \Xi$ the function
$$\Theta \ni \theta \rightarrow P(\theta, \xi)$$
is smooth.

(2) $\forall \theta \in \Theta$ the triple
$$(\Xi, \Omega, P(\theta, -))$$
is a probability space.

(3) $\forall \theta, \theta^* \in \Theta$ with $\theta \neq \theta^*$ there exists $\xi \in \Xi$ such that
$$P(\theta, \xi) \neq P(\theta', \xi).$$
(with the requirement (3) (Θ, P) is called identifiable.)

(4) The differentiation d_θ commutes with the integration $\int_\Xi$. The Fisher information of a model (Θ, P) is the symmetric bi-linear form g which is defined by
$$g(X,Y)(\theta) = \int_\Xi P(\theta, \xi)[[d_\theta log(P)]^{\otimes 2}(X,Y)](\theta, \xi)d\xi.$$
Here d_θ stands for the differentiation with respect to the argument $\theta \in \Theta$.

(5) The Fisher information is positive definite.

Remark 3. *The Fisher information g can be defined using any Koszul connection ∇ according to the following formula*
$$g(X,Y)(\theta) = -\int_\Xi P(\theta, \xi)[(\nabla^2 log(P))(X,Y)](\theta, \xi)]d\xi.$$

The right member of the last equality does not depend on the choice of the Koszul connection ∇.

From now on, we deal with a generic statistical model. This means that we do not assume the Fisher information g is definite.

Let $\theta = (\theta_1, ..., \theta_m)$ be a system of Euclidean coordinate functions in $\mathbb{R}^m$. To every real number α is assigned the torsion free Koszul connection ∇^α whose Christoffel symbols in the coordinate (θ_j) are

$$\Gamma^\alpha_{ij,k} = \int_\Xi P(\theta, \xi)[(\frac{\partial^2 log(P(\theta, \xi))}{\partial\theta_i\partial\theta_j} + \frac{1-\alpha}{2}\frac{\partial log(P(\theta, \xi))}{\partial\theta_i}\frac{\partial log(P(\theta, \xi))}{\partial\theta_j})\frac{\partial log(p(\theta, \xi))}{\partial\theta_k}]d\xi.$$

This definition agrees with any affine coordinate change. We put $\partial_i = \frac{\partial}{\partial\theta_i}$. We have

$$\nabla^\alpha_{\partial_i}\partial_j = \sum_k \Gamma^\alpha_{ij,k}\partial_k.$$

Now we assume a model (Θ, P) is regular. Then the Christoffel symbols and the Fisher information are related by the formula

$$\Gamma^{\alpha}_{ij,k} = g(\nabla^{\alpha}_{\partial_i}\partial_j, \partial_k).$$

Further every quadruple $(\Theta, g, \nabla^{\alpha}, \nabla^{-\alpha})$ is a statistical manifold [18,48].

Thus we have a family of splitting short exact sequences

$$0 \to \Omega^{\nabla^{\alpha}}(\Theta) \to \mathcal{M}(\nabla^{-\alpha}, \nabla^{\alpha}) \to \mathcal{S}_2^{\nabla^{\alpha}}(\Theta) \to 0.$$

So the machinery we have developed in the preceding sections can be performed to explore the differential topology of regular local statistical models. For that purpose the crucial tool is the family of vector space

$$\mathcal{S}_2^{\alpha}(\Theta) = \mathcal{S}_2^{\nabla^{\alpha}}(\Theta).$$

We consider the abstract trivial bundle

$$\cup_{\alpha}[\mathcal{S}^{\alpha} \times \{\alpha\}] \to \mathbb{R}$$

whose fiber over $\alpha \in \mathbb{R}$ is $\mathcal{S}^{\alpha}(\Theta)$. To every $B \in \mathcal{S}^{\alpha}(\Theta)$ we assign the unique $\psi^{+} \in \Sigma(g)$ defined by

$$g(\psi^{+\alpha}(X), Y) = B(X, Y).$$

The machinery in the preceding subsection leads to the following proposition.

Proposition 8. *We assume (Θ, P) is regular.*

(1) Every non zero singular section

$$\mathbb{R} \ni \alpha \to B_{\alpha} \in \mathcal{S}^{\alpha}(\Theta)$$

gives rise the family of (g-orthogonal) 2-web

$$T\Theta = Ker(\psi^{+\alpha}) \oplus im(\psi^{+\alpha}).$$

Further according to the notation used previously (B_{α}) is a family of Riemannian foliations as in [39,40].

(2) By replacing $\mathcal{S}^{\alpha}(\Theta)$ by $\Omega_2^{\nabla}(\Theta)$ every non zero singular section

$$\mathbb{R} \ni \alpha \to \omega_{\alpha} \in \Omega_2^{\nabla}(\Theta)$$

yields a family of symplectic foliations ω_{α}.

Reminder.

(i) $\alpha \to B_{\alpha}$ is called a singular section if each B_{α} is non inversible.
(ii) $\alpha \to \omega_{\alpha}$ is called a simple section if each ω_{α} is simple.

We have used some gauge morphisms to construct Riemannian submersions of statistical manifolds over symplectic manifolds. The notions we just introduced lead to similar situations.

Theorem 15. *Let (Θ, P) be a regular statistical model whose Fisher information is denoted by g. Every simple non zero singular section*

$$\mathbb{R} \ni \alpha \to \omega_{\alpha} \in \Omega^{\alpha}(\Theta)$$

defines an α-family of Riemannian submersions of (Θ, g) onto symplectic manifolds.

Proposition 9. *We assume a model* (Θ, P) *is regular. For every nonzero real number* α *one has*

$$\mathcal{M}(\nabla^\alpha, \nabla^{-\alpha}) \cap \mathcal{M}(\nabla^{-\alpha}, \nabla^\alpha) = \mathcal{M}(g, \nabla^{-\alpha}, \nabla^\alpha) + \mathcal{M}(g, \nabla^*, \nabla).$$

The Sketch of Proof. The proof is based on the short exact sequences

$$0 \to \mathcal{M}(g, \nabla^\alpha, \nabla^{-\alpha}) \to \mathcal{M}(\nabla^\alpha, \nabla^-\alpha) \to \mathcal{S}_2^\nabla(\Theta) \to 0,$$

$$0 \to \mathcal{M}(g, \nabla^{-\alpha}, \nabla^\alpha) \to \mathcal{M}(\nabla^{-\alpha}, \nabla^\alpha) \to \mathcal{S}_2^{-\nabla}(\Theta) \to 0.$$

Let us suppose that the conclusion of the the proposition fails. Then there is a nonzero 2-form $B \in \mathcal{S}_2^{\nabla^\alpha}(\Theta) \cap \mathcal{S}_2^{\nabla^{-\alpha}}(\Theta)$.

(1) If $Ker(B) = 0$, then both ∇^α and $\nabla^{-\alpha}$ coincide with the Levi-Civita connection of B. This implies $\alpha = 0$, this contradicts our choice of α.
(2) If $Ker(B) \neq 0$ then $Ker(B)$ and $Ker(B)^\perp$ are geodesic for both ∇^α and $\nabla^{-\alpha}$. Thus the pair $(Ker(B), Ker(B)^\perp)$ defines a g-orthogonal 2-web.

At one side, by the virtue of the reduction theorem as in [18] every leaf F of $Ker(B)^\perp$ inherits a dual pair $(F, g_F, \nabla_F^\alpha, \nabla_F^{-\alpha})$.

At another side, B gives rise to the Riemannian structure (F, B). Furthermore both ∇_F^α and $\nabla_F^{-\alpha}$ are torsion free metric connections in (F, B). Thereby one gets

$$\nabla_F^\alpha = \nabla_F^{-\alpha}$$

The last equality holds if and only if $\alpha = 0$. This contradicts our assumption. The proposition is proved. □

The proposition above is a separation criterion for α-connections in the following sense. For every nonzero real number α the vector subspace $\mathcal{S}_2^{\nabla^\alpha}(\Theta)$ is transverse to $\mathcal{S}_2^{\nabla^{-\alpha}}(\Theta)$ in the vector space $\mathcal{S}_2(\Theta)$ of symmetric forms in Θ.

5.4. The Exponential Models and the Hyperbolicity

A challenge is the search of a criterion for deciding whether a model (Θ, P) is an exponential family. That is the challenge in [22]. Still, nowadays, this problem is open.

The Fisher information of a regular exponential model is a Hessian Riemannian metric. We are going to demonstrate that the converse is globally true. Our proof is based on cohomological arguments.

In the Appendix A to this paper we introduce a new numerical invariant $r^b(\Theta, P)$ which measures how far from being an exponential family is a model (Θ, P).

The invariant r^b derives from the global analysis of differential operators

$$D_\alpha = D_{\nabla^\alpha},$$

$$D^\alpha = D^{\nabla^\alpha}.$$

Now we are going to provide a cohomological characterization of exponential models.

Before pursuing we recall a definition.

Let $\theta_j, j := 1, ..., m$ be a system of Euclidean coordinate functions of $\mathbb{R}^m$.

Definition 40. *[18] An m-dimensional statistical model* (Θ, P) *is called an exponential model for* (Ξ, Ω) *if there exist a map*

$$\Xi \ni \xi \to [C(\xi), F_1(\xi), ..., F_m(\xi)] \in \mathbb{R}^{m+1}$$

and a smooth function

$$\Theta \ni \theta \to \psi(\theta) \in \mathbb{R}$$

such that

$$P(\theta,\xi) = exp(C(\xi) + \sum_1^m F^j(\xi)\theta_j - \psi(\theta_1,...,\theta_m)).$$

Theorem 16. *Let (Ξ,Ω) be a measurable set and let (Θ,P) be an m-dimensional statistical model for (Ξ,Ω). The Fisher information of (Θ,P) is denoted by g. The following statements are equivalent.*

(1) There exists $\nabla \in \mathcal{LF}(\Theta)$ such that

$$\delta_{KV}g = 0,$$

(2) The model (Θ,P) is an exponential model.

Demonstration.

$(2) \Rightarrow (1)$.

We assume that (2) holds. Then we fix a system of affine coordinate functions

$$\theta = (\theta_1,...,\theta_m).$$

By the virtue of (2) we have

$$P(\theta,\xi) = exp(C(\xi) + \sum_1^m F^j(\xi)\theta_j - \psi(\theta)).$$

Here $\psi \in C^\infty(\Theta)$ and $(C,F)(\xi) = (C(\xi),F^1(\xi),...,F^m(\xi)) \in \mathbb{R}^{m+1}$. Therefore, one has

$$\frac{\partial^2 log(P(\theta,\xi))}{\partial\theta_i\partial\theta_j} = -\frac{\partial^2\psi}{\partial\theta_i\partial\theta_j}.$$

Thereby one can write

$$-\int_\Xi P(\theta,\xi)\frac{\partial^2 log(P(\theta,\xi))}{\partial\theta_i\partial\theta_j} = \frac{\partial^2\psi(\theta,\xi)}{\partial\theta_i\partial\theta_j}.$$

This shows that we have

$$g = \delta_{KV}(d\psi) \in B^2_{KV}(\mathcal{A},\mathbb{R}).$$

The implication $(2) \to (1)$ is proved.

$(1) \Rightarrow (2)$.

We use a strategy similar to that used in [52]. However our arguments do not depend on rank(g). Let $\nabla \in \mathcal{LF}(\Theta)$ whose KV algebra is denoted by $\mathcal{A}$. We assume

$$g \in Z^2_{KV}(\mathcal{A},C^\infty(\Theta)).$$

Thus we have

$$\delta_{KV}g = 0.$$

In (Θ,∇) we fix a system of local affine coordinate functions

$$\{\theta_1,...,\theta_m\}$$

whose domain is a convex open subset U. We write the matrix of Fisher information g in the basis

$$\left\{\frac{\partial}{\partial\theta_j}\right\},$$

namely

$$g = \sum g_{ij} d\theta_i d\theta_j.$$

Here

$$g_{ij} = g(\frac{\partial}{\partial\theta_i}, \frac{\partial}{\partial\theta_j}).$$

The assumption

$$\delta_{KV} g = 0$$

is equivalent to the system

$$\frac{\partial g_{ij}}{\partial\theta_k} - \frac{\partial g_{kj}}{\partial\theta_i} = 0$$

for all i, j, k.

We use a notation which is used in [52]. We consider the differential 1-forms

$$h_j = \sum_i g_{ij} d\theta_i.$$

Every differential 1-form h_j is a de Rham cocycle. By the Lemma of Poincaré the convex open set U supports smooth functions $\phi_j, j =: 1, ..., m$ which have the following property

$$d\phi_j = h_j.$$

We put

$$\omega = \sum_j \phi_j d\theta_j.$$

Then we have

$$(\delta_{KV}\omega)(\frac{\partial}{\partial\theta_i}, \frac{\partial}{\partial\theta_j}) = g_{ij}.$$

Thus the differential 1-form $\sum_j \phi_j d\theta_j$ is de Rham closed.

Since U is convex it supports a local smooth function Ψ such that

$$d\Psi = \sum \phi_j d\theta_j.$$

So we get

$$g(\frac{\partial}{\partial\theta_i}, \frac{\partial}{\partial\theta_j}) = \frac{\partial^2\psi}{\partial\theta_i\partial\theta_j}.$$

To continue we fix $\theta_0 \in U$ and we consider the function

$$\theta \rightarrow a(\theta)$$

which is defined in U by

$$a(\theta) = \int_\Xi P(\theta_0, \xi)[\psi(\theta) + log(P(\theta, \xi))]d\xi.$$

Now recall that the integration $\int_\Xi$ commutes with the differentiation $\frac{d}{d\theta}$. Therefore, $\forall i, j \leq dim(\Theta)$ one has

$$\frac{\partial^2 a}{\partial\theta_i\partial\theta_j}(\theta) = \int_\Xi P(\theta_0, \xi)\frac{\partial^2(\psi + log(P))}{\partial\theta_i\partial\theta_j}(\theta, \xi)d\xi.$$

The identities above show that

$$\frac{\partial^2 a}{\partial\theta_i\partial\theta_j}(\theta_0) = 0$$

for all $\theta_0 \in U$. Thereby the function

$$\theta \to \psi(\theta) + log(P(\theta,\xi))$$

depends affinely on $\theta_1, ..., \theta_m$. So there exists a $\mathcal{R}^{m+1}$-valued function

$$\Xi \ni \xi \to (C(\xi), F_1(\xi), ..., F_m(\xi)) \in \mathcal{R}^{m+1}$$

such that

$$\psi(\theta) + log(P(\theta,\xi)) = C(\xi) + \sum_1^m F_i(\xi)\theta_i.$$

In final we get

$$P(\theta,\xi) = exp(C(\xi) + \sum F_i(\xi)\theta_i - \psi(\theta))$$

for all $(\theta,\xi) \in U \times \Xi$. So (Θ, P) is locally an exponential family.

Since the exponential function is injective this local property of (Θ, P) is a global property, in other words the model is globally an exponential model. In final assertion (1) implies assertion(2). This ends the demonstration of the theorem

Some Comments.

(i) It must be noticed that the demonstration above is independent of the rank of the Fisher information g. Therefore, the theorem holds in singular statistical models.

(ii) In regular statistical models the theorem above leads to the notion of e-m-flatness as in [18].

(iii) When the Fisher information g is semi-definite the dualistic relation is meaningless. However data $(\Theta, g, \nabla, \nabla^*)$ may be regarded as data depending on the transversal structure of the distribution $Ker(g)$.

(iv) In the analytic category the Fisher information is a Riemannian foliation. Therefore, both the information geometry and the topology of information are transversal concepts. This may be called the transversal geometry and the transversal topology of Fisher-Riemannian foliations.

(v) The theorem above does not solve the question as how far from being an exponential family is a given model. It only tells us that exponential families are objects of the Hessian geometry.

The framework for addressing the challenge just mentioned is the theory of invariants. That is the purpose of a forth going work. Some new results are anounced in the Appendix A to this paper.

6. The Similarity Structure and the Hyperbolicity

We consider a dually flat pair (M, g, ∇, ∇^*). Both (M, ∇, g) and (M, g, ∇^*) are locally hyperbolic in the sense of [2]. So they locally support the geometry of Koszul. That is a consequence of the classical Lemma of Poincare.

Every point of M has an open neighborhood U supporting a local de Rham closed differential 1-forms

$$\omega \in C^1_{KV}(\mathcal{A}, \mathbb{R})$$

and

$$\omega^* \in C^1_{KV}(\mathcal{A}^*, \mathbb{R})$$

subject to the following requirements

$$g|U = \delta\omega,$$

$$g|U = \delta^*\omega^*.$$

By the virtue of Theorem 3 as in [2], for both (M, g, ∇) and (M, g, ∇^*) being globally hyperbolic it is necessary that

$$[g] = 0 \in H^2_{KV}(\mathcal{A}, \mathbb{R})$$

and

$$[g] = 0 \in H^2_{KV}(\mathcal{A}^*, \mathbb{R}).$$

Every choice of local differential 1-forms ω and ω^* gives rise to a unique pair of local similarity vector fields (H, H^*), viz

$$\nabla_X H = X,$$

$$\nabla^*_X H^* = X$$

for all vector fields X. The vector fields H and H^* are Riemannian gradients of ω^* and of ω respectively. This means that those differential 1-forms are defined by

$$\omega = \iota_H g,$$

$$\omega^* = \iota_{H^*} g.$$

Here ι_H stands for the inner product by H, viz

$$\iota_H g(X) = g(H, X).$$

This short discussions lead to the following statement

Theorem 17. *Let (M, g, ∇, ∇^*) be a compact dually flat pair whose KV algebras are denoted by $\mathcal{A}$ and by $\mathcal{A}^*$. The following assertions are equivalent*

(1) *The locally flat manifold (M, ∇) is hyperbolic,*
(2) *the locally flat manifold (M, ∇^*) admits a global similarity vector field H^*.*

Definition 41. *Let $\nabla \in \mathcal{LC}(M)$.*

(1) *The gauge structure (M, ∇) is called a similarity structure if ∇ admits a global similarity vector field $H \in \mathcal{X}(M)$.*
(2) *A dual pair (M, g, ∇, ∇^*) is a similarity dual pair if either (M, ∇) or (M, ∇^*) is a similarity structure.*

The following proposition is a straightforward consequence of our definition.

Proposition 10. *If a gauge structure (M, ∇) is flat and is locally a similarity structure, then (M, ∇) is a locally flat manifold*

7. Some Highlighting Conclusions

In this Part A our aim has been to address various purposes involving the theory of KV homology. Doing that we have pointed significant relationships between some major topics in mathematics and the local information geometry. Those relationships might be sources of new investigations.

We summarize some relevant relationships we have been concerned with.

7.1. The Total KV Cohomology and the Differential Topology

We have addressed the existence problem for a few major objects of the differential topology. Riemannian foliations and symplectic foliations. Riemannian webs and their linearization problem.

To those questions we have obtained substantial solutions in the category of locally flat manifolds. The cohomological methods we have used are based on the split exact short cohomology sequence

$$0 \rightarrow H^2_{dR}(M, \mathbb{R}) \rightarrow H^2_{\tau}(\mathcal{A}, \mathbb{R}) \rightarrow \mathcal{S}^{\mathcal{A}}_2(M) \rightarrow 0.$$

7.2. The KV Cohomology and the Geometry of Koszul

The Hessian Geometry is a byproduct local vanishing Theorems in the theory of KV cohomology. The geometry of Koszul is a byproduct of global vanishing Theorem in the same sitting.

7.3. The KV Cohomology and the Information Geometry

The category of finite dimensional statistical models for a measurable set (Ξ, Ω) contains the subcategory of finite dimensional Hessian manifolds. From this viewpoint the Hessian information geometry is nothing but the exponential information geometry (i.e., the geometry of exponential families and their genelarizations). The framework for those purposes is closely related to vanishing Theorems in the theory of KV cohomology.

At another side cotangent bundles of Hessian manifolds are Kaehlerian manifolds. This aspect has been discussed by many authors, see [52] and the bibliography ibidem.

7.4. The Differential Topology and the Information Geometry

A lot of outstanding links between the differential topology and the information geometry are based on the dualistic relation of Amari. This approach leads to significant results in the category of statistical manifolds. In a statistical manifold (M, g, ∇, ∇^*) we have introduced the splitting short exact sequence

$$0 \rightarrow \Omega^{\nabla}_2(M) \rightarrow \mathcal{M}(\nabla, \nabla^*) \rightarrow \mathcal{S}^{\nabla}_2(M) \rightarrow 0.$$

Here (i) $\Omega^{\nabla}_2(M)$ is the space of ∇-geodesic symplectic foliations in M; (ii) $\mathcal{S}^{\nabla}_2(M)$ is the space of ∇-geodesic Riemannian foliations in M.

The numerical invariant $n(\nabla)$ has outstanding impacts on the differential topology of M. See our results on orthogonal 2-webs and on Riemannian submersions on symplectic manifolds.

7.5. The KV Cohomology and the Linearization Problem for Webs

In a locally flat pair (M, g, ∇, ∇^*) we consider the short exact sequence

$$O \rightarrow \Omega^{\nabla}_2(M) \rightarrow \mathcal{M}(\nabla, \nabla *) \rightarrow \mathcal{S}^{\nabla}_2(M) \rightarrow O.$$

The linearization of webs of is a difficult outstanding problem in the differential topology. $\mathbb{G}^k[\Omega^{\nabla}_2(M)]$ stands for the family formed by

$$[\omega_1, ..., \omega_k] \subset \Omega^{\nabla}_2(M)$$

such that

$$dim[\Sigma_j Ker(\omega_j)] = \min[dim(M), \Sigma_j dim(Ker(\omega_j))].$$

$\mathbb{G}^k[\mathcal{S}^{\nabla}_2(M)]$ stands for the family formed by

$$[B_1, ..., B_k] \subset \mathcal{S}^{\nabla}_2(M)$$

such that

$$dim[\Sigma_j Ker(B_j)] = \min[dim(M), \Sigma_j dim(Ker(B_j))].$$

(i) Elements of $\mathbb{G}^p[\Omega^{\nabla}_2(M)]$ are LINEARIZABLE symplectic k-webs.
(ii) Elements of $\mathbb{G}^p[\mathcal{S}^{\nabla}_2(M)]$ are LINEARIZABLE Riemannian k-webs.

We have introduced the double complex

$$\left\{C: \quad C^{ij} = C^i_\tau(\mathcal{A}^*, \mathcal{A}^*) \otimes C^j_\tau(\mathcal{A}, C^\infty(M)), \quad \delta_{ij}\right\}.$$

It gives rise to spectral sequences which may be useful for computing the KV cohomologies $H *_\tau (\mathcal{A}^*, \mathcal{A}^*)$ and $H *_{KV} (\mathcal{A}, C^\infty(M))$. That is not the purpose of this paper. However this double complex is useful for replacing the first order differential equation

$$D^{\nabla\nabla^*}(\psi) = 0$$

by the homological equation

$$\delta_{1,2} q_\psi = 0.$$

We have proved the homological nature of the space of gauge homomorphisms $\mathcal{M}(\nabla, \nabla^*)$. This is useful for relating the image of $\mathcal{M}(\nabla, \nabla^*)$ in $H^1_\tau(\mathcal{A}^*, \mathcal{A}^*)$ and the pair $H^2_{dR}(M), H^{1,2}(C)$.

8. B. The Theory of StatisticaL Models

In the introduction of this paper we have recalled the problem raised by Peter McCullagh.

What is a statistical model [30]?

By the way we have recalled a variant request of Misha Gromov.

In a Search for a Structure. The Fisher Information [15,16].

McCullagh and Gromov choose the same framework for addressing their purpose, The theory of category. This Part B is devoted to the same purpose.

Further the moduli space of isomorphism class of objects of a category $\mathcal{C}$ is denoted by $[\mathcal{C}a]$. In general it is difficult to find an invariant inv_a which encodes $[\mathcal{C}a]$. Subsequently to the questions raised by McCullagh and by Gromov the moduli space of isomorphism class of statistical models is discussed in this Part B. Nowadays, there exists a well established theory of statistical models. The classical references are Amari [17], Amari and Nagaoka [18]. Other remarkable references are Barndorff-Nielsen (Indian Journal of Mathematics 29, Ramanujan Centenary Volume) [21,24], Kass and Vos [37], Murray-Rice (Chapter 1, Section 15 in [22]). In Part A of this paper we have been dealing with this current theory. It has been called the local theory. We suggest reading the attempt by McCullagh to establish a conceptually consistent theory of statistical models [30]. In its time, the paper of McCullagh had been the object of controversy and questions.

We are aimed at re-establishing the theory of statistical models. Our motivations have emerged from some criticisms.

The current theory presents some deficiencies that we plan outlining. (i) A weakness of the current theory is its lacking in geometry; (ii) In the literature on the information geometry many references define an m-dimensional statistical model as an open subset of an Euclidean space $\mathbb{R}^m$. Though this definition may be useful for dealing with coordinate functions, it is topologically and geometrically useless. Let Γ be the group of measurable isomorphisms of a measurable set (Ξ, Ω). The information geometry of a statistical model $\mathbb{M}$ includes the geometry in the sense of Erlangen program of the pair $[\mathbb{M}, \Gamma]$.

Let $\mathbb{M}$ and $\mathbb{M}^*$ be m-dimensional statistical models for the same measurable set (Ξ, Ω). An isomorphism of $\mathbb{M}$ on $\mathbb{M}*$ looks like an sufficient statistic. The geometries $[\mathbb{M}, \Gamma]$ and $[\mathbb{M}*, \Gamma]$ provide the same information. So the impact on the applied information geometry of the theory of moduli space is notable. Subsequently the search for characteristic invariants presents a challenge. An invariant is called characteristic if it determine a model up to isomorphism. So a characteristic invariant encodes the moduli space. That increases the interest in the search of both McCullagh and Gromov.

The Fisher information of widely used models are Hessian metrics [52]. This observation is relevant. However the Fisher information is not a characteristic invariant.

We intend to face the following challenges.
Challenge 1. Revisit the theory of geometric statistical models for measurable sets.
Challenge 2. The Search for a geometric characteristic invariant for statistical models. We recall that such an invariant will encode the points the moduli space of models. Before continuing we recall some definitions.

Definition 42. *A geometric invariant of a model for* (Ξ, Ω) *is a datum which is invariant under the action of the symmetry group* $Aut(\Xi, \Omega)$.

The framework which is useful for re-establishing the theory of statistical models is the category of locally trivial fiber bundles.

As we have mentioned the need for introducing a new theory of statistical model emerges from some criticisms. We recall the definition a statistical model [18,22,24].

Definition 43. *An m-dimensional statistical model for a measurable set* (Ξ, Ω) *is a pair* (Θ, P) *having the properties which follow.*

(1) *The manifold* Θ *is an open subset of the m-dimensional Euclidean space* $\mathbb{R}^m$.
(2) *P is a positive real valued function*

$$\Theta \times \Xi \ni (\theta, \xi) \to P(\theta, \xi) \in \mathbb{R}$$

subject to the requirements which follow.
(3) *The function* $P(\theta, \xi)$ *is differentiable with respect to* $\theta \in \Theta$.
(4) *For every fixed* $\theta \in \Theta$ *one set*

$$P_\theta = P(\theta, -)$$

then the triple

$$(\Xi, \Omega, P_\theta)$$

is a probability space, viz

$$\int_\Xi P_\theta(\xi) d\xi = 1$$

Furthermore the operation of differentiation

$$d_\theta = \frac{d}{d\theta}$$

commutes with the operation of integration $\int_\Xi$.
(5) (Θ, P) *is identifiable, viz for* $\theta, \theta^* \in \Theta$

$$P_\theta = P_{\theta^*}$$

if and only if

$$\theta = \theta^*$$

(6) *The Fisher information*

$$g_\theta(X, Y) = \int_\Xi P(\theta, \xi)[d_\theta log(P(\theta, \xi))]^{\otimes 2}(X, Y) d\xi$$

is positive definite.

Some Criticisms.

The First Critique

The first critique arises from requirement (5).

From the viewpoint of fiber bundles the requirement (5) is useless. Consider the Cartesian product

$$\mathcal{E} = \Theta \times \Xi.$$

That is the same thing as the trivial fiber bundle

$$\mathcal{E} \ni (\theta, \xi) \to \pi(\theta, \xi) = \theta \in \Theta.$$

Therefore P_θ is the restriction to the fiber $\mathcal{E}_\theta$ of the function P.

The Second Critique

The second critique emerges from the requirement (1).

This requirement (1) is too restrictive. It excludes many interesting compact manifolds such as flat tori, euclidean sphere, compact Lie groups.

The Third Critique

From the viewpoint of the differential topology the requirement (6) may be damage to the topology of Θ. When the Fisher information g is singular its kernel is in involution. Thus the topological-geometrical information that are contained in g are transverse to the distribution $Ker(g)$. If $Ker(g)$ is completely integrable then topological and geometrical informations which are contained in g are transversal to the foliation $Ker(g)$. See Part A of this paper. This ends the criticisms.

To motivate for deleting the requirement (1) we construct a compact statistical model which satisfies all of the requirements except the requirement (1).

Let $\mathcal{E}$ be the tangent bundle of the circle S^1. E is the trivial line bundle

$$S^1 \times \mathbb{R} \ni (\theta, t) \to \theta \in S^1.$$

We consider the fonctions f, F and P defined by

$$f(\theta, t) = [\sin^2(\frac{t^2\theta}{1+t^2})\cos^2(\frac{\theta}{4})e^{-t^2} + \frac{\pi}{e^2}t^2],$$

$$F(\theta) = \int_{-\infty}^{+\infty} e^{-f(\theta,t)}dt,$$

$$P(\theta, t) = \frac{e^{-f(\theta,t)}}{F(\theta)}.$$

The function $P(\theta, t)$ has the following properties

(i) $(i) : P(\theta, t)$ is smooth,

(ii) $P(0, t) = P(2\pi, t) \quad \forall t \in \mathbb{R}$,

(iii) the $\frac{d}{d\theta}$ commutes with $\int_{\mathbb{R}}$,

(iv) $P(\theta, t) \leq 1 \quad \forall (\theta, t) \in S^1 \times \mathbb{R}$,

(v) if $0 < \theta, \theta^* < 2\pi$ then $P_\theta = P^*_\theta$ if and only if $\theta = \theta^*$,

(vi) $\int_{-\infty}^{+\infty} P(\theta, t)dt = 1$.

These properties show that there is a one to one correspondence between the circle S^1 and a subset of probability densities in $\mathbb{R}$. Thus S^1 is a compact 1-dimensional manifold of probabilities in the measurable set $(\mathbb{R}, \beta(\mathbb{R}))$. Here $\beta(\mathbb{R})$ is the family of Borel subsets of $\mathbb{R}$.

So (S^1, P) is a compact parametric model for $(\mathbb{R}, \beta(\mathbb{R}))$.

A Digression.

Let $\{(\Theta_j, P_j), j := 1,2\}$ *be statistical models for measurable sets* $\{(\Xi_j, \Omega_j), j := 1,2\}$. *We put*

$$\Theta = \Theta_1 \times \Theta_2,$$

$$(\Xi, \Omega) = (\Xi_1 \times \Xi_2, \Omega_1 \times \Omega_2),$$

$$P = P_1 \otimes P_2.$$

The function P *is defined in* $\Theta \times \Xi$ *by*

$$P((\theta_1, \theta_2), (\xi_1, \xi_2)) = P_1(\theta_1, \xi_1) P_2(\theta_2, \xi_2)$$

The integration on Ξ *is defined by*

$$\int_{\Xi} f((\theta_1, \theta_2), (\xi_1, \xi_2)) d(\xi_1, \xi_2) = \int_{\Xi_1 \times \Xi_2} f((\theta_1, \theta_2), (\xi_1, \xi_2)) d\xi_1 d\xi_2.$$

Thus we get

$$\int_{\Xi} P[(\theta_1, \theta_2), (\xi_1, \xi_2)] d\xi_1 d\xi_2 = \int_{\Xi_1 \times \Xi_2} P_1(\theta_1, \xi_1) P_2(\theta_2, \xi_2) d\xi_1 d\xi_2 = 1.$$

So (Θ, P) *is a statistical model for* $[\Xi_1 \times \Xi_2, \Omega_1 \times \Omega_2]$.
One is in position to prove that every Euclidean torus $\mathbb{T}^m$ *is a statistical model for* $(\mathbb{R}^m, \beta(\mathbb{R}^m))$.

Another Construction.

For every positive integer m *we consider positive real numbers*

$$\alpha_1 < \alpha_2 < ... < \alpha_m$$

and the real functions which are defined by

$$f_j(\theta, t) = \sin^2(\frac{t^2\theta}{1+t^2}) \cos^2(\frac{\theta}{4}) e^{-t^2} + \alpha_j t^2 \quad (\theta, t) \in E,$$

$$F_j(\theta) = \int_{-\infty}^{+\infty} e^{-f_j(\theta,t)} dt,$$

$$P_j(\theta, t) = \frac{e^{-f_j(\theta,t)}}{F_j(\theta)}.$$

Now we consider the tangent bundle of the m-dimensional flat torus $T\mathbb{T}^m$,

$$\mathbb{T}^m = S^1 \times S^1 \times ... \times S^1.$$

Let

$$(\theta, t) = [(\theta_1, t_1), (\theta_2, t_2), ..., (\theta_m, t_m)] \in T\mathbb{T}^m.$$

We put

$$F(\theta) = \int_{\mathbb{R}^m} e^{-\sum_1^m f_j(\theta_j, t_j)} dt_1 dt_2 ... dt_m,$$

$$P(\theta, t) = \frac{e^{-\sum_1^m f_j(\theta_j, t_j)}}{F(\theta)}.$$

The function $P(\theta, t)$ *satisfies the following requirements*

(1) *If $\theta \neq \theta^*$ there exists $t^{**} \in \mathbb{R}^m$ such that*

$$P(\theta, t^{**}) \neq P(\theta^*, t^{**}),$$

(2) $P(\theta, t) \leq 1 \forall (\theta, t) \in T\mathbb{T}^m$,
(3) $\int_{\mathbb{R}^m} P(\theta, t) dt = 1$.

We deduce that the pair $(\mathbb{T}^m, P)$ is an m-dimensional manifold of probability densities in the measurable set $(\mathbb{R}^m, \beta(\mathbb{R}^m))$.

The image of every local chart of $\mathbb{T}^m$ is a local statistical model in the classical sense [17,18,22]. This ends the Digression.

We are motivated for introducing a new theory of statistical models whose localization yields the current theory. The theory we introduce is an answer to McCullagh and to Gromov.

8.1. The Preliminaries

In this Part B we face three major challenges.
Challenge 1. Taking into account the criticisms we have raised our aim is to introduce a new theory of statistical model whose localization leads to the classical theory of statistical models.
Challenge 2. The second challenge is the Search for an invariant which encodes the point of the moduli space of isomorphism class of statistical models.
Challenge 3. We introduce the theory of homological statistical model and we explore the links between this theory and the challenge 2.
Challenge 4. The fourth challenge is to explore the relationships between "challenge 1, challenge 2, challenge 3" and "Vanishing Theorems in the theory of KV homology".

The theory of KV cohomology and the geometry of Koszul play important roles. We introduce the needed definitions.

Let (Ξ, Ω) be a measurable set. Let $Aut(\Xi, \Omega)$ be the group of measurable isomorphisms of Ξ. Let (M, D) be a locally flat manifold whose KV algebra is denoted by $\mathcal{A}$.

We keep the notation used in Part A of this paper. For instance $\mathcal{S}_2(M)$ is the vector space of differentiable symmetric bi-linear forms in M.

Definition 44. *A random Hessian metric in (M, D) is a map*

$$M \times \Xi \ni (x, \xi) \rightarrow Q(x, \xi) \in S^2[T_x^* M],$$

which has the following properties

(1) *for every vector field X the real number $Q(x, \xi)[X, X]$ is non negative, furthermore $\forall v \in T_x M \setminus \{0\}\ \exists \xi \in \Xi$ such that*

$$Q(x, \xi)(v, v) > 0,$$

(2) *for every $\xi \in \Xi$, the random KV cochain*

$$(X, Y) \rightarrow Q_\xi(X, Y)(x)$$

with

$$Q_\xi(X, Y)(x) = Q(x, \xi)(X_x, Y_x)$$

is a random cocycle of the KV complex $[C_{KV}^*(\mathcal{A}, C^\infty(M)), \delta_{KV}]$.

Let (Ξ, Ω, p) be a probability space. A random Hessian metric Q generates a Hessian structure (M, g_Q, D) whose tensor metric g_Q is defined by

$$g_Q(X,Y)(x) = \int_{\Xi} Q(x,\xi)(X,Y)p(\xi)d\xi.$$

The group $Aut(\Xi, \Omega)$ of measurable isomorphisms of (Ξ, Ω) is denoted by Γ.

Warning.

When Ξ is a topological space elements of Γ are continuous maps. When Ξ is a differentiable manifold elements of Γ are differentiable maps. Let $\mathbb{P}(\Xi)$ be the Boolean algebra of all subsets of Ξ. The abstract group $Aut(\Xi, \mathbb{P}(\Xi))$ is a subgroup of the group $Isom(\Xi)$ of isomorphisms of the set Ξ.

Definition 45. *A measurable set (Ξ, Ω) is called homogeneous if the natural action of Γ in Ξ is transitive.*

Throughout this paper we will be dealing with homogeneous measurable sets. Below we introduce the framework of the theory of statistical models.

8.2. The Category $\mathcal{FB}(\Gamma, \Xi)$

8.2.1. The Objects of $\mathcal{FB}(\Gamma, \Xi)$

Definition 46. *An object of the category $\mathcal{FB}(\Xi,)$ is a datum $[\mathcal{E}, \pi, M, D]$ which is composed as it follows.*

(1) *M is a connected m-dimensional smooth manifold. The map*

$$\pi : \mathcal{E} \to M$$

is a locally trivial fiber bundle whose fibers $\mathcal{E}_x$ are isomorphic to the set Ξ.

(2) *The pair (M, D) is an m-dimensional locally flat manifold.*

(3) *There is a group action*

$$\Gamma \times [\mathcal{E} \times M] \times \mathbb{R}^m \ni (\gamma, [e, x, \theta]) \to [[(\gamma \cdot e), \gamma \cdot x], \tilde{\gamma} \cdot \theta] \in [\mathcal{E} \times M] \times \mathbb{R}^m.$$

That action is subject to the compatibility requirement

$$\pi(\gamma \cdot e) = \gamma \cdot \pi(e) \quad \forall e \in \mathcal{E}.$$

(4) *Every point $x \in M$ has an open neighborhood U which is the domain of a local fiber chart*

$$\Phi_U \times \phi_U : [\mathcal{E}_U \times U] \ni (e_x, x) \to [\Phi_U(e_x), \phi_U(x)] \in [\mathbb{R}^m \times \Xi] \times \mathbb{R}^m.$$

The local charts are subject to the following compatibility relation

- *(U, ϕ_U) is an affine local chart of the locally flat manifold (M, D),*
- *$\phi_U(\pi(e)) = p_1(\Phi_U(e))$.*

(5) *We set*

$$\Phi_U(e) = (\theta_U(e), \xi_U(e)) \in \mathbb{R}^m \times \Xi.$$

Let $(U, \Phi \times \phi)$ and $(U^*, \Phi^* \times \phi^*)$ be two local charts with

$$U \cap U^* \neq \emptyset,$$

then there exists a unique $\gamma_{UU^*} \in \Gamma$ *such that*

$$[\gamma_{UU^*} \cdot \Phi](e) = \Phi^*(e) \quad \forall e \in \mathcal{E}_{U \cap U^*}.$$

Comments. *Requirements (3) and (4) mean that*

$$[\Phi_U(e), \phi_U(\pi(e)] = [[\theta_U(e), \xi_U(e)], \theta_U(e)]$$

Both requirements (4) and (5) yield the following remarks: the following action is differentiable

$$\Gamma \times M \ni (\gamma, x) \rightarrow \gamma \cdot x \in M,$$

the following action is an affine action

$$\Gamma \times \mathbb{R}^m \ni (\gamma, \theta) \rightarrow \tilde{\gamma} \cdot \theta,$$

both the left side member and the right side member of (5) have the following meaning.

$$\gamma_{UU^*} \cdot [\theta_U(e), \xi_U(e)] = [\theta_{U^*}(e), \xi_{U^*}(e)].$$

Consequently (5) implies that for all $x \in U \cap U^*$ *one has*

$$\tilde{\gamma}_{UU^*} \cdot \phi(x) = \phi^*(x).$$

Therefore we get

$$\gamma_{UU^*} = \phi^* \circ \phi^{-1}.$$

Suppose that U, U^* and U^{**} are domains of local chart with

$$U \cap U^* \cap U^{**} \neq \emptyset$$

then

$$\gamma_{U^*U^{**}} \circ \gamma_{UU^*} = \gamma_{UU^{**}}.$$

The requirement (3) means that the fibration π is Γ equivariant.
The Figure 2 expresses the requirement property (3).

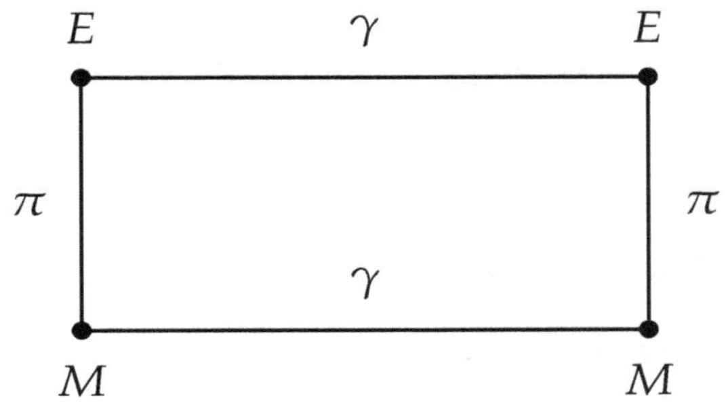

Figure 2. Fibration.

We recall that the group Γ acts in both $\mathcal{E}$ and M. Figure 2 expresses that the projection π of $\mathcal{E}$ on M is Γ-equivariant.

8.2.2. The Morphisms of $\mathcal{FB}(\Gamma, \Xi)$

Let $[\mathcal{E}, \pi, M, D]$ and $[\mathcal{E}^*, \pi^*, M^*, D^*]$ be two objects of $\mathcal{FB}(\Gamma, \Xi)$.
Let $\Psi \times \psi$ be a map

$$[\mathcal{E} \times M] \ni (e, x) \to (\Psi(e), \psi(x)) \in [\mathcal{E}^* \times M^*].$$

Definition 47. *A pair $(\Psi \times \psi)$ is a morphism of the category $\mathcal{FB}(\Gamma, \Xi)$ if the following conditions are satisfied*
(m.1): $\pi^* \circ \Psi = \psi \circ \pi$,
(m.2): both Ψ and ψ are Γ-equivariant isomorphism, that is to say

$$\Psi(\gamma \cdot e) = \gamma \cdot \Psi(e),$$

$$\psi(\gamma \cdot x) = \gamma \cdot \psi(x),$$

(m.3): ψ is an affine map of (M, D) in (M^*, D^*).

The Figure 3 represents the properties (m.1) and (m.2). We are going to define the category of statistical model for (Ξ, Ω). The framework is the category $\mathcal{FB}(\Gamma, \Xi)$.

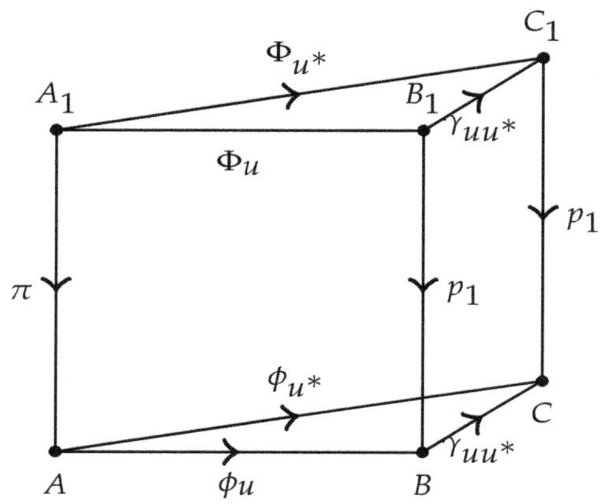

Figure 3. Equivariance.

At one side we recall that the group Γ also acts in $\mathbb{R}^m \times \Xi$. At another side the localizations are made coherent thanks to Cech cocycles γ_{UU^*}. Figure 3 tells two informations. Firstly localizations are Γ-equivariant, secondly thanks to Cech cocycles localizations are coherent.

8.3. The Category $\mathcal{GM}(\Xi, \Omega)$

We keep the notation used in the previous subsections. Our purpose is the category of statistical models $\mathcal{GM}(\Xi, \Omega)$.

8.3.1. The Objects of $\mathcal{GM}(\Xi, \Omega)$

Definition 48. *An m-dimensional statistacal model for (Ξ, Ω) is an object of $\mathcal{FB}(\Gamma, \Xi)$, namely*

$$\mathbb{M} = [\mathcal{E}, \pi, M, D]$$

which has the following properties (ρ_*).

$[\rho_1]$: For every local chart $(U, \Phi_U \times \phi_U)$ the subset

$$[\Theta_U \times \Xi] = \Phi_U(\mathcal{E}_U)$$

supports a non negative real valued function P_U *subject to the following requirements.*
$[\rho_{1.1}]$: *For every fixed* $\xi \in \Xi$ *the function*

$$\Theta_U \ni \theta \to P_U(\theta, \xi)$$

is differentiable.
$[\rho_{1.2}]$: *For every fixed* $\theta \in \Theta_U$ *the triple*

$$(\Xi, \Omega, P_U(\theta, -))$$

is a probability space. Further the operation of integration $\int_\Xi$ *commutes with the operation of differentiation* $d_\theta = \frac{d}{d\theta}$.
$[\rho_{1.3}]$: *Let* $(U, \Phi_U \times \phi_U, P_U)$ *and* $(U^*, \Phi_{U^*} \times \phi_{U^*}, P_{U^*})$ *be as in* $[\rho_{1.1}]$ *and in* $[\rho_{1.2}]$.
If $U \cap U^* \neq \emptyset$ *then* P_U, P_{U^*} *and* γ_{UU*} *are related by the formula*

$$P_{U^*} \circ \gamma_{UU^*} = P_U.$$

$[\rho_{1.4}]$: *Let* $U \subset M$ *be an open subset and let* $\gamma \in \Gamma$. *Let us assume that both* U *and* $\gamma \cdot U$ *are domains of local charts*

$$(U, \Phi_U \times \phi_U, P_U)$$

and

$$(\gamma \cdot U, \Phi_{\gamma \cdot U} \times \phi_{\gamma \cdot U}, P_{\gamma \cdot U}).$$

We assume that those local charts satisfy $\rho_{1.1}$, $\rho_{1.2}$ *and* $\rho_{1.3}$. *Then the relations*

$$\Phi_{\gamma \cdot U} \circ \gamma = \gamma \circ \Phi_U,$$

$$\phi_{\gamma \cdot U} \circ \gamma = \gamma \circ \phi_U,$$

implies the equality

$$P_{\gamma \cdot U} \circ \gamma = P_U.$$

A Comment.

Actually, $([\rho_{1.3}])$ *has the following meaning:*

$$P_{U^*}[\tilde{\gamma}_{UU^*} \cdot \theta_U(e), \gamma_{UU^*} \cdot \xi_U(e)] = P_U(\theta_U(e), \xi_U(e))$$

$\forall e \in \mathcal{E}_{U \cap U^*}$.

The Figure 4 represents $(\rho_{1.3})$

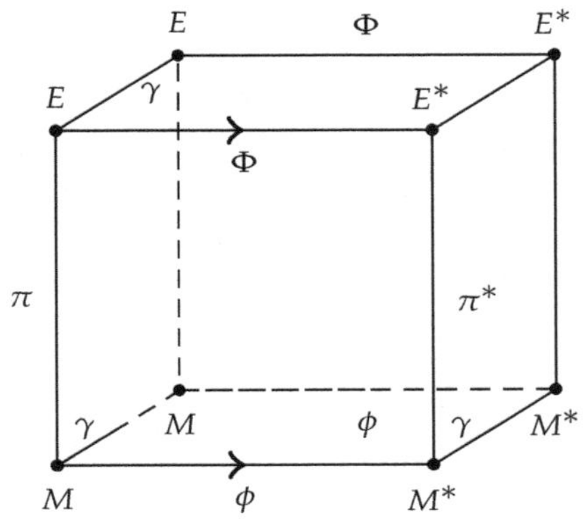

Figure 4. Moduli.

This ends the comment.

Definition 49. *A datum* $[U, \Phi_U \times \phi_U, P_U, \gamma_{UU^*}]$ *as in the last definition is called a local statistical chart of* $[\mathcal{E}, \pi, M, D]$.

Figure 4 is represents what are crucial steps toward the serach of characteristic invariants, viz invariants encoding the points of the moduli space of statistical models. At the present Figure 4 describes the moduli space of the category $\mathcal{FB}(\Gamma, \Xi)$

Before dealing with morphisms of the category $\mathcal{GM}(\Xi, \Omega)$ we introduce a relevant global geometrical invariant.

8.3.2. The Global Probability Density of a Statistical Model

We consider a COMPLETE (or maximal statistical) atlas of an object $[\mathcal{E}, \pi, M, D]$ (of the category $\mathcal{GM}(\Xi, \Omega)$), namely

$$A_\Phi = [U_j, \Phi_j, \phi_j, P_j, \gamma_{ij}].$$

The family U_j is an open covering of M. The pair $\mathcal{E}_j \times U_j$ is the domain of the local chart $(\Phi_j \times \phi_j)$. We have

$$\mathcal{E}_j = \mathcal{E}_{U_j}.$$

If $U_i \cap U_j \neq \emptyset$ then one has

$$\phi_j(x) = \tilde{\gamma}_{ji} \cdot \phi_i j(x) \quad \forall x \in U_i \cap U_j.$$

In particular $A = (U_j, \phi_j)$ is an affine atlas of the locally flat manifold (M, D). We have

$$\Phi_j(\mathcal{E}_{y^*}) = \phi_j(y^*) \times \Xi \quad \forall y^* \in U_j.$$

Therefore we set

$$[\mathcal{E}_{y^*}, \Omega_{y^*}] = \Phi_j^{-1}[[\phi_j(y^*) \times \Xi], \Omega].$$

The atlas A_Φ satisfies requirements $(\rho_{1.1})$, $(\rho_{1.2})$ and $(\rho_{1.3})$. In $\mathcal{E}_{U_j}$ the local function p_j is defined by

$$p_j = P_j \circ \Phi_j.$$

We suppose that

$$U_i \cap U_j \neq \emptyset.$$

By the virtue of of $[\rho_{1.3}]$ one has

$$p_i(e) = p_j(e)$$

for all $e \in \mathcal{E}_{U_i \cap U_j}$.

Thereby there exists a unique function

$$\mathcal{E} \ni e \to p(e) \in \mathbb{R}$$

whose restriction to $\mathcal{E}_j$ coincides with p_j. The restriction to $\mathcal{E}_x$ is denoted by p_x. The triple

$$(\mathcal{E}_x, \Omega_x, p_x)$$

is a probability space.

Definition 50. *The function*

$$\mathcal{E} \ni e \to p(e) \in \mathbb{R}$$

is called the probability density of the model $[\mathcal{E}, \pi, M, D]$.

Comments.

(i) We take into account the global probability density p. Then an object of the category $\mathcal{GM}(\Xi,\Omega)$ is denoted by

$$[\mathcal{E},\pi,M,D,p].$$

(ii) The function p is Γ-equivariant. THIS IS THE GEOMETRY in the sense of Erlangen program.
(iii) We have not used any argument depending the dimension of manifolds.

The Figure 5 expresses coherence to local probability densities We are in position to define the morphisms of the category $\mathcal{GM}(\Xi,\Omega)$.

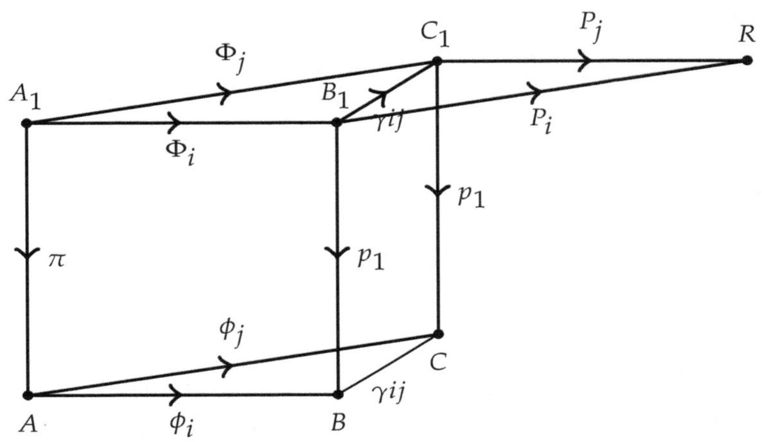

Figure 5. Localisation.

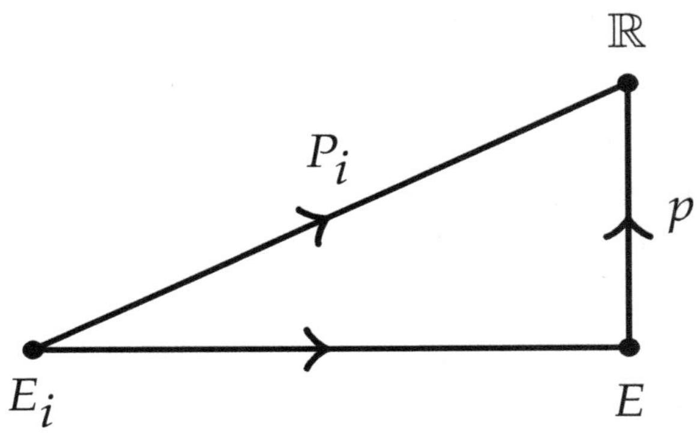

Figure 6. Probability Density.

In Figure 5 one sees that modulo the dynamics of the group Γ in $\mathbb{R}^m \times \Xi$ all localizations look alike. Figure 6 show that local probability densities $\{p_i\}$ are but localizations of a global probability density p

8.3.3. The Morphisms of $\mathcal{GM}(\Xi,\Omega)$

Definition 51. *Let $\mathbb{M} = [\mathcal{E},\pi,M,D,p]$ and $\mathbb{M}* = [\mathcal{E}^*,\pi^*,M^*,D^*,p^*]$ be two objects of the category $\mathcal{GM}(\Xi,\Omega)$. A $\mathcal{FB}(\Gamma,\Xi)$-morphism*

$$(\Psi\times\psi):[\mathcal{E},\pi,M,D]\to[\mathcal{E}^*,\pi^*,M^*,D^*]$$

is a morphism of $[\mathcal{E}, \pi, M, D, p]$ *in* $[\mathcal{E}^*, \pi^*, M^*, D^*, p^*]$ *if*

$$p^* \circ \Psi = p \cdot$$

A Comment.

Let $[\mathcal{E}, \pi, M, D]$ *be an object of the category* $\mathcal{FB}(\Gamma, \Xi)$. *Let G be the group of isomorphisms of* $[\mathcal{E}, \pi, M, D]$. *If M is finite dimensional then G is a finite dimensional Lie group. The group G acts in the category M whose objects are probability densities in* $[\mathcal{E}, \pi, M, D,]$.

Definition 52. *The orbit space space*

$$m = \frac{M}{G}$$

is called the moduli space of M.

A Comment.

Every trivialization of

$$\mathbb{M} = [\mathcal{E}, \pi, M, D, p]$$

is a statistical model in the classical sense [18,22]. So we have taken up the challenge 1.

8.3.4. Two Alternative Definitions

We introduce two other presentations of the category $\mathcal{GM}(\Xi, \Omega)$. Those presentations highlight the connection with the searches of McCullagh and Gromov. Those presentation is useful in both the theoretical statistics and the applied statistics [17,18,21,24,55,57,58].

We consider the category $\mathcal{MSE}$ whose objects are a probability spaces (Ξ, Ω, p).

Definition 53. *A morphism of a probability space* (Ξ, Ω, p) *in another probability space* (Ξ^*, Ω^*, p^*) *is a measurable map* Ψ *of* (Ξ, Ω) *in* (Ξ^*, Ω^*) *such that*

$$p = p^* \circ \Psi \cdot$$

Remark 4. *A morphism as in the last definition has a statistical nature. An isomorphisms of* (Ξ, Ω, p) *on* $(\Xi*, \Omega*, p*)$ *is an sufficient statistic. The category* $\mathcal{MSE}$ *is useful for introducing two variant descriptions of the category* $\mathcal{GM}(\Xi, \Omega)$.

Definition 54. *We use the previous notation.*

(1) *A statistical model is a locally trivial fiber bundle over a locally flat manifold*

$$\pi : \mathcal{E} \to M.$$

The fibers of π *are probability spaces.*

(2) *The functor*

$$[\mathcal{E}, p] \to [M, D]$$

is called a $\mathcal{MSE}$*-fibration.*

The category of $\mathcal{MSE}$-fibrations is denoted by $\mathcal{FB}(\mathcal{MSE})$. The morphisms the category $\mathcal{GM}(\Xi, \Omega)$ are called $\mathcal{MSE}$-morphisms.

Definition 55. *A statistical model for a measurable set* (Ξ, Ω) *is a functor of the category* $\mathcal{FB}(\Gamma, \Xi)$ *in the category* $\mathcal{FB}(\mathcal{MSE}$, *namely*

$$[\mathcal{E}, \pi, M, D] \to [\mathcal{E}, \pi, M, D, p]$$

At the present step it is clear that the information geometry is structured.

8.3.5. Fisher Information in $\mathcal{GM}(\Xi, \Omega)$

We consider a $\mathcal{MSE}$-fibration

$$\mathbb{M} := [\mathcal{E}, p] \to (M, D).$$

The Fisher information to be defined is an elementg of $\Gamma(S^2(T^*M))$.
We recall that every $\mathcal{MSE}$-fiber $\mathbb{M}_x, x \in M$ has a structure of probability space

$$\mathbb{M}_x := [\mathcal{E}_x, \Omega_x, p_x].$$

Let X, Y be local vector fields which are defined in a open neighbourhood of $x \in M$.

Definition 56. *The Fisher information at x is defined by*

$$g_x(X, Y) = -\int_{\mathcal{E}_x} p(e)[D^2 log(p(e))](X, Y)d(e)$$

We recall that the horizontal differentiation commutes with the integration along the $\mathcal{MSE}$-fibers, viz

$$d_\theta \circ \int_F = \int_F \circ \frac{\partial}{\partial \theta}.$$

So the Fisher information g is well defined. It has the following properties

(1) g is positive semi-definite,
(2) g is an invariant of the Γ-geometry in $[\mathcal{E}, \pi, M, D, p]$.

8.4. Exponential Models

Let $[\mathcal{E}, \pi, M, D, p]$ be an object of $\mathcal{GM}(\Xi, \Omega)$. We recall that data which are defined in $\mathcal{E}$ are called random data in the base manifold M. The operation of integration along the $\mathcal{MSE}$-fibers is denoted by $\int_F$. Thus a random datum μ is called smooth if its image $\int_F(\mu)$ is smooth.

Conversely every datum θ^* which is point-wise defined in M is the image of the random datum

$$\theta = \theta^* \circ \pi.$$

So we get

$$\theta^* = \int_F [\theta^* \circ \pi].$$

Thus at every $x \in M$ one has

$$\theta^*(x) = \int_{\mathcal{E}_x} \theta^*(\pi(e)) p_x(e) de.$$

A random affine function is a function

$$\mathcal{E} \ni e \to a(e) \in \mathbb{R}$$

subject to the requirement

$$D^2 \int_F a = 0.$$

Definition 57. *An $\mathcal{MSE}$-fibration*

$$[\mathcal{E}, p] \to [M, D]$$

is called an exponential model if the following conditions are satisfied.

(1) *The base manifold M supports a locally flat structure (M, ∇) and a real valued function $\psi \in C^\infty(M)$.*
(2) *The total space $\mathcal{E}$ supports a real valued random function a.*
(3) *The triple $[a, \nabla, \psi]$ is subject to the following requirement*
(4) $\nabla^2 \int_F(a) = O,$
(5) $p(e) = exp[a(e) - \psi(\pi(e))].$

Remark 5. *At one side Localizations of exact Hessian homological statistical models yield the classical Koszul Information Geometry [55]. That is but the classical Hessian Information Geometry. At another side the KV homology learns that the Hessian information Geometry is the same think as the geometry of exponential famillies see Part A Section 5, Theorem 16.*

Reminder.

In the Appendix A to this paper the reader will find a new invariant $r^b(p)$ measuring how far from being an exponential model is an $\mathcal{MSE}$-fibration

$$[\mathcal{E}, p] \to [M, D].$$

By the virtue of results in Part A, to be an exponential model depends on homological conditions.

8.4.1. The Entropy Flow

We are going to introduce the notion of local entropy flow. Subsequently we will show that the Fisher information of a model $[\mathcal{E}, \pi, M, D, p]$ is the Hessian of the local entropy flow.

To start we consider a $\mathcal{MSE}$-fibration

$$[\mathcal{E}, p] \to [M, D].$$

That is another presentation of the statistical model $[\mathcal{E}, \pi, M, D, p]$.

Let $[U_j, \Phi_j \times \phi_j, \gamma_{ij}, P_j]$ be an atlas of $[\mathcal{E}, \pi, M, D, p]$. We put

$$[\Theta_j, P_j] = [\Phi_j(\mathcal{E}_j), p \circ \Phi_j^{-1}].$$

Then every $[\Theta_j, P_j]$ is a local statistical model for (Ξ, Ω).

Le X, Y be two vector fields defined in U_j and let $\psi_X(t)$ and $\psi_Y(s)$ be their local flows defined in U_j. Then we set

$$\Phi_j(e) = [\theta_j(e), \xi_j(e)] = [\phi_j(\pi(e)), \xi_j(e)], \quad e \in \mathcal{E}_j,$$

$$\tilde{\psi}_j(t)[\theta_j(e), \xi_j(e)] = ([\phi_j \psi_X(t) \phi_j^{-1}][\theta_j(e)], \xi_j(e)),$$

$$\tilde{\psi}_j(s)[\theta_j(e), \xi_j(e)] = ([\phi_j \psi_Y(s) \phi^{-}1_j][\theta_j(e)], \xi_j(e)).$$

Definition 58. *The local entropy flow of the pair (X, Y) is the function $Ent^j_{X,Y}$ defined by*

$$Ent^j_{X,Y}(s,t)(\pi(e)) = \int_\Xi \{P_j[\tilde{\psi}_X(s)(\Phi(e))] log[P_j[\tilde{\psi}_Y(t)(\Phi_j(e))]]\} \, d\xi(e).$$

To pursue we use the compatibility of local charts of the atlas$[U_j, \Phi_j \times \phi_,, \gamma_{ij}, P_j]$. If

$$U_i \cap U_j \neq \emptyset$$

then forall $e \in \mathcal{E}_{U_i \cap U_j}$ we have

$$\Phi_j(e) = \gamma_{ij} \cdot \Phi_i(e),$$

$$\phi_j(\pi(e)) = \gamma_{ij} \cdot \phi_i(\pi(e)).$$

We recall that $\tilde{\psi}(t)$ and $\tilde{\psi}(s)$ are defined by

$$\tilde{\psi}_j(t) = \phi_j \psi_X(t) \phi_j^{-1},$$

$$\tilde{\psi}_j(s) = \phi_j \psi_Y(s) \phi_j^{-1}.$$

Those reminders are useful for concluding that whenever

$$U_i \cap U_j \neq \emptyset$$

we have

$$Ent^i_{X,Y}(s,t)(\pi(e)) = Ent^j_{X,Y}(s,t)(\pi(e)).$$

So the local entropy flow does not depend on local charts.

If both X and Y are complete vector fields then their entropy flow is globally defined. A notable consequence is the following statement.

Theorem 18. *Every $\mathcal{MSE}$-fibration over a compact manifold M admits a globally defined entropy map*

$$\mathcal{X}(M) \times \mathcal{X}(M) \ni (X,Y) \to Ent_{X,Y} \in C^\infty(\mathbb{R}^2).$$

8.4.2. The Fisher Information as the Hessian of the Local Entropy Flow

we consider the function

$$H_j(s,t,\xi) = P_j[\tilde{\psi}_X(s)(\phi(e))]log[P_j[\tilde{\psi}_j(t)(\Phi_j(e))]].$$

Direct calculations yield

$$[\frac{\partial^2(H_j(s,t))}{\partial s \partial t}][(s,t) = (0,0)] = P_j[\phi_j(e)](X \cdot log[P_j(\Phi_j(e))])(Y \cdot log[P_j(\phi_j(e))]).$$

We know that $\frac{\partial^2}{\partial s \partial t}$ commutes with $\int_\Xi$. Therby we conclude that

$$g\pi(e)(X,Y) = \frac{\partial^2 Ent_j(s,t)(\pi(e))}{\partial s \partial t}[(s,t) = (0,0)].$$

Theorem 19. *We consider an $\mathcal{MSE}$-firation over a compact manifold*

$$\mathbb{M} := [\mathcal{E}, p] \to [M, D].$$

The Fisher information of $\mathbb{M}$ is the Hessian of the entropy map.

8.4.3. The Amari-Chentsov Connections in $\mathcal{GM}(\Xi, \Omega)$

Let

$$\mathbb{M} = [\mathcal{E}, \pi, M, D, p]$$

be an m-dimensional statistical model for a measurable set (Ξ, Ω). To define the family of α-connections we work in a local chart (Φ_U, Φ_U).

We set

$$\Theta_U = \phi_U(U),$$

$$\Theta \times \Xi = \Phi_U(\mathcal{E}_U).$$

In the base manifold (M, D) the local chart $[U, \phi_u)$ yields a system of local affine coordinate functions

$$\theta = (\theta_1, ..., \theta_m).$$

We use the notation as in [18]. Given a real number α we define the α-connection ∇^α by its Christoffel symbols in the local coordinate functions θ. Those Christoffel are denoted by $\Gamma^\alpha_i j : k$. We proceed as it follows.

Step 1: In the open subset $\Theta_U \subset \mathbb{R}^m$ we put

$$\tilde{\Gamma}^{\alpha,U}_{ij:k}(\theta) = \int_\Xi P_U(\theta,\xi)\left\{[\frac{\partial^2 l_n(\theta,\xi)}{\partial\theta_i\partial\theta_j} + \frac{1+\alpha}{2}\frac{\partial l_n(\theta,\xi)}{\partial\theta_i}\frac{\partial l_n(\theta,\xi)}{\partial\theta_j}]\frac{\partial l_n(\theta,\xi)}{\partial\theta_k}\right\} d\xi.$$

This local definition of $\tilde{\Gamma}^\alpha_{ij:k}$ agrees with affine coordinate change in Θ_U.

Step 2: In the open subset U $\Gamma^{\alpha,U}_{ij:k}$ is defined by

$$\Gamma^{\alpha,U}_{ij:k} = \tilde{\Gamma}^{\alpha,U}_{ij:k} \circ \phi_U.$$

Since the definition of $\tilde{\Gamma}^\alpha_{ij:k}$ agrees with an affine coordinate change we can use an atlas

$$\mathbb{A} = [U_j, \Phi_j \times \phi_j, \gamma_{ij}]$$

for constructing a Koszul connection $\nabla^\alpha(\mathbb{A})$. Since the construction of $\nabla^\alpha(\mathbb{A})$ agrees with affine coordinate change the connection $\nabla^\alpha(\mathbb{A})$ is independent from the choice of $\mathbb{A}$. Every α-connection is torsion free. So an $\mathcal{MSE}$-fibration

$$[\mathcal{E}, p] \to [M, D]$$

gives rise to a ma

$$\mathbb{R} \ni \alpha \to \nabla^\alpha \in \mathcal{SLC}(M).$$

If the Fisher information g is definite then $(M, g, \nabla^\alpha, \nabla^{-\alpha})$ is a dual pair [17,48].

By the virtue of the definition of the Fisher information g a local section of section of $Ker(g)$ is a local vector field $X \in \mathcal{X}(M)$ such that

$$X \cdot p = 0.$$

Therefore, it is easily seen that

$$L_X g = 0.$$

So if data are analytic then g is a stratified Riemannian foliation.

8.4.4. The Homological Nature of the Probability Density

We consider a $\mathcal{MSE}$-fibration

$$\mathbb{M} := [\mathcal{E}, p] \to [M, D].$$

We recall that a random differential q-form in $\mathcal{E}$ is a mapping

$$\mathcal{E} \ni e \to \omega(e) \in \wedge^q(T^*_{\pi(e)})M$$

such

$$\int_F \omega \in \Omega^q(M).$$

The vector space of random differential q-forms is denoted by $\Omega^q(\mathcal{E})$.

Let $\Phi_U \times \phi_U$ be a local chart of $\mathbb{M}$. We set

$$(\Theta_U, \Xi) = (\phi_U(U), \Xi) = \Phi_U(\mathcal{E}_U).$$

We recal that in $\Phi_U(\mathcal{E}_U)$ the partial differentiation $\frac{\partial}{\partial\theta}$ is called the horizontal differentiation in $\mathcal{E}_U$. Therefore we use the relation

$$\int_F \circ \frac{\partial}{\partial\theta} = \frac{d}{d\theta} \circ \int_F$$

for setting the de Rham complex of random differential forms. Namely

$$\Omega(\mathcal{E}): \quad 0 \to \mathbb{R} \to \Omega^0(\mathcal{E}) \to ...\Omega^q(\mathcal{E}) \to \Omega^{q+1}(\mathcal{E})... \to \Omega^m(\mathcal{E}) \to 0.$$

The complex $\Omega(\mathcal{E})$ is a complex of Γ-modules. Here

$$\Gamma = Aut(\Xi, \Omega).$$

Then the cohomology space $H*(\Gamma, \Omega(\mathcal{E}))$ is bigraded,

$$H^{p,q}(\Gamma, \Omega(\mathcal{E})) = H^p(\Gamma, \Omega^q(\mathcal{E})).$$

The probability density p is Γ-invariant. It is an element of $H^{0,0}(\Gamma, \Omega(\mathcal{E}))$.

8.4.5. Another Homological Nature of Entropy

One of main purpose of [14] is the homological nature of the entropy. The classical entropy function of a statistical model $[\mathcal{E}, \pi, M, D, p]$ is defined by

$$E(\pi(e)) = \int_{\mathcal{E}_{\pi(e)}} p(e^*)log(p(e^*)).$$

In the complex $\Omega(\mathcal{E})$ we perform the machinery of Eilenberg [59]. That yields the exact sequence (of random cohomology spaces)

$$\to H^{q-1}_{res}(\mathcal{E}, \mathbb{R}) \to H^q_e(\mathcal{E}, \mathbb{R}) \to H^q_{dR}(\mathcal{E}, \mathbb{R}) \to H^q_{res}(\mathcal{E}, \mathbb{R}) \to$$

We take into account the identities

$$p(\gamma \cdot e) = p(e),$$

$$\gamma \cdot (\pi(e)) = \pi(\gamma \cdot e).$$

Then we have

$$\begin{aligned} E(\gamma \cdot \pi(e)) &= \int_{\mathcal{E}_{\gamma\cdot\pi(e)}} p(\gamma \cdot e^*)log(p(\gamma \cdot e^*)) \\ &= \int_{\mathcal{E}_{\pi(\gamma\cdot e)}} p(e^*)log(p(e^*)) \\ &= E(\pi(e)). \end{aligned}$$

Thus the entropy $E(\pi(e))$ is Γ-equivariant. Therefore, it defines an equivariant cohomology class

$$[E] \in H^0_e(M, \mathbb{R}).$$

This is another topological nature of the entropy. For another viewpoint see [16,31].

Our purpose is to show the theory of statistical models has a homological nature in the category $\mathcal{FB}(\Gamma, \Xi)$.

Definition 59. *A statistical model for a measurable set* (Ξ, Ω) *is couple* $(\mathbb{M}, [p])$ *formed by an object of the category* $(\mathcal{FB} - \Gamma, \Xi)$*, namely*

$$\mathbb{M} = [\mathcal{E}, \pi, M, D]$$

and a smooth Γ*-equivariant random cohomology class*

$$[p] \in H_e^0(\mathcal{E}, \mathbb{R}).$$

Further the to every fiber $p|\mathcal{E}_x$ *is a probability density.*

A Comment.

Let $(U, \Phi \times \phi)$ *be a local chart of* $[\mathcal{E}, \pi, M, D]$ *and let* $x* \in U$*. We set*

$$\Theta_U = \phi(U),$$

$$(\mathcal{E}_{x^*}, \Omega_{x^*}) = \Phi^{-}1[\{\phi(x^*)\} \times (\Xi, \Omega)].$$

The last definition above says that we obtain the probability space

$$(\mathcal{E}_{x^*}, \Omega_{x^*}, [p]).$$

This property does not depend on the choice of the local chart $(U, \Phi \times \phi)$*. Thus we can regard* $[\mathbb{M}, p]$ *as a special type of homological map*

$$\mathcal{FB}(\Gamma, \Xi) \ni \mathbb{M} \to [p] \in H_e^0(\mathcal{E}, \mathbb{R}).$$

9. The Moduli Space of the Statistical Models

We are going to face another major open problem. The challenge is the search for an invariant which encodes the points of the orbit space

$$m = \frac{M}{G}.$$

That is what is called the problem of moduli space. This problem of moduli space is a major challenge in both the differential geometry and the algebraic geometry (see the theory of Teichmuller). The problem is rather confusedly addressed in [30]. Subsequently it provoked controversies and criticisms.

The Hessian Functor

We consider the category $\mathcal{BF}$ whose objects are pairs $\{M, B\}$ formed by a manifold M equipped bilinear forms $B \in \Gamma(T^{*\otimes 2M})$.

In Part A we have defined the Hessian differential operator of a Koszul connection ∇, namely

$$D_\nabla = \nabla^2.$$

Those operators are useful for addressing the problem of moduli spaces. For our purpose four categories are involved,

(1) The category $\mathcal{LC}$ whose objects are gauge structures (M, ∇),
(2) The category $\mathcal{GM}$ whose objects are statistical models for measurable sets,
(3) the category $\mathcal{BF}$ whose objects are manifolds equipped bilinear forms,

(4) the category $\mathcal{F}(\mathcal{LC}, \mathcal{BF})$ whose objects are functors

$$\mathcal{GM} \to \mathcal{BF}.$$

Definition 60. *The Hessian functor is the functor*

$$\mathcal{GM} \ni \mathbb{M} = [\mathcal{E}, \pi, M, D, p] \to q_{\mathbb{M}} \in \mathcal{F}(\mathcal{LC}, \mathcal{BL})$$

Here $q_{\mathbb{M}}$ is defined by

$$q_{\mathbb{M}}[\nabla] = \nabla^2 log(p).$$

Reminder.

We recall that for vector fields X, Y the bilinear form $q_{\mathbb{M}}[\nabla](X, Y)$ is defined by

$$q_{\mathbb{M}}[\nabla](X, Y) = X \cdot (Y \cdot log(p)) - \nabla_X Y \cdot log(p)$$

The functor $q_{\mathbb{M}}$ is called the Hessian functor of the model

$$\mathbb{M} = [\mathcal{E}, \pi, M, D, p].$$

Our aim is to demonstrate the following claim. Up to isomorphism a statistical model $\mathbb{M}$ is defined by its Hessian functor $q_{\mathbb{M}}$. The functor $q_{\mathbb{M}}$ is an significant contribution to the information geometry.

We fix an object of $\mathcal{FB}(\Gamma, \Xi)$, namely $[\mathcal{E}, \pi, M, D]$. Let $\mathcal{P}(\mathcal{E})$ be the convex set of probability densities in $[\mathcal{E}, \pi, M, D]$. The multiplicative group of positive real valued functions defined in Ξ is denoted by $\mathbb{R}^{\Xi}_{+}$. The quotient of $\mathcal{P}(\mathcal{E})$ modulo $\mathbb{R}^{\Xi}_{+}$ is denoted by

$$\mathcal{PRO}(\mathcal{E}) = \frac{\mathcal{P}(\mathcal{E})}{\mathbb{R}^{\Xi}_{+}}.$$

Lemma 6. *For every $p \in \mathcal{P}(\mathcal{E})$ the image of*

$$\mathbb{M} = [\mathcal{E}, p],$$

namely $q_{\mathbb{M}}$ depends only on the class $[p] \in \mathcal{PRO}(\mathcal{E})$

Proof. We consider

$$\mathbb{M} = [\mathcal{E}, \pi, M, D, p],$$
$$\mathbb{M}^* = [\mathcal{E}, \pi, M, D, p^*].$$

We assume that

$$q_{\mathbb{M}} = q_{\mathbb{M}^*}.$$

Thus in every local trivialization $\Theta_U \times \Xi$ one has the identity

$$X(Ylog(\frac{p^*(x,\xi)}{p(x,\xi)}) - \nabla_X Ylog(\frac{p^*(x,\xi)}{p(x,\xi)})) = 0$$

forall $X, Y \in \Gamma(T\Theta)$, forall $\nabla \in \mathcal{LC}(\Theta)$. That identity holds if and only if the function

$$(x, \xi) \to \frac{p^*(x,\xi)}{p(x,\xi)}$$

belongs to $\mathbb{R}^{\Xi}_{+}$. This ends the idea. $\square$

A Comment.

The mapping

$$\mathbb{M} \to q_{\mathbb{M}}$$

is a global geometrical invariant in the sense of Erlangen. In other words it is an invariant of the Γ-geometry in $[\mathcal{E}, \pi, M, D, p]$.

Our aim is to demonstrate that

$$\mathbb{M} \to q_{\mathbb{M}}$$

is a characteristic invariant in the category $\mathcal{GM}(\Xi, \Omega)$. In other words the isomorphism class of the model

$$\mathbb{M} = [\mathcal{E}, \pi, (M, D), p]$$

is encoded by the functor

$$\nabla \to q_{\mathbb{M}}[\nabla].$$

The first step is the following lemma.

Lemma 7. *In the same object* $[\mathcal{E}, \pi, M, D]$ *we consider two statistical models*

$$\mathbb{M}_1 = [\mathcal{E}, \pi, M, D, p_1],$$

$$\mathbb{M}_2 = [\mathcal{E}, \pi, M, D, p_2].$$

The following assertions are equivalent

(1) $q_{\mathbb{M}_1} = q_{\mathbb{M}_2}$,

(2) $p_1 = p_2$.

Proof. We work in the domain of a local trivialization of $[\mathcal{E}, \pi, M, D]$. By the virtue of Lemma 6 above we know that

$$q_{p_1} = q_{p_2}$$

if and only if

$$p_1(x, \xi) = \lambda(\xi) p_2(x, \xi)$$

with $\lambda \in \mathbb{R}^{\Xi}_{+}$. Since both p_1 and p_2 are Γ-equivariant the function

$$\Xi \ni \xi \to \lambda(\xi)$$

is Γ-invariant too. Now we take into account that the natural action of Γ in Ξ is transitive. Therefore the Γ-equivariant function $\lambda(\xi)$ is a constant function. Therefore

$$p_1(x, \xi) = \lambda p_2(x, \xi)$$

The operation of integration along a fiber of π yields

$$\lambda = 1$$

This ends the proof. $\square$

We consider two m-dimensional statistical models for (Ξ, Ω), namely

$$\mathbb{M}_j = [\mathcal{E}_j, \pi_j, M_j, D_j, p_j], \quad j := 1, 2.$$

To simlplify we use the following notation.

$$q_{p_j} = q_{\mathbb{M}_j}.$$

In the category $\mathcal{FB}(\Gamma, \Xi)$ we consider an isomorphism

$$[\mathcal{E}_1 \times M_1] \in (e, x) \to [\Psi(e), \psi(x)] \in \mathcal{E}_2 \times M_2.$$

(1) Let ψ_* be the differential of ψ. For $\nabla \in \mathcal{LC}(M_1)$ the image $\psi_*(\nabla) \in \mathcal{LC}(M_2)$ is defined by

$$[\psi_*(\nabla)]_{X^*} Y^* = \psi_*[\nabla_{\psi_*^{-1}(X^*)} \psi_*^{-1}(Y^*)]$$

for all vector fields $X^*, Y^* \in \mathcal{X}(M_2)$.

(2) It is clear that the datum $[\mathcal{E}_1, \pi, M_1, D_1, p_2 \circ \Psi]$ is an object of the category $\mathcal{GM}(X, \Omega)$. Then for vector fields X, Y in M_1 we calculate (at X, Y) the right hand member of the following equality

$$[q_{p_2 \circ \Psi}(\nabla)] = \nabla^2[log(p_2 \circ \Psi)].$$

Direct calculations yield

$$\begin{aligned}
\nabla^2[log(p_2 \circ \Psi)](X, Y) &= X \cdot [Y \cdot log(p_2 \circ \Psi)] - \nabla_X Y \cdot log(p_2 \circ \Psi) \\
&= X \cdot [Y \cdot log(p_2) \circ \Psi] - \nabla_X Y \cdot [log(p_2) \circ \Psi] \\
&= \psi_*(X) \cdot [\psi_*(Y) \cdot log(p_2)] - \psi_*(\nabla_X Y) \cdot log(p_2) \\
&= [\psi_*(\nabla)^2 log(p_2)](\psi_*(X), \psi_*(Y)).
\end{aligned}$$

Thus for all $\nabla \in \mathcal{LC}(M_1)$ we have

$$q_{[p_2 \circ \Psi]}(\nabla) = q_{p_2}(\psi_*(\nabla)).$$

We summarize the calculations just carried out as it follows

Lemma 8. *Keeping the notation we just used namely p_2 and $\Psi \times \psi$ we have the following equality*

$$q_{[p_2 \circ \Psi]} = q_{p_2} \circ \psi_*$$

We are in position to face the problem of moduli space in the category $\mathcal{GM}(\Xi, \Omega)$.

Theorem 20. *We consider two m-dimensional statistical models*

$$\mathbb{M}_j = [\mathcal{E}_j, \pi_j, M_j, D_j, p_j], \quad j := 1, 2.$$

In the category $\mathcal{FB}(\Gamma, \Xi)$ let $\Psi \times \psi$ be an isomorphism of $[\mathcal{E}_1, \pi_1, M_1, D_1]$ onto $[\mathcal{E}_2, \pi_2, M_2, D_2]$. The following assertions are equivalent.

$$(1) \quad q_{p_2} \circ \psi_* = q_{p_1},$$

$$(2) \quad p_2 \circ \Psi = p_1.$$

Demonstration.

The demonstration is based on Lemmas 7 and 8.

According to our construction one has the following functor

$$q_\Psi = q_{p_2} \circ \psi_*.$$

This functor q_Ψ is the Hessian functor of the model

$$\mathbb{M}_\Psi = [\mathcal{E}_1, \pi, M_1, D_1, p_2 \circ \Psi].$$

Further $\Psi \times \psi$ is an isomorphism of $\mathbb{M}_\Psi$ onto $\mathbb{M}_2$.

Let us prove that assertion (2) implies assertion (1).

By of the definition of morphism of models, the pair $\Psi \times \psi$ is an isomorphism of $\mathbb{M}_1$ onto $\mathbb{M}_2$ if and only if

$$p_2 \circ \Psi = p_1.$$

Here we set the explicit formulas. Let $\nabla \in \mathcal{LC}(M_1)$. For all vector fields X, Y in M_2 we have

$$X \cdot (Y \cdot log(p_1)) - \nabla_X Y \cdot log(p_1) = X \cdot (Y \cdot log(p_2 \circ \Psi)) - \nabla_X Y \cdot log(p_2 \circ \Psi)$$
$$= \psi_*(X) \cdot (\psi_*(Y) \cdot log(p_2)) - \psi_*(\nabla_X Y) \cdot log(p_2).$$

now we observe that

$$\psi_*(\nabla_X Y) = [\psi_*(\nabla)]_{[\psi_*(X)]} \psi_*(Y).$$

Therefore (2) impies the equality

$$\psi^*[q_{[p_2]}(\psi_*(\nabla))] = q_{[p_2 \circ \Psi]}(\nabla) = q_{p_1}.$$

This shows the implication (2) $\rightarrow$ (1).

Let us prove that assertion (1) implies assertion (2).

Now we assume that that (1) holds, viz

$$q_{[p_2 \circ \Psi]} = q_{p_1}.$$

Then both $\mathbb{M}_1$ and $\mathbb{M}_\Psi$ have the same Hessian functor. By the virtue of Lemma 8 above we deduce that

$$p_2 \circ \Psi = p_1.$$

This ends the demonstration.

Reminder.

(i) *Objects of $\mathcal{GM}(\Gamma, \Xi)$ are quintuplets*

$$\mathbb{M} = [\mathcal{E}, \pi, M, D, p].$$

They are called statistical models for the measurable set (Ξ, Ω).

(ii) *Objects of $\mathcal{FB}(\mathcal{MSE})$ are functors*

$$[\mathcal{E}, p] \rightarrow [M, D].$$

They are called $\mathcal{MSE}$-fibrations.

Those categories are canonically equivalent. Further the actions the group G in those categories lead to the same moduli space

$$m = \frac{\mathcal{GM}(\Xi,\Omega)}{G} = \frac{\mathcal{FB}(\mathcal{MSE})}{G}.$$

We rephrase the theorem on moduli space.

Theorem 21. *The functor*

$$[\mathcal{E}, p] \to q_p \in \mathcal{BF}(M)$$

parametrizes the moduli space m.

This ends the challenge 2.

10. The Homological Statistical Models

In this section we introduce the theory of homological statistical models (HSM in short). We address the links between this theory and the local theory as in [17].

The theory of homological statistical models is useful for strengthening the central role played by the theory of KV homology in the information geometry and in the topology of the information [14,16–18,22,30,37,60,61].

We introduce the theory of localization of homological models. We use it to highlight the role played by local cohomological vanishing theorems as well as the role played by global cohomological vanishing theorems.

The framework is the category $\mathcal{FB}(\Gamma,\Xi)$.

Let $[\mathcal{E},\pi,M,D]$ be an m-dimensional object of the category $\mathcal{FB}(\Gamma,\Xi)$, viz $m = dim(M)$. The KV algebra of (M,D) is denoted by $\mathcal{A}$. The smooth manifold $\mathbb{R}^m$ supports a sheaf of KV algebras $\tilde{\mathcal{A}}$. This sheaf is locally isomorphic to $\mathcal{A}$. The vector space $C^\infty(\mathbb{R}^m)$ is a left module of $\tilde{\mathcal{A}}$. The affine action of Γ in $\mathbb{R}^m$ is $\tilde{\mathcal{A}}$-preserving.

Let $(U,\Phi_U \times \phi_U)$ be a local chart of $[\mathcal{E},\pi,M,D]$. We recall that $d\phi_U$ is the differential of ϕ_U. We have

$$d\phi_U(\mathcal{A}) = \tilde{\mathcal{A}}(\phi_U(U)).$$

Definition 61. *A homological model consists of the following data. The datum $[\mathcal{E},\pi,M,D]$ is an object of the category $\mathcal{FB}(\Gamma,\Xi)$. Every $x \in M$ has an open neighborhood U which is the domain of a local chart of $[\mathcal{E},\pi,M,D]$, namely $(\Phi_U \times \phi_U)$. We set*

$$\Theta_U \times \Xi = \Phi_U(\mathcal{E}_U).$$

Those data are subject to the following requirements.
HSM.1 : $\Theta \times \Xi$ supports a non negative random symmetric 2-cocycle

$$\Theta_U \times \Xi \ni (\theta,\xi) \to Q_U(\theta,\xi) \in Z^2_{KV}(\tilde{\mathcal{A}},\mathbb{R}).$$

HSM.2 : Let $[U,\Phi_U \times \phi_U, Q_U]$ and $[U^*,\Phi_{U^*} \times \phi_{U^*}, Q_{U^*}]$ as in HSM.1.
If we assume that

$$U \cap U^* \neq \emptyset$$

then there exists $\gamma_{UU^*} \in \Gamma$ such that

$$HSM.2.1 \quad \Phi_{U^*}(e) = \gamma_{UU^*} \cdot \Phi_U(e) \quad \forall e \in \mathcal{E}_{U\cap U^*},$$

$$HSM.2.2 \quad Q_U(\Phi_U(e)) = \gamma^*_{UU^*} \cdot [Q_{U^*}(\Phi_{U^*}(e))].$$

Comments.

The equality

$$Q_U(\Phi_U(e)) = \gamma^*_{UU^*} \cdot [Q_{U^*}(\Phi_{U^*}(e))]$$

has the following meaning. For $v, w \in T_{\phi_U(e)}\Theta_U$ one has

$$Q_U[\theta(e), \xi(e)](v, w) = Q_{U^*}[\gamma_{UU^*} \cdot (\theta(e), \xi(e))](d[\gamma_{UU^*}] \cdot v, d[\gamma_{UU^*}] \cdot w).$$

Morphisms of homological models are defined by replacing the probability P_U by the random cocycle Q_U.

Definition 62. *The category of homological statistical models for a measurable set* (Ξ, Ω) *is denoted by* $\mathcal{HSM}(\Xi, \Omega)$.

10.1. The Cohomology Mapping of $HSM(\Xi, \Omega)$

We consider an m-dimensional object of $\mathcal{HSM}(\Xi, \Omega)$ which is defined by a complete atlas

$$\mathbb{A} = [U_j, \Phi_j \times \phi_j, \gamma_{ij}, Q_j]$$

The underlying object of the atlas $\mathbb{A}$ is denoted by $[\mathcal{E}, \pi, M, D]$. We set

$$\Theta_j = \phi_j(U_j) \subset \mathbb{R}^m.$$

We are not making any difference between $(U_j, \mathcal{A})$ and $(\Theta_j, \tilde{\mathcal{A}})$. We put set

$$\mathcal{E}_{ij} = \mathcal{E}_{U_i \cap U_j}.$$

If we assume that

$$U_i \cap U_j \neq \emptyset$$

then we have

$$\Phi_j(e) = \gamma_{ij} \cdot \Phi_i(e), \quad \forall e \in \mathcal{E}_{ij}$$

and

$$Q_i(\Phi_i(e)) = \gamma^*_{ij} \cdot Q_j(\Phi_j(e)) \quad \forall e \in \mathcal{E}_{ij}.$$

We put

$$q_j(e) = Q_j(\Phi_j(e)) \quad \forall e \in \mathcal{E}_j.$$

Here

$$\mathcal{E}_j = \mathcal{E}_{U_j}.$$

If

$$U_i \cap U_j \neq \emptyset$$

then we know that

$$[\Phi_j \circ \Phi_i^{-1}](\theta_i(e), \xi_i(e)) = \gamma_{ij}(\theta_i(e), \xi_i(e)) \quad \forall e \in \mathcal{E}_{ij}.$$

Therefore we get

$$q_i(e) = q_j(e) \quad \forall e \in \mathcal{E}_{ij}.$$

Therefore q_j is the restriction to $\mathcal{E}_j$ of a (globally defined) map

$$\mathcal{E} \ni e \to Q(e) \in Z^2_{KV}(\mathcal{A}, \mathbb{R}).$$

It is clear that

$$Q_j = Q \circ \Phi_j^{-1}.$$

The action of Γ in $\mathcal{E}$ is Q-preserving. Thus a homological statistical model is a quintuplet

$$[\mathcal{E}, \pi, M, D, Q].$$

Here Q is a map

$$\mathcal{E} \ni e \rightarrow Q(e) \in Z^2_{KV}(\mathcal{A}, \mathbb{R}).$$

Thus we get random cohomological map

$$\mathcal{E} \ni e \rightarrow [Q](e) = [Q(e)] \in H^2_{KV}(\mathcal{A}, \mathbb{R}).$$

Definition 63. *The mapping* $[Q]$ *is called cohomology mapping of the homological model* $[\mathcal{E}, \pi, M, D, Q]$.

10.2. An Interpretation of the Equivariant Class [Q]

We intend to interpret the cohomology class $[Q]$ as an obstruction class.

Definition 64. *(1) A homological statistical model whose cohomological map vanishes is called an EXact Homological Statistical Model, (EXHSM); (2) A homological statistical model whose cocycle is a random Hessian metric is called a HEssian Homological Statistical Model (HEHSM); (3) An exact Hessian homological statistical model is called a HYperbolic Homological Statistical Model (HYHSM).*

Given a Hessian Homological model

$$\mathbb{M} = [\mathcal{E}, \pi, M, D, Q]$$

the cohomology map $[Q]$ is the obstruction for $\mathbb{M}$ being an Hyperbolicity model.

The following proposition leads to impacts on the differential topology.

Proposition 11. *The kernel of an exact homological statistical model is in involution. Further if M and all data depending on M are analytic then Q is a stratified transversally Riemannian foliation in M.*

If $[\mathcal{E}, \pi, M, D, Q]$ is exact then there exists a random differential 1-form θ such that

$$Q = \delta_{KV}\theta,$$

viz

$$Q(X,Y) = X \cdot \theta(Y) - \theta(D_X Y) \quad \forall X, Y \in \mathcal{X}(M).$$

That useful for seing that $Ker(Q)$ is in involution.

10.3. Local Vanishing Theorems in the Category $\mathcal{HSM}(\Xi, \Omega)$

Reminder.

The category whose objects are homological statistical models (for (Ξ, Ω)*) is denoted by* $\mathcal{HSM}(\Xi, \Omega)$. *Henceforth we fix an auxiliary structure of probability space* (Ξ, Ω, p^*).

Definition 65. *We are interested in random functions defined in* $\mathbb{R}^m \times \Xi$.

(1) *A random function f has the property $p^* - EXP$ if*

$$exp(f(x,\xi)) \leq \int_{\Xi} exp(f(x,\xi))dp^*(\xi) \quad \forall x \in \mathbb{R}^m.$$

(2) *A random closed differential 1-form θ has the property $p^* - EXP$ if every $x \in \mathbb{R}^m$ has an open neighbourhood U satisfying the following conditions, $U \times \Xi$ support a random function f subject to two requirements:*

- $\theta = df$,
- *f has the property $p^* - Exp$.*

(3) *An exact homological statistical model $[\mathcal{E}, \pi, M, D, Q]$ has property $p^* - EXP$ if there exists a random differential 1-form θ satisfying the following conditions*

- *θ has the property $p^* - EXP$,*
- $Q = \delta_{KV}\theta$.

Localization

Our purpose is to explore the relationships between the theory of homological statistical models and the theory of local statistical model as in [18,22], Barndorff-Nielsen 1987

Our aim is to show that the current (local) theory is a byproduct of the localization of homological models. The notion of localization of homological models is but the notion of local vanishing theorem.

Theorem 22. *Let $[\mathcal{E}, \pi, M, D, Q]$ be a homological statistical model.*

(1) *$[\mathcal{E}, \pi, M, D, Q]$ is locally exact.*
(2) *If the $[\mathcal{E}, \pi, M, D, Q]$ has the property $p^* - EXP$ then $[\mathcal{E}, \pi, M, D, Q]$ is locally isomorphic to a classical statistical model (Θ, P) as in [18].*

The Sketch of Proof of (1). Let $(U, \Phi \times \phi)$ be a local chart of $[\mathcal{E}, \pi, M, Q]$. We set

$$\Theta_U = \phi(U).$$

We assume that Θ_U is an open convex subset of $\mathbb{R}^m$. Θ supports a system of affine coordinate functions

$$\theta = (\theta_1, ..., \theta_m).$$

We have

$$Q(\theta,\xi) = \sum Q_{ij}(\theta,\xi)d\theta_i d\theta_j.$$

Since $Q(\theta,\xi)$ is a random KV cocycle of $\tilde{\mathcal{A}}$ we have

$$\delta_{KV}Q = 0.$$

The last equality is equivalent to the following system

$$\frac{\partial Q_{jk}}{\partial \theta_i} - \frac{\partial Q_{ik}}{\partial \theta_j} = 0.$$

We fix $\xi \in \Xi$. For every j the random differential 1-form β_j is defined by

$$\beta_j(\theta,\xi) = \sum_i Q_{ij}d\theta_i.$$

Every $\beta_j(\theta,\xi)$ is a cocycle of the de Rham complex of Θ_U.

By the virtue of the Lemma of Poincaré there exists a local function $h_j(\theta,\xi)$ such that

$$\beta_j = dh_j.$$

Now the differential 1-form $\tilde{\theta}$ is defined by

$$\tilde{\theta} = \sum_j h_j(\theta,\xi)d\theta_j.$$

Direct calculations lead to the following equality

$$Q = \delta_{KV}\tilde{\theta}.$$

This ends the proof of (1). □

The proof of (2). We assume that $\mathbb{M} = [\mathcal{E},\pi,M,D,Q]$ has the property $p^* - Exp$. We keep the notation we just used.

The random differential 1-form $\tilde{\theta}$ is a (de Rham) cocycle. Therefore $\Theta \times \Xi$ supports a random function $h(\theta,\xi)$ such that

$$\tilde{\theta} = dh.$$

So we have the following conclusion

$$Q(\theta(e),\xi(e)) = D^2h(\theta(e),\xi(e)) \quad \forall e \in \mathcal{E}_U.$$

Equivalently one gets

$$\frac{\partial^2 h}{\partial\theta_i\partial\theta_j} = Q_{ij}.$$

Since $\mathbb{M}$ has the property $p * -Exp$ we choose a function h has the property $p^* - EXP$. The functions $F(\theta)$ and $P(\theta,\xi)$ are defined by

$$F(\theta) = \int_\Xi exp(h(\theta,\xi))dp^*(\xi),$$

$$P_Q(\theta,\xi) = \frac{exp(h(\theta,\xi))}{F(\theta)}.$$

By the virtue of the property $p^* - Exp$ the function $P(\theta,\xi)$ satisfies the following requirements

(i) $P_Q\theta,\xi)$ is differentiable with respect to θ,
(ii) P_Q satisfies the following inequalities

$$0 \le P_Q(\theta,\xi) \le 1,$$

(iii) P_Q satifies the following identity

$$\int_\Xi P_Q(\theta,\xi)d\xi = 1.$$

Thus the pair (Θ_U, P_Q) is a local statistical model for (Ξ,Ω). This ends the proof of (2). The theorem is demonstrated. □

The pair (Θ_U, P_Q) is called a localization of $\mathbb{M}$.

Definition 66. *A localization (Θ_U, P_Q) is called a Local Vanishing Theorem of $[\mathcal{E},\pi,M,Q]$.*

Let $[\mathcal{E}, \pi, M, D, p]$ be an object of the category $\mathcal{GM}(\Xi, \Omega)$. We set

$$Q_p = D^2 log(p).$$

Thefore we get the exact homological statistical model

$$\mathbb{M}_p = [\mathcal{E}, \pi, M, D, Q_p]$$

So the notion of vanishing theorem has significant impacts on the information geometry. To simplify an exact models which having the property $p^* - Exp$ (for some probabily space) are is called $Exp - models$.

Theorem 23. *The notation is that used previously.*

(1) *The category $\mathcal{GM}(\Xi, \Omega)$ is a subcategory of the category $\mathcal{EXHSM}(\Xi, \Omega)$.*
(2) *Ojects of $\mathcal{GM}(\Xi, \Omega)$ but homological $Exp - models$.*

Reminder: New Insigts.

(1.1) *The Information GEometry is the geometry of statistical models.*
(1.2) *The Information topology is the topology of statistical models.*
(2.1) *The homological nature of the Information Geometry.*
(2.2) *What is a statistical model? The answer to the question raised by McCullagh should be: A statistical model is a Global Vanishing Theorem in the theory of homological models.*
(2.3) *A local statistical model is a Local Vanishing Theorem in the theory of homological models.*

11. The Homological Statistical Models and the Geometry of Koszul

Our purpose is to to relate the category of homological statistical models and the geometry of Koszul. This relationship is based on the localization of homological statistical models.

Proposition 12. *$\mathcal{EXPHEHSM}(\Xi, \Omega)$ stands for the subcategory whose objects are Hessian Exp-models.*

(1) *The holomogical map leads to the functor of $\mathcal{EXPHEHSM}(ß, \Omega)$ in the category of Hessian structures in (M, D)*

$$[\mathcal{E}, \pi, M, D, Q] \to (M, D, \tilde{Q}).$$

(2) *If M is compact then the subcategory of exact Hessian homological Exp-models $\mathcal{EXPHYHSM}(\Xi, \Omega)$ is sent in the category of hyperbolic structure in in (M, D).*

12. Examples

This section is devoted to a few examples. The construction involves some basic notions of the differential topology.

Example 1: Dynamics

We consider a triple $[M \times H, p_1, M, \nabla]$. Here (M, ∇) is a compact locally flat manifold, $(H, d\mu)$ is an amenable group. There is an effective affine action

$$H \times (M, \nabla) \to (M, \nabla).$$

Let $f \in C^\infty(M)$ and $x \in M$. The function

$$f_x : H \to \mathbb{R}$$

is defined by

$$f_x(h) = f(h \cdot x).$$

Let $L^\infty_H(M)$ be the set of $f^* \in C^\infty(M)$ such that

$$f^*_x \in L^\infty(H),$$

viz

$$\sup_{[h\in H]} |f_x(h)| < \infty \quad \forall x.$$

Now $EXP(L^\infty_H(M)$ stands for the set of $f^* \in L^\infty_H(M)$ such that

$$exp(f^*_x(h)) \leq \mu(exp(f^*_x)) \quad \forall x.$$

The function $P_{f^*}(x,h)$ is defined by

$$P_{f^*}(x,h) = \frac{exp(f^*(x \cdot h))}{\mu(exp(f^*_x))}.$$

The pair (M, P_{f^*}) is a probability density in H. Now set

$$\tilde{f}^*(x,h) = f^*(x \cdot h)$$

Therefore the datum $[M \times H, p_1, M, \nabla, P_{f^*}]$ is a statistical model for $(H, \mathcal{P}(H))$. Here $\mathcal{P}(M)$ is the boolean algebra of subsets of H and p_1 is the trivial fibration of $M \times H$ over M.

Example 2: Geometry

We focus on an example which plays a significant role in global analysis (and geometry) in some type of bounded domains [2,3]. This example relates the geometry of Koszul and Souriau Lie groups thermodynamics [4] and bibliography therein.

Let $\mathcal{C} \subset \mathcal{R}^m$ be a convex cone and let $\mathcal{C}^*$ be its dual. The characteristic function of $\mathcal{C}$ is defined by

$$\mathcal{C} \ni v \rightarrow \int_{\mathcal{C}^*} exp(- < v, w^* >) dw^*.$$

This gives rise to the following function

$$\mathcal{C} \times \mathcal{C}^* \ni (v, v^*) \rightarrow P(v, v^*) = \frac{exp(- < v, v^* >)}{\int_{\mathcal{C}^*} exp(- < v, w^* >) dw^*}$$

So $(\mathcal{C}, P)$ is a statistical model for $(\mathcal{C}^*, dw^*)$. Here dw^* is the standard Borel measure.

Stratified Analytic Riemannian Foliations

Reminder.

We recall that a (regular) Riemannian foliation M is a symmetric bilinear form $g \in \mathcal{S}_2(M)$ having the following properties

(1) $rank(g) = constant,$

(2) $L_X g = 0 \forall X \in \mathcal{G}(Ker(g)).$

From (2) one easily deduces that $Ker(g)$ is in involution. By the virtue of Theorem of Frobenius (1) and (2) imply that $Ker(g)$ is completely integrable.

In the category of differentiable manifolds, not all involutive singular distributions are completely integrable. Nevertheless, that is true in the category of analytic manifolds [62].

This subsection is mainly devoted to examples of stratified Riemannian foliations in analytic manifolds. For more details about those object the readers are referred to [46,63].

Theorem 24. *Let M be an orientable compact analytic manifold. Let $C^\omega(M^2)$ be the space real valued analytic functions defined in M^2. There exists a canonical map of $C^\omega(M \times M)$ in the family of analytic stratified Riemannian foliations in M.*

The Idea of Construction.

Let dv be an analytic volume element in M. In M we fix an analytic torsion free Koszul connection ∇. To a function $f \in C^\omega(M^2)$ we assign the function $P \in C^\omega(M^2)$

$$P_f(x, x^*) = \frac{exp[f(x, x^*)]}{\int_M exp[f(x, x^{**})]dv(x^{**})}.$$

We make the following identification

$$\mathcal{X}(M) = \mathcal{X}(M) \times 0 \subset \mathcal{X}(M^2).$$

The analytic bilinear form $g_f \in \mathcal{S}_2(M)$ is defined by

$$[g_f(x)](X, Y) = -\int_M P_f(x, x^*)[\nabla^2(log(P_f))(X, Y)](x, x^*)dv(x^*).$$

The form g_f has the following properties.

(a) *g_f does not depend on the choice of ∇,*
(b) *g_f is symmetric and positive semi-definite,*
(c) *If X is a section of $Ker(g_f)$ then $L_X g_f = 0$.*

Conclusion.

If

$$rank(g_f) = constant$$

then g_f is a Riemannian foliation as in [38–40,46].

If $rank(g_f)$ is not constant we apply [62]. Thereby g_f is an analytic stratified Riemannian foliation.

Reminder.

The idea of the strafication of g_f.

Step 0

The open subset $U_0 \subset M$ is defined by

$$x \in U_0 \quad iff \quad rank(g_f(x)) = \max_{[x^* \in M]} rank(g_f(x^*)).$$

The closed analytic submanifold $F_1 \subset M$ is defined by

$$F_1 = M \setminus U_0.$$

Step 1

In the pair (F_1, g_f) the open subset $U_1 \subset F_1$ is defined by

$$x \in U_1 \quad iff \quad rank(g_f(x)) = \max_{[x* \in F_1]} rank(g_f(x*)).$$

Step 2

We iterate this construction. Then we have get a filtration of M

$$... \subset F_n \subset F_{n-1} \subset ... \subset F_1 \subset F_O = M.$$

This filtration has the following properties

(1) $F_{j-1} \setminus F_j$ *is a analytic submanifold of M.*
(2) g_f *defines a regular Riemannian foliation in* $(F_{j-1} \setminus F_j, g_f)$,

Remark 6. *The extrinsic geometry of submanifolds is a particular case of the geometry of singular foliation [25].*

13. Highlighting Conclusions

13.1. Criticisms

In Part B we have raised some criticisms. We have constructed structures of statistical models in flat tori. An m-dimensional flat torus is not homeomorphic to an open subset of $\mathbb{R}^m$. The second criticism is the lack of dynamics. Subsequently, the problem of moduli space is absent from the classical theory. That deficiency is filled in by the characteristic functor

$$\mathbb{M} = [\mathcal{E}, \pi, M, D, p] \to q_{\mathbb{M}}.$$

The current theory requires a model to be identifiable. From the viewpoint of locally trivial fiber bundles, that requirement is useless.

13.2. Complexity

In both the theoretical information geometry and the applied information the exponential models and their generalizations play notable roles. What we call the complexity of a model $[\mathcal{E}, \pi, M, D, p]$ is its distance from the category of exponential models. Up to today there does not exist any INVARIANT which measures how far from being an exponential is a given model. This problem has a homological nature. We have produced a function r^b which fills in that lack. (See the Appendix A below).

13.3. KV Homology and Localization

We have introduced the theory of homological model. Among the notable notions that we have studied is the localization of homological statistical models. It links the theory of homological models and the current theory as in [22]. It may be interpreted as a functor from the theory of homological models to the classical theory of statistical models.

13.4. The Homological Nature of the Information Geometry

$\mathcal{GM}(\Xi, \Omega)$ and $\mathcal{HSM}(\Xi, \Omega)$ are introduced in this Part B. The category of local statistical models for (Ξ, Ω) is denoted by $\mathcal{LM}(\Xi, \Omega)$. On one side, the right arrows below mean subcategory. Then we have

$$\mathcal{LM}(\Xi, \Omega) \to \mathcal{GM}(\Xi, \Omega) \to \mathcal{HSM}(\Xi, \Omega).$$

On another side, the notion of Vanishing Theorem is useful in linking $\mathcal{HSM}(\Xi,\Omega)$ with both $\mathcal{GM}(\Xi,\Omega)$ and $\mathcal{LM}(\Xi,\Omega)$.

(1) The Global Vanishing Theorem is the functor

$$\mathcal{HSM}(\Xi,\Omega) \to \mathcal{GM}(\Xi,\Omega).$$

(2) The Local Vanishing Theorem is the functor

$$\mathcal{HSM}(\Xi,\Omega) \to \mathcal{LM}(\Xi,\Omega).$$

13.5. Homological Models and Hessian Geometry

In the category $\mathcal{HSM}(\Xi,\Omega)$ the Hessian functor is the functor from $\mathcal{HEHSM}(\Xi,\Omega)$ to the category of randon Hessian manifolds.

Furthermore, every structure of probability space (Ξ,Ω,p^*) gives rise to a canonical functor from $\mathcal{HEHSM}(\Xi,\Omega)$ to the category of Hessian manifolds. The canonical functor is defined by

$$[\mathcal{E},\pi,M,D,Q] \to \int_F p^*Q$$

Acknowledgments: The author gratefully thanks the referees for number of comments and suggestions. Their criticisms have been helpful to improve parts of the original manuscript.

Appendix A

Usually the appendix is devoted to overview the notions which are used in a paper. In this appendix we announce a few outstanding impacts of Hessian differential operators of Koszul connections.

In the introduction a pair of Koszul connections (∇,∇^*) is used for defining three differential operators

$$X \to D^{\nabla}(X) = \iota_X R^{\nabla} - L_X\nabla \quad \forall X \in \Gamma(TM).$$

The differential operator D^{∇} is elliptic and involutive in the sense of the global analysis [50,51,64]. Let $\mathcal{J}^{\nabla}$ be the sheaf of germ of solutions to the equation

$$FE^{**}(\nabla) : D^{\nabla}(X) = 0.$$

If ∇ torsion free then $FE^{**}(\nabla)$ is a Lie equation.

The non negative integers $r^b(\nabla)$ and $r^b(M)$ are defined by

$$r^b(\nabla) = \min_{[x\in M]} \left\{dim(\mathcal{J}^{\nabla}(x)\right\},$$

$$r^b(M) = \min_{[\nabla\in\mathcal{SLC}(M)]} \left\{dim(M) - r^b(\nabla)\right\}.$$

Here $\mathcal{SLC}(M)$ is the convex set of torsion free Koszul connections in M. We set the following notation: $\mathcal{R}ie(M)$ is the set of Riemannian metric tensors in M. $\mathcal{LF}(M)$ is the set of locally flat Koszul connection in M. At one side every $g \in \mathcal{R}ie(M)$ gives rise to the map

$$\mathcal{LF}(M) \ni \nabla \to \nabla^* \in \mathcal{LC}(M)$$

which is defined by

$$g(Y,\nabla^*_X Z) = Xg(Y,Z) - g(\nabla_X Y, Z).$$

At another side every $\nabla \in \mathcal{LF}(M)$ gives rise to the map

$$\mathcal{R}i(M) \ni g \to \nabla^g \in \mathcal{LC}(M)$$

which is defined by

$$g(Y, \nabla^g_X Z) = Xg(Y,Z) - g(\nabla_X Y, Z).$$

In every Riemannian manifold (M, g) we define the following numerical invariants

$$r^b(M,g) = \min_{[\nabla \in \mathcal{LF}(M)]} \left\{ dim(M) - r^b(\nabla^*) \right\},$$

$$r^B(M) = \min_{[g \in \mathcal{R}ie(M)]} \left\{ r^b(M,g) \right\}.$$

In every locally flat manifold (M, ∇) we define the following numerical invariant

$$r^b(M,\nabla) = \min_{[g \in \mathcal{R}ie(M)]} \left\{ r^b(\nabla^g) \right\}$$

The numerical invariants we just defined have notable impacts.

Appendix A.1. The Affinely Flat Geometry

Theorem A1. *In a smooth manifold M the following assertions are equivalent*

(1) $r^b(M) = 0$,
(2) *the manifold M admits locally flat structures.*

Appendix A.2. The Hessian Geometry

Theorem A2 (Answer a hold questions of [65]). *In a Riemannian manifold (M, g) the following assertions are equivalent*

(1) $r^b(M,g) = 0$,
(2) *the Riemannian manifold (M, g) admits Hessian structures (M, g, ∇)*

A Comment.

Assertion (2) has the following meaning.

(i) (M, ∇) *is a locally flat manifold.*
(ii) *every point has an open neighborhood U supporting a system of affine coordinate functions $(x_1, ..., x_m)$ and a local smooth function $h(x_1, ..., x_m)$ such that*

$$g(\frac{\partial}{\partial x_i}, \frac{\partial}{\partial x_j}) = \frac{\partial^2 h}{\partial x_i \partial x_j}.$$

Appendix A.3. The Geometry of Koszul

Theorem A3. *In a locally flat manifold (M, ∇) whose KV algebra is denoted by $\mathcal{A}$ the following assertions are equivalent*

(1) $r^b(M, \nabla) = 0$,
(2) *the KV cohomology space $H^2_{KV}(\mathcal{A}, \mathbb{R})$ contains a metric class $[g]$,*
(3) *the locally flat manifold (M, ∇) admits Hessian structures (M, ∇, g).*

Appendix A.4. The Information Geometry

Let (Ξ, Ω) be a transitive measurable and let

$$\mathbb{M} = [\mathcal{E}, \pi, M, D, p]$$

be an object of $\mathcal{GM}(\Xi, \Omega)$. Let g be the Fisher information of $\mathbb{M}$. Let

$$\{\nabla^{\alpha}, \quad \alpha \in \mathbb{R}\}$$

be the family of α-connections of $\mathbb{M}$. We define the following numerical invariant

$$r^b(\mathbb{M}) = \min_{[\alpha \in \mathbb{R}]} \left\{ dim(M) - r^b(\nabla^{\alpha}) \right\}.$$

Theorem A4. *In $\mathbb{M}$ the following assertions are equivalent.*

(1) $r^b(\mathbb{M}) = 0$,
(2) *$\mathbb{M}$ is an exponential family.*

Corollary A1. *Assume that $\mathbb{M}$ is regular, viz g is positive definite, then the following assertions are equivalent*

(1) $r^b(\mathbb{M}) = 0$,
(2) $r^b(M, g) = 0$,

Appendix A.5. The Differential Topology of a Riemannian Manifold

A Riemannian manifold (M, g) , (whose Levi-Civita connection is denoted by ∇^*), is called special if

$$\mathcal{J}^{\nabla^*} \neq 0$$

Theorem A5. *A special positive Riemannian manifold(M, g) has the following properties*

(1) *(M, g) admits a geodesic flat Hessian foliation*

$$[\mathcal{F}, g|\mathcal{F}, \nabla^*].$$

(2) *The leaves of $\mathcal{F}$ are the orbits of a bi-invariant affine Cartan-Lie group $(\tilde{G}, \tilde{\nabla})$.*
(3) *The bi-invariant affine Cartan-Lie group $(\tilde{G}, \tilde{\nabla})$ is generated by an effective infinitesimal action of a simply connected bi-invariant affine Lie group (G, ∇).*

References

1. Faraut, J.; Koranyi, A. *Analysis on Symmetric Cones*; Oxford University Press: Oxford, UK, 1994.
2. Koszul, J.-L. Déformation des connexions localement plates. *Ann. Inst. Fourier* **1968**, *18*, 103–114. (In French)
3. Vinberg, E.B. The theory of homogeneous convex cones. *Trans. Moscow. Math. Soc.* **1963**, *12*, 303–358.
4. Barbaresco, F. Geometric Theory of Heat from Souriau Lie Groups Thermodynamics and Koszul Geometry: Applications in Information Geometry. *Entropy* **2016**, doi:10.3390/e18110386.
5. Nguiffo Boyom, M.; Wolak, R. Foliations in affinely flat manifolds: Information Geometric. Science of Information. In *Geometric Science of Information*; Springer: Berlin/Heidelberg, Germany, 2013; pp. 283–292.
6. Gindikkin, S.G.; Pjateckii, I.I.; Vinerg, E.B. Homogeneous Kahler manifolds. In *Geometry of Homogeneous Bounded Domains*; Cremonese: Roma, Italy, 1968; pp. 3–87.
7. Kaup, W. Hyperbolische komplexe Rume. *Ann. Inst. Fourier* **1968**, *18*, 303–330.
8. Koszul, J.-L. Variété localement plate et convexité. *Osaka J. Math.* **1965**, *2*, 285–290. (In French)
9. Nguiffo Boyom, M. The Cohomology of Koszul-Vinberg Algebras. *Pac. J. Math.* **2006**, *225*, 119–153.
10. Nguiffo Boyom, M. Réductions Kahlériennes dans les groupes de Lie Résolubles et Applications. *Osaka J. Math.* **2010**, *47*, 237–283. (In French)

11. Shima, H. Homogeneous hessian manifolds. *Ann. Inst. Fourier* **1980**, *30*, 91–128.
12. Vey, J. Une notion d'hyperbolicité sur les variétés localement plates. *C. R. Acad. Sci. Paris* **1968**, *266*, 622–624. (In French)
13. Barbaresco, F. Information geometry of covariance matrix: Cartan-Siegel homogeneous bounded domains. Mostov/Berger fibration and Frechet median. In *Matrix Information Geometry*; Springer: Berlin/Heidelberg, Germany, 2013; pp. 199–255.
14. Baudot, P.; Bennequin, D. Topology forms of informations. *AIP Conf. Proc.* **2014**, *1641*, 213–221.
15. Gromov, M. In a Search for a Structure, Part I: On Entropy. Available online: www.ihes.fr/gromov/PDF/structure-serch-entropy-july5-2012.pdf (accessed on 28 November 2016).
16. Gromov, M. On the structure of entropy. In Proceedings of the 34th International Workshop on Bayesian Inference and Maximum Entropy Methods in Science and Engineering MaxEnt 2014, Amboise, France, 21–26 September 2014.
17. Amari, S.-I. *Differential Geometry Methods in Statistics*; Lecture Notes in Statistics; Springer: Berlin/Heidelberg, Germany, 1990.
18. Amari, S.-I.; Nagaoka, H. *Methods of Information Geometry, Translations of Mathemaical Monographs*; American Mathematical Society: Providence, RI, USA, 2007; Volume 191.
19. Arnaudon, M.; Barbaresco, F.; Yan, L. Medians and Means in Riemannian Geometry: Existence, Uniqueness and Computation. In *Matrix Information Geometry*; Springer: Berlin/Heidelberg, Germany, 2013; pp. 169–197.
20. Armaudon, M.; Nielsen, F. Meadians and means in Fisher geometry. *LMS J. Comput. Math.* **2012**, *15*, 23–37.
21. Barndorff-Nielsen, O.E. Differential geometry and statistics: Some mathematical aspects. *Indian J. Math.* **1987**, *29*, 335–350.
22. Murray, M.K.; Rice, J.W. Monographs on statistics and applied probability. In *Differential Geometry and Statistics*; Chapman and Hall/CRC: Boca Raton, FL, USA, 1993; Volume 48.
23. Nguiffo Boyom, M.; Byande, P.M. KV cohomology in information geometry. In *Matrix Information Geometry*; Springer: Berlin/Heidelberg, Germany, 2013; pp. 69–92.
24. Barndorff-Nielsen, O.E. *Information and Exponential Families in Stattistical Theory*; Wiley: New York, NY, USA, 1978.
25. Nguiffo Boyom, M.; Jamali, M.; Shahid, M.H. Multiply CR warped product Statistical submanifolds of holomorphic statistical space form. In *Geometric Science of Information*; Springer: Berlin/Heidelberg, Germany, 2015; pp. 257–268.
26. Milnor, J. The fundamental groups of complete affinely flat manifolds. *Adv. Math.* **1977**, *25*, 178–187.
27. Gerstenhaber, M. On deformations of Rings and Algebras. *Ann. Math.* **1964**, *79*, 59–103.
28. Nijenhuis, A. Sur une classe de propriétés communes à quelques types différents d'algèbres. *Enseign. Math.* **1968**, *14*, 225–277. (In French)
29. Nguiffo Boyom, M.; Byande, P.M.; Ngakeu, F.; Wolak, R. KV Cohomology and diffrential geometry of locally flat manifolds. Information geometry. *Afr. Diaspora J. Math.* **2012**, *14*, 197–226.
30. McCullagh, P. What is statistical model? *Ann. Stat.* **2002**, *30*, 1225–1310.
31. Baudot, P.; Bennequim, D. The homological nature of Entropy. *Entropy* **2015**, *17*, 3253–3318.
32. Hochschild, G. On the cohomology groups of an associative algebra. *Ann. Math.* **1945**, *46*, 58–67.
33. Hochschild, G.; Serre, J.-P. Cohomology of Lie algebras. *Ann. Math.* **1953**, *57*, 591–603.
34. McCleary, J. *A User's Guide to Spectral Sequences*; Cambridge University Press: Cambridge, UK, 2001.
35. Eilenberg, S. Cohomology of space with operators group. *Trans. Am. Math. Soc* **1949**, *65*, 49–99.
36. Koszul, J.-L. Homologie des complexes de formes différentielles d'ordre supérieur. *Ann. Sci. Ecole Norm. Super.* **1974**, *7*, 139–153. (In French)
37. Kass, R.E.; Vos, P.W. *Geometrical Foundations of Asymptotic Inference*; Wiley: New York, NY, USA, 1997.
38. Moerdijk, I.; Mrcun, J. *Introduction to Foliations and Lie Groupoids*; Cambridge Studies in Advanced Mathematics; Cambridge University Press: Cambridge, UK, 2003.
39. Molino P. *Riemannian, Foliations*; Birkhauser: Boston, MA, USA, 1988.
40. Reinhardt, B.L. Foliations with bundle-like metrics. *Ann. Math.* **1959**, *69*, 119–132.
41. Akivis, M.; Goldberg, V.; Lychagin, V. Linearizability of d-webs, on two-dimensional manifolds. *Sel. Math.* **2004**, *10*, 431–451.
42. Grifone, J.; Saab, J.; Zoltan, M. On the linearization of 3-webs. *Nonlinear Anal.* **2001**, *47*, 2643–2654.
43. Henaut, A. Sur la linéarisation des tissus de C^2. *Topology* **1993**, *32*, 531–542.

44. Hess, H. Connectionson symplectic manifolds and geometric quantization. In *Differential Geometry Methods in Maththematical Physics*; Lectures Notes 836; Springer: Berlin/Heidelberg, Germany, 1980; pp. 153–166
45. Nguiffo Boyom, M. Structures localement plates dans certaines variétés symplectiques. *Math. Scand.* **1995**, *76*, 61–84. (In French)
46. Kamber, F.; Tondeur, P.H. De Rham-Hodge Theory for Riemannian foliations. *Math. Ann.* **1987**, *277*, 415–431.
47. Byande, P.M. *De Structures Affines à la Géométrie de L'information*; Éditions Universitaires Européennes: Saarbrücken, Germany, 2012. (In French)
48. Chentsov, N.N. *Statistical Decision Rules and Optimal Inference*; American Mathematical Society: Providence, RI, USA, 1982.
49. Petrie, T.; Handal, J. *Connections, Definite Forms and Four-Manifolds*; Oxford Mathematcal Monograph, Oxford Science Publications; Clarendon Press: Oxford, UK, 1990.
50. Guillemin, V.; Sternberg, S. An algebraic model for transitive differential geometry. *Bull. Am. Math. Soc.* **1964**, *70*, 16–47.
51. Singer, I.M.; Sternberg, S. The infinte groups and lie and Cartan. *J. Anal. Math. Jerus.* **1965**, *15*, 1–144.
52. Shima, H. *The Geometry of Hessian Structures*; World Scientific: Singapore, 2007.
53. Barbaresco, F. *Symplectic Structure of Information Geometry: Fisher Information and Euler-Poincare Equation of Souriau Lie Group Thermodynamics*; Nielsen, F., Barbaresco, F., Eds.; Springer: Berlin/Heidelberg, Germany, 2015; pp. 529–540.
54. Matsuzoe, H.; Henmi, M. *Hessian Structures on Deformed Exponential Families*; Nielsen, F., Barbaresco, F., Eds.; Springer: Berlin/Heidelberg, Germany, 2013; Volume 8085, pp. 275–282.
55. Barbaresco, F. Koszul Information Geometry and Souriau Geometric Temperature/Capacity of Lie Group Thermodynamics. *Entropy* **2014**, *16*, 4521–4565.
56. Nijenhuis, A.; Richardson, W. Commutative algebra cohomology and deformation of Lie algebras and associative algebras. *J. Algebra* **1968**, *9*, 42–53.
57. Chevallier, E.; Barbaresco, F.; Angulo, J. Kernel Density Estimation on Siegel Space Applied to Radar Processing. *Entropy* **2016**, *18*, 396.
58. Chevallier, E.; Barbaresco, F.; Angulo, J. Probability density estimation on hyperbolic space applied to radar processing. In *Geometric Science of Information*; Springer: Berlin/Heidelberg, Germany, 2015; pp. 753–761.
59. Eilenberg, S. Homology of spaces with operators group I. *Trans. Am. Math. Soc.* **1947**, *62*, 378–417.
60. Nguiffo Boyom, M.; Wolak, R. Transverse Hessian metrics information geometry. In Proceedings of the 34th International Workshop on Bayesian Inference and Maximum Entropy Methods in Science and Engineering, Amboise, France, 21–26 September 2014.
61. Nguiffo Boyom, M.; Wolak, R.A. Transversely Hessian foliations and information geometry. *Int. J. Math.* **2016**, *27*, doi:10.1142/S0129167X16500920.
62. Zeghib, A. On Gromov theory of rigid transformation groups, a dual approach. *Ergod. Theorey Dyn. Syst.* **2000**, *20*, 935–946.
63. Kamber, F.; Tondeur, P.H. *Invariant Operators and the Cohomology of Lie Algebra Sheaves*; Memoirs of the American Matheamtical Society: Providence, RI, USA, 1971.
64. Kumpera, A.; Spencer, D. *Lie Equation*; Princeton University Press: Princeton, NJ, USA, 1972.
65. Guts, A.K. (Omsk State University). Private communication.

Permissions

All chapters in this book were first published in MDPI; hereby published with permission under the Creative Commons Attribution License or equivalent. Every chapter published in this book has been scrutinized by our experts. Their significance has been extensively debated. The topics covered herein carry significant findings which will fuel the growth of the discipline. They may even be implemented as practical applications or may be referred to as a beginning point for another development.

The contributors of this book come from diverse backgrounds, making this book a truly international effort. This book will bring forth new frontiers with its revolutionizing research information and detailed analysis of the nascent developments around the world.

We would like to thank all the contributing authors for lending their expertise to make the book truly unique. They have played a crucial role in the development of this book. Without their invaluable contributions this book wouldn't have been possible. They have made vital efforts to compile up to date information on the varied aspects of this subject to make this book a valuable addition to the collection of many professionals and students.

This book was conceptualized with the vision of imparting up-to-date information and advanced data in this field. To ensure the same, a matchless editorial board was set up. Every individual on the board went through rigorous rounds of assessment to prove their worth. After which they invested a large part of their time researching and compiling the most relevant data for our readers.

The editorial board has been involved in producing this book since its inception. They have spent rigorous hours researching and exploring the diverse topics which have resulted in the successful publishing of this book. They have passed on their knowledge of decades through this book. To expedite this challenging task, the publisher supported the team at every step. A small team of assistant editors was also appointed to further simplify the editing procedure and attain best results for the readers.

Apart from the editorial board, the designing team has also invested a significant amount of their time in understanding the subject and creating the most relevant covers. They scrutinized every image to scout for the most suitable representation of the subject and create an appropriate cover for the book.

The publishing team has been an ardent support to the editorial, designing and production team. Their endless efforts to recruit the best for this project, has resulted in the accomplishment of this book. They are a veteran in the field of academics and their pool of knowledge is as vast as their experience in printing. Their expertise and guidance has proved useful at every step. Their uncompromising quality standards have made this book an exceptional effort. Their encouragement from time to time has been an inspiration for everyone.

The publisher and the editorial board hope that this book will prove to be a valuable piece of knowledge for researchers, students, practitioners and scholars across the globe.

List of Contributors

Frank Nielsen
Sony Computer Science Laboratories, Tokyo 141-0022, Japan

Akio Fujiwara
Department of Mathematics, Osaka University, Toyonaka, Osaka 560-0043, Japan

Koichi Yamagata
Graduate School of Informatics and Engineering, The University of Electro-Communications, Chofu, Tokyo 182-8585, Japan

David C. de Souza
Instituto Federal do Ceará, Campus Maracanaú, Fortaleza 61939-140, Brazil

Rui F. Vigelis
Computer Engineering School, Campus Sobral, Federal University of Ceará, Sobral 62010-560, Brazil

Charles C. Cavalcante
Department of Teleinformatics Engineering, Federal University of Ceará, Fortaleza 60455-900, Brazil

Emmanuel Chevallier
Department of Computer Science and Applied Mathematics, Weizmann Institute of Science, Rehovot 7610001, Israel

Frédéric Barbaresco
Thales Air Systems, Surface Radar Business Line, Advanced Radar Concepts Business Unit, Voie Pierre-Gilles de Gennes, Limours 91470, France

Jesus Angulo
CMM-Centre de Morphologie Mathématique, MINES ParisTech, PSL-Research University, Paris 75006, France

Thibault Forget
Thales Air Systems, Surface Radar Business Line, Advanced Radar Concepts Business Unit, Voie Pierre-Gilles de Gennes, Limours 91470, France
CMM-Centre de Morphologie Mathématique, MINES ParisTech, PSL-Research University, Paris 75006, France

Hiroshi Matsuzoe
Department of Computer Science and Engineering, Nagoya Institute of Technology, Nagoya 466-8555, Japan

Stéphane Puechmorel and Florence Nicol
Laboratoire de Mathématiques Appliquées, Informatique et Automatique pour l'Aérien (MAIAA), Département Sciences et Ingénierie de la Navigation Aérienne (SINA), École Nationale de l'Aviation Civile (ENAC), 7 avenue Edouard Belin CS 54005, 31055 Toulouse, France

Michel Nguiffo Boyom
ALEXANDER GROTHENDIECK INSTITUTE, IMAG-UMR CNRS 5149-c.c.051, University of Montpellier, PL. E. Bataillon, F-34095 Montpellier, France

Index

Q

R

S

T

V

www.ingramcontent.com/pod-product-compliance
Lightning Source LLC
LaVergne TN
LVHW081038210926
848485LV00004B/1050